_____ Drinking Water Quality

Drinking Water Quality

Problems and Solutions

N. F. Gray

Trinity College, University of Dublin, Ireland

JOHN WILEY & SONS

Chichester · New York · Brisbane · Toronto · Singapore

Copyright © N.F. Gray, 1994

Published 1994 by John Wiley & Sons Ltd,
Baffins Lane, Chichester,
West Sussex PO19 1UD, England
Telephone National Chichester (0243) 779777
International (+44) 243 779777

Reprinted March 1995 (paper)

Other Wiley Editorial Offices

John Wiley & Sons, Inc., 605 Third Avenue,
New York, NY 10158-0012, USA

Jacaranda Wiley Ltd, 33 Park Road, Milton,
Queensland 4064, Australia

John Wiley & Sons (Canada) Ltd, 22 Worcester Road,
Rexdale, Ontario M9W 1L1, Canada

John Wiley & Sons (SEA) Pte Ltd, 37 Jalan Pemimpin #05-04,
Block B, Union Industrial Building, Singapore 2057

Library of Congress Cataloging-in-Publication Data

A catalog record for this book is available from the Library of Congress

British Library Cataloguing in Publication Data

A catalogue record for this book is available from the British Library

ISBN 0-471-94817-9 (case)
0-471-94818-7 (paper)

Typeset in 10/12pt Times by Vision Typesetting, Manchester
Printed and bound in Great Britain by Bookcraft (Bath) Avon

Contents

Preface

Every day each one of us uses large quantities of water. It is vital not only for cooking and of course drinking, but for washing clothes, dishes and ourselves, for flushing the toilet, watering the garden and a hundred other uses about the home. Industry uses water in vast quantities to manufacture everything from paper to motor cars, electricity to computers, and even to build our homes. It is the single most important commodity that we all require, and yet it is the one we take most for granted. In many arid and semi-arid areas water has to be collected each day, often by women and children, walking many miles to the nearest well and then back carrying a heavy jar or bucket full of water. A common practice in many African and Asian countries, yet in the developed world we expect to be able to turn on the tap and have as much water as we want, when we want. Not only that, but we expect it all to be clean and safe to drink, even though only a small fraction will be actually consumed or used for food preparation.

To enable us to have the facility of clean water piped to our homes requires an enormous investment both in money and expertise. It also requires a remarkable commitment from those whose job it is to supply our water. Countries which have a piped water system are in the minority in the world. There are even fewer where one expects to be able to drink the water that comes from the tap.

Running water is considered a luxury in many countries. In the third world the lack of water has contributed to the death of 100 million children in the past 20 years. Water is the single most important commodity that each and everyone of us needs, without it society and life itself comes to an end.

This book gives a broad overview of drinking water quality by examining the passage of water from its source to when it comes out of the consumer's tap. How safe is water to drink? What are the risks to health? Who is at risk? Surprisingly, while there are a number of problem areas, the overall quality of water in the United Kingdom and the Republic of Ireland is amongst the highest in Europe, if not the world, and yet consumers are generally unhappy with the quality of the water they are expected to drink. The book examines why this should be the case by identifying all the problems with drinking water and explaining where and why they occur. Alternatives such as the use of bottled water and home treatment systems, such as water filters, are also discussed.

This text is designed to be an introduction to the water supply industry and in particular to the problems associated with drinking water quality. It is aimed primarily at undergraduates and postgraduates studying environmental science or environmental engineering. However, I hope that hydrologists, geographers, chemists, microbiologists or those simply interested in water quality problems will find it equally useful.

Nick Gray
Trinity College, Dublin
April, 1994

Acknowledgements

There are very many people who have kindly provided me with information. I am particularly grateful to the water companies and water undertakers throughout the British Isles for sending me so much information about their water supply policies and operations.

I would like to thank the following publishers and organizations for permission to reproduce copyright material, or to modify copyright material, in this text.

Academic Press, Figure 2.10; American Water Works Association, Table 6.2; British Geological Survey, Figures 2.4, 4.1, Bruner Ltd, Table 8.4; Commission of the European Union, Table 2.6; Doulton Water Care Ltd, Figure 8.3; Dublin Regional Public Analyst, Table 5.2; Elsevier Science Ltd, Table 3.2; English Nature, Table 4.1; Foundation for Water Research, Table 4.8; Figure 2.8; The Controller of Her Majesty's Stationery Office, Table 1.11, 4.3, 4.4, 4.6, 7.3, 7.4; Figures 4.3, 5.6, 6.5; Institution of Water and Environmental Management, Figures 3.1, 3.8, 3.10, 3.11; Journal of the Institution of Water Engineers and Scientists, Table 7.5, 7.6; Figure 7.6; Journal of the Institution of Water and Environmental Management, Figure 4.4; Longman Education, Figures 6.2, 6.3, 6.4; Methuen Publishers Ltd, Table 2.1; National Rivers Authority, Table 1.2; National Water Council, Table 1.5; Open University, Figures 2.3, 2.5, 2.6, 3.7; Oxford University Press, Figures 2.9, 5.5; Penguin Books Ltd, Figure 2.11; Severn Trent Water, Table 1.4; US Environmental Protection Agency, Tables 1.12, 1.13, 1.14, 4.15; Water Engineering and Management, Table 4.13; Water Research Centre, Figures 1.1, 2.7, 4.5, 5.1, 5.2, 5.3, 6.1, 7.5; Water Services Association, Tables 1.1, 1.3, 1.7, 1.8, 1.9, 2.3, 2.4, 5.1, 5.4; Figures 1.2, 1.3, 1.4, 1.5; World Health Organization Tables 1.15, 1.16, 1.17, 1.18, 2.2; WRc Evaluation and Testing Centre, Figures 7.1, 7.2, 7.3, 7.4.

Glossary

All the terms, and abbreviations, listed below have been explained fully in the text. A brief summary or explanation is given here as an *aide-mémoire*.

Absorption
The process by which a substance is taken into the body of another substance (normally a biological cell).

Activated carbon
Made from materials such as coal or coconut shells, it has a very highly porous structure which is able to adsorb dissolved organic matter and certain dissolved gases from water.

Acute toxicity
A toxic effect coming speedily to a crisis, usually caused by a large dose of poison of short duration.

ADAS
Agricultural Development Advisory Service.

ADI
Acceptable Daily Intake of a substance to ensure no ill effect on consumer.

Adsorption
The process by which a gas, vapour, dissolved material or small suspended particle is attracted to, and attached to, the surface of another material either by physical or chemical forces.

Aeration
The vigorous mixing of water to put oxygen into solution to strip out carbon dioxide, remove odorous compounds, and to facilitate oxidative reactions.

Algae
Small, microscopic plant forms found as single cells, colonies or as filaments.

Algal bloom
A prolific growth of algae due to nutrient enrichment resulting in a serious reduction in water quality.

Alum
The common name for aluminium sulphate which is widely used as a coagulant.

Anaerobic
A process (normally biological) performed in the complete absence of oxygen.

Anion
Negatively charged ions.

AOC
Assimilible organic carbon.

Aquifer
An underground water-bearing layer of porous rock.

Artificial recharge
An artificial method of replacing the water abstracted from an aquifer at rates in excess of those that occur naturally.

Backwashing
The process in which the flow is reversed through a slow sand filter or ion exchange resin to loosen the bed and to flush out any suspended matter collected.

Breakpoint chlorination
Chlorinating water until the chlorine demand is satisfied, with the excess chlorine remaining in the water as free chlorine.

Cation
Positively charged ions.

CEN
Comité Européen de Normalisation publish European standards.

Chloramines
Compounds formed by the reaction of chlorine with ammonia in water.

Chlorination
The use of chlorine gas or solutions to disinfect drinking water.

Chlorine demand
The amount of chlorine consumed by organic matter and other oxidizable compounds in the water without leaving a chlorine residue.

Chronic toxicity
A toxic effect which continues for a long time and may be either lethal or sub-lethal, generally caused by a low dose of poison over a long time.

Coagulant
Normally a salt of aluminium or iron added to water to form a hydroxide precipitate.

Coagulation
The process in which small particles of floc begin to agglomerate into larger particles which settle out of solution more readily.

Coagulant aid
A compound added to water to promote coagulation. These are normally large organic ions whose charges help to form larger flocs.

Conductivity
The power of a liquid to transmit an electrical charge which is related to the concentration of ions.

Coliforms
A group of bacteria found in vast numbers in faeces, used as indicator organisms for microbial pathogens in water.

CTC
Tetrachloromethane or carbon tetrachloride – an industrial solvent.

DCM
Methylene chloride – an industrial solvent.

Denitrification
The reduction of nitrate to nitrogen gas.

Desalination
The removal of dissolved salts from brackish water in order to make it potable.

Diffuse pollution
Pollution emanating from a large undefined area.

Disinfection
Reduction of the microbial contamination of water.

DWI
Drinking Water Inspectorate

Filtration
Process to remove particles from water by passing it through a porous layer.

Flocculation
A process in which small suspended particles agglomerate to form large fluffy flocs. Aluminium and iron salts are generally added to water to bring this about.

Fluoride
A general compound containing fluorine added to water to reduce the incidence of dental caries within the community (fluoridation).

Fluorosis
A condition caused by excess fluoride in drinking water resulting in pitting and discoloration of teeth

FTU
Formazin Turbidity Unit. Widely used unit for expressing turbidity.

GAC
Granular activated carbon.

Groundwater
Water from aquifers or other natural underground sources.

GV
Guide value for substances and parameters listed in the World Health Organization drinking water guidelines.

Hardness
Caused by the dissolved salts of calcium and magnesium. Hard waters cause scaling and reduce the effectiveness of soap, while soft waters are generally acidic and corrosive.

Humic acids
This is a general name for a wide variety of organic compounds which are the breakdown products of vegetable matter, including peat.

Hydrogencarbonate
Bicarbonate (HCO_3).

Hydrological cycle
The continuous circulation of water between the atmosphere, land and sea by precipitation, transpiration and evaporation.

Ions
Charged particles in water. They are either atoms (e.g. Ca^{2+}, Cl^-) or groups of atoms (e.g. HCO_3^-).

Ion exchange
A process in which ions of like charge are exchanged between a solid resin and the water. Water softeners replace the calcium ions in the water for sodium ions thus reducing the water hardness.

Iron (II)
Ferrous iron.

Iron (III)
Ferric iron.

Langelier index
An equation for calculating the corrosiveness of water.

Lime
The common name for calcium oxide (CaO).

Methaemoglobinemia
A condition in babies and infants where the oxygen carrying capacity of haemoglobin is reduced due to the uptake of nitrite.

mg/l
Milligram per litre (1000 mg/l = 1 g/l).

MAC
Maximum admissible concentration. The maximum concentration of a substance listed in the EC Drinking Water Directive.

MCL
Maximum contaminant levels are enforceable drinking water standards set by the Office of Drinking Water of the USEPA.

MCLG
Maximum contaminant level goals are non-enforceable health based goals for drinking water set by the Office of Drinking Water of the USEPA.

Ml/d
Megalitre per day (1 Ml/d $= 1000$ m^3/d $= 1\,000\,000$ l/d).

ng/l
Nanogram per litre (1000 ng/l $= 1$ μg/l).

NRA
National Rivers Authority.

OFWAT
Office of Water Services.

Oocyst
The dormant stage of a protozoan parasite such as *Cryptosporidium*.

Oxidation
A chemical reaction in which the oxygen content of a compound is increased, or in which electrons are removed from an ion or compound. Soluble iron becomes insoluble when oxidized and precipitates out of solution.

Ozone
A gas which is an unstable form of oxygen, with the chemical formula O$_3$ used as an aggressive disinfectant.

PAC
Powdered activated carbon.

PAH
Polycyclic aromatic hydrocarbons.

Pathogen
An organism which is capable of causing disease.

PCBs
Polychlorinated biphenyls.

PCE
Tetrachloroethene or perchloroethylene – an industrial solvent.

PCTs
Polychlorinated terphenyls.

PCV
Prescribed concentration or value. This is the maximum admissible concentration allowed under the Water Supply (Water Quality) Regulations for England and Wales, 1989.

Pesticide
A pesticide is any chemical used to control animal and plant pests. These include fungicides, herbicides and insecticides.

pH
A measure of how acid or alkaline water is in relation to the concentration of hydrogen ions present.

Plumbosolvency
Solubility of lead in water.

Point pollution
Pollution from a clearly defined source, usually an outlet pipe.

Primary National Drinking Water Regulations
Standards set by the USEPA for parameters considered to be potentially harmful to health.

Private supply
Water taken from a private source or supplied by a non-licensed supplier.

Public supply
Water supplied by a company licensed or authorized to do so.

Rapid sand filter
A filter containing coarse sand or other filtration media through which water passes through at high rates, often under pressure.

Reduction
A chemical process in which electrons are added to an ion or compound, or in which the oxygen content is reduced.

Reverse osmosis
A process for the removal of dissolved ions and organic compounds from water which is passed under pressure through a semi-permeable membrane.

Service reservoir
A water tower or reservoir used for storage of treated water within the distribution system.

Secondary National Drinking Water Regulations
Standards set by the USEPA for parameters intended to protect the aesthetic quality of water rather than concerned with public health.

Sedimentation
Process for the removal of settleable solids within a tank under semi-quiescent conditions. The settled particles form a sludge which is removed from the tank at regular intervals.

Slow sand filters
Water passes through a layer of fine sand on top of a layer of gravel. Solids are removed by filtration and nutrients are removed by biological activity.

Softening
Process to reduce the hardness of water either by precipitation or ion-exchange processes.

Surface water
General term for any water body which is found flowing or standing on the surface such as rivers, lakes or reservoirs.

Taste threshold
Minimum concentration at which a compound in water can be tasted.

TDS
Total dissolved solids is a crude measure of the total concentration of inorganic salts in water. Measured by evaporation.

TCA
Methyl chloroform or 1,1,1-trichloroethane – an industrial solvent.

TCE
Trichloroethylene – an industrial solvent.

TCM
Trichloromethane or chloroform – an industrial solvent.

THM
Trihalomethanes – organic chemicals formed during chlorination of water containing natural organic compounds such as humic and fulvic acids.

Turbidity
Due to colloidal and suspended matter which imparts a cloudiness to water. It is determined by measuring the degree of scattering of a beam of light which is passed through the water.

µg/l
Microgram per litre (1000 µg/l = 1 mg/l).

USEPA
US Environmental Protection Agency.

UV
Ultraviolet radiation used to disinfect water.

Water supply zone
These are either a discrete area served by a single source or an area supplying no more than 50 000 people. Used as the basic unit for monitoring water quality in England and Wales.

Water undertaker
A company holding a licence to supply drinking water. In England and Wales these are the water service companies and the water-only companies.

Wholesomeness
A concept of water quality which is defined by reference to the standards and other

requirements listed in the Water Supply (Water Quality) Regulations for England and Wales, 1989.

WRc
Water Research Centre.

WSC
Water supply company.

Chapter 1
The Water Business

1.1 WATER CONSUMPTION

1.1.1 The Demand for Water

During 1990–1 the water undertakers (i.e. those who supply water) in the UK supplied a total of 20 361 million litres of water each day (Ml/d) to consumers. This included 17 381 Ml/d to about 50 243 000 people in England and Wales alone. When this is broken down by region, the greatest demand is in the Thames and North West Regions, which have large populations (Table 1.1). As we shall see in Chapter 2, the areas of highest demand do not normally correspond to the areas where adequate water resources are to be found, so shortages occur. The current demand for potable water in England and Wales is increasing and is estimated by the National Rivers Authority (NRA) to increase by 13.5% to 19 617 Ml/d by the year 2011 (Table 1.2).

The demand for water varies over the 24 hour period. This is known as the diurnal variation, with peak usage occurring between 08.00–12.00 and from 18.00–19.00 each day (Figure 1.1). Demand is greater during weekends (by about 12%) and seasonal trends also occur, with demand being higher in the summer than in the winter.

1.1.2 Demand in the Home

In the UK the typical household water consumption, typical here meaning a family of two adults and two children, is 475 l/d. This is equivalent to a per capita water consumption rate of 140 l/d.

Toilet flushing is the single major use of water at 32% of the total consumption per person per day, followed by bathing and showering at 17%, then washing clothes at 12%. Dishwashers are becoming increasingly popular, but currently only account for 1% of the average water usage. It is expected that dishwasher ownership may reach 40% of the population by the end of the century. If this happens it would add an extra 6 l/d to the per capita consumption, increasing the overall demand by about 5%. Table 1.3 gives some idea of the ownership of water-using appliances, the number of times on average they are used daily and how much water they use. At the top of the list are automatic washing machines, which can use a staggering 110 l every time they are used.

Table 1.1. Size of water supply area, population and amount of water supplied by water companies and water-only companies in England and Wales in 1991 and totals for other parts of the UK. Reproduced by permission of the Water Services Association

Water service company or area	Area (km²)		Resident population		Unmetered water (Ml/d)	Metered potable water (Ml/d)	Non-potable water (Ml/d)	Total water (Ml/d)
	Water supply	Sewerage	Water supply	Sewerage				
Anglian	22 000	27 000	3.8	5.4	860	331	40	1231
Northumbrian	3850	9400	1.2	2.6	279	157	234	670
North West	14 415	14 445	6.8	6.8	1770	638	77	2485
Severn Trent	18 960	21 650	6.9	8.3	1429	561	0	1990
Southern	4450	10 450	2.1	4.3	517	181	7	705
South West	10 300	10 800	1.5	1.5	381	117	0	498
Thames	8200	13 750	7.3	11.7	2248	547	0	2795
Dŵr Cymru	20 400	21 300	2.8	3.1	807	240	148	1195
Wessex	7350	10 000	1.1	2.5	292	131	6	429
Yorkshire	13 900	13 600	4.4	4.6	1043	382	0	1425
Total (water service companies)	124 000	152 000	37.9	50.8	9626	3285	512	13 423
Water-only companies	28 000	—	12.9	—	2926	968	64	3958
England and Wales	152 000		50.8		12 552	4253	576	17 381
Scotland	77 750		5.1		1651	630	19	2300
Northern Ireland	14 150		1.6		537	143	0	680
UK	244 000		57.5		14 740	5026	595	20 361

Table 1.2. Current estimated water demand in the year 2011 for all water companies in England and Wales. Reproduced by permission of the National Rivers Authority (1991)

National Rivers Authority region	Demand in 1990 (Ml/d)	Estimated demand in 2011 (Ml/d)
Anglian	1839	2343
Northumbria	1114	1218
North West	2550	2446
Severn Trent	2398	2573
Southern	1324	1577
South West	493	598
Thames	4032	4723
Dŵr Cymru	1216	1417
Wessex	861	1135
Yorkshire	1457	1587
Total for England and Wales	17 284	19 617

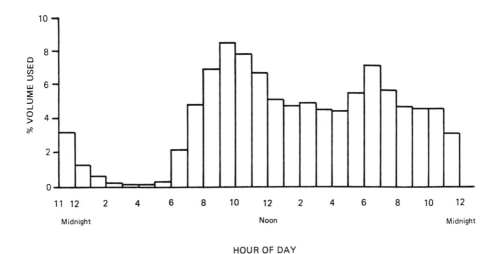

Figure 1.1. Diurnal variation in household water consumption. Reproduced from Bailey *et al.* (1986) by permission of the Water Research Centre

A bath uses on average 80 l a time compared with a shower which only uses 5 l/min. Garden sprinklers use about a 1000 l/h, which is the average daily water usage for seven or eight people. Clearly not all the public supplies are utilized for domestic purposes. This is illustrated by the analysis of daily water usage in 1984–5 by the former Severn Trent Water Authority. Of the total 1942 Ml supplied each day, 840 Ml was used for domestic purposes, 530 Ml for industrial, 50 Ml for agricultural purposes and a remarkable 522 Ml (26.9%) was lost every day through a leaky distribution system (Table 1.4). Losses from leaks are a widespread problem as water mains not only deteriorate with age, but are often damaged by heavy vehicles, building work or

Table 1.3. Estimated cost of water usage in the home (1991–2) in England and Wales. Reproduced by permission of the Water Services Association

Purpose of water use	Ownership (%)	Cost unit	Amount of water use (l)	Cost (pence)
Cooking, drinking, washing up and personal hygiene	—	Daily per person	27	1.3
Bath	95	One	90	4.3
Shower	25	One	20	1.0
Toilet	99	One	9	0.4
Automatic washing machine	85	Per cycle	100	5.2
Non-automatic washing machine	—	Per wash	40	1.9
Dishwashing machine	10	Per wash cycle	50	2.4
Hose pipe/sprinkler	—	Per minute	18	0.9

Table 1.4. Estimated daily use of water supplied by the former Severn Trent Water Authority during 1984–5. Reproduced from Archibald (1986) by permission of Severn Trent Water

Type of use	Amount used (Ml/d)	Purpose	Amount used (Ml/d)
Domestic	840	Basic	288
		Toilet flushing	242
		Bathing	155
		Washing machine	114
		External use	27
		Luxury appliances	14
Industrial and commercial	530	Processing	256
		Domestic	153
		Cooling: direct	77
		Cooling: recycled	44
Agricultural	50	Livestock	35
		Domestic	10
		Protected crops	3
		Outdoor irrigation	2
Unaccounted for	522	Distribution system	287
		Consumers' service pipes	167
		Trunk mains	52
		Service reservoirs	16
Total	1942		

subsidence. Leakage control is a vital method of conserving water. Detecting and repairing leaks is both labour intensive and time consuming, which means that it is very expensive. However, if leaks are not controlled then water demand will escalate, with most of the extra demand seeping away into the ground instead of making its way to the consumer.

Table 1.5. Comparison of average water consumption per person in three water company areas with respect to household size and socioeconomic group. Reproduced by permission of the National Water Council (1982)

	Average water consumption (l/d)		
	South West	Severn Trent	Thames
Household size			
1	126	116	136
2	124	118	151
3	118	109	123
4	110	92	116
5	103	96	103
6	97	75	97
7+	92	69	64
Social group			
A	138	126	134
B	124	117	126
C1	122	100	124
C2	111	93	113
D	103	93	120
E	86	76	141

Studies of water usage suggest that the figures for the average consumption rates currently used by the water industry may be a little high. A more reliable guide is given in Table 1.5. The National Water Council (1982) and the Water Research Centre (WRc) (Bailey *et al.*, 1986) have carried out studies on the pattern of domestic water usage. Both studies reported that the per capita consumption decreased slightly with an increase in household size, and that social grouping also appeared to have an influence, with Social Group A using about 50 l/d more than Social Group E. The WRc study was conducted throughout the UK and found that the daily volume of water consumed per household for non-potable purposes is dependent on the size of the household. So on a national basis about 3% of the total volume of domestic water consumed each day is used for potable purposes, which is equivalent to about 10 l for the average household. Interestingly, 25% of the first draws of water taken from the system each day are for potable purposes, a habit that may have significant consequences in those areas where the water is corrosive and lead or galvanized plumbing is used (Section 7.2).

1.1.3 Droughts

Since the Second World War there have been a number of major droughts in the UK, in 1949, 1959, 1975–6, 1984 and 1989–90. There is no clear lead up to a drought. Sometimes they can be caused by a dry winter which fails to fill reservoirs and replenish groundwaters, as happened in the 1975–6 drought, and in fact the winter of 1988 was also exceptionally dry. Alternatively they may be due to a very dry spring and early

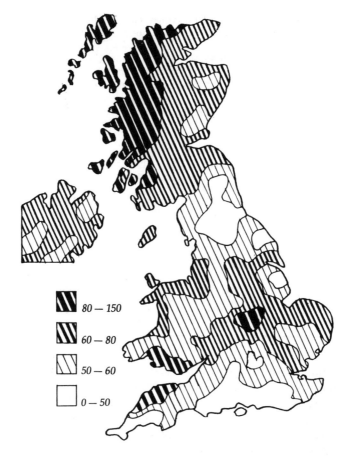

Figure 1.2. Percentage of mean annual rainfall that fell in the UK during the drought period May to September 1989. Reproduced by permission of the Water Services Association

summer, as was the case in the 1984 drought which affected the west of the country in particular. In the case of the 1959 drought, it was a long dry summer with the autumn and winter rains failing to come as expected.

The year of 1989 was the hottest and driest period experienced in Britain for 13 years. It was preceded by a very dry November in 1988. The 1988–9 winter—that is, December 1988, January and February 1989—was incredibly dry up to the first weeks of February, certainly drier than anything previously recorded since the official records began in 1727. However, in the last weeks of February it poured with rain, resulting in flooding. May was extremely sunny and so the remarkable weather of 1989, which ran from May to August, began. Figure 1.2 shows that the south of England and also the north-east of the country received less than 60% of their normal rainfall during May to September 1989. Even a 20% reduction in rainfall can result in a much higher reduction in the volume of water in a river, perhaps by as much as 50%, and if this is combined with a dry winter, then aquifers will not be recharged, resulting in the water-table falling (Sections 2.2 and 2.3). So by 31 December 1989, the UK had probably had the finest weather over a 12 month period this century, and England and Wales had received about three-quarters of its normal rainfall. This low rainfall scenario is expected to occur once every 40 years, and has a serious effect on water resources.

The situation between May and September 1989 was exceptional and should not recur in England and Wales, according to the experts, for another 300 years, whereas the very dry spring of 1990 was a once in a 100 years phenomenon. Water availability can vary over fairly short distances. For example, in 1990 the reservoirs were full in the north whereas in the south river beds were dry. This is in contrast with the 1984 drought when the Lake District and north-west were very dry and the south-east had sufficient water. In that year 20 million people were subject to water restrictions.

Is our weather changing? Some are predicting a mini-ice age, which occurs every 180 years or so. Certainly there have been two exceptionally cold winters in 1962–3 and 1981–2, but the general trend appears to be milder, drier winters. Most of the global climate models used to predict long-term changes suggest that over the next 30–50 years there will be a 1.5–3.0°C increase in temperature, with temperatures increasing relatively more at the poles than at the equator. Rainfall patterns will change with some areas getting more, others less. Any change in climate will alter the way we live, the crops we grow and eat, in fact our very social make-up. Researchers working on consequences of the greenhouse effect on precipitation in Ireland are gloomy about the future; they estimate that summers will become significantly warmer and that more vigorous storm activity will ensue. They are recommending civil engineers not to plan long-term structures using old climate averages for either precipitation or wind. Overall they expect a 15% decrease in precipitation in the summer, with greater soil moisture deficits and longer spells of drought in the early summer becoming common. This will significantly affect both our surface water flows and our groundwater recharge characteristics (Section 2.3) and is going to mean a major redesign of current water resource management strategies. Recent estimates have suggested that these longer warmer summers could reduce the yields from reservoirs by up to 15% due to increased evaporation alone.

Clearly it costs a great deal to supply adequate volumes of water all the year around without imposing restrictions. Meters allow those who want to use it to pay for it, and many believe that special tariffs such as off-peak use, seasonal tariffs and water shortage penalties for exceeding minimum volumes should be used to control water usage rather than imposing restrictions. Reducing leaks could significantly increase the volumes of water available. Water is going to cost more due to the significant increase in quality which we are currently seeing, which has involved a massive capital investment and an increase in the operational cost of treatment. To overcome the shortages experienced during droughts, which should in theory only occur every 40–300 years, then water is going to cost even more to pay for new reservoir construction and exploitation of other resources (Section 2.5). Conservation of supplies, including metering, will be an important factor in keeping these costs down (Section 2.4). The problems of water resources are explored in Chapter 2.

1.2 THE UK WATER INDUSTRY

1.2.1 Water Undertakers

The Water Act of 1973 established 10 authorities to be responsible for water services throughout England and Wales. These regional water authorities came into being on 1 April 1974, taking over from a host of local authorities and river boards the whole management of water resources. It was an innovative idea to have those responsible for river pollution, land drainage, fisheries, water supply, sewerage and sewage treatment, water conservation, recreation and amenity, and nature conservation all under the same organizational umbrella, and often under the same roof. This allowed water resources to be managed for the first time in a holistic manner, rather than in the previous fragmented way. It also meant that money could be spent in priority areas, that large schemes could be financed and major problems tackled.

On 5 February 1986 the UK Government announced its intention to privatize the regional water authorities. This only affected England and Wales, with the situation in Scotland and Northern Ireland remaining unaltered. Seventeen months later in July 1987, plans for a public body to regulate the privatized water industry were announced, the NRA. The draft Water Bill was presented to the UK Parliament on 11 November 1988 and enacted on 6 July 1989. This transferred the functions of the 10 regional water authorities relating to water supply and the provision of sewage treatment to 10 new water service companies. The transfer actually took place on 1 September 1989, so overnight the responsibility for drinking water became that of the new water service companies. The existing private statutory water companies retained their functions, although there were minor boundary changes. With this transfer went most of the property, rights and, of course, liabilities of the old water authorities. However, the entire infrastructure, the reservoirs, treatment plants, distribution mains, sewers and waste water treatment plants, was transferred to the new companies overnight. Certain other functions of the regional water authorities were not transferred to the new water service companies, such as pollution control monitoring, water resource management, fisheries, flood protection and alleviation and land drainage. These functions were transferred along with related property, rights and any liabilities to the NRA.

The new water service companies cover approximately the same geographical areas as the old regional water authorities, and they have also retained their name, except that the term 'authority' has been dropped, so the Southern Water Authority is now known simply as Southern Water (Table 1.6).

The water service companies are the principal operating subsidiaries of 10 holding (parent) companies, although each holding company has other subsidiaries which it runs as separate companies. It is, however, the water service companies that are the core businesses, which hold the statutory responsibilities for water and sewerage services and are licensed to provide those services within their geographical boundaries.

The Water Act 1989 sets out regulatory provisions to water and sewerage undertakers (the water service companies and the statutory water companies). These provisions do not, on the whole, apply to the business activities of the holding

Table 1.6. Water undertakers by region. Each region has sewerage and waste water treatment facilities provided by a single water service company, although there may be several water supply companies

Anglia Region
  Anglian Water Services Ltd
  Cambridge Water Company
  East Anglian Water Company
  Essex Water Company
  Tendring Hundred Water Company

Northumbria Region
  Northumbrian Water Ltd
  Hartlepool's Water Company
  Newcastle and Gateshead Water Company
  Sunderland and South Shields Water
    Company

North West Region
  North West Water Ltd

Severn Trent Region
  Severn Trent Water Ltd
  East Worcestershire Waterworks Company
  South Staffordshire Waterworks Company

Southern Region
  Southern Water Services Ltd
  Eastbourne Water Company
  Folkestone and District Water Company
  Mid-Kent Water Company
  Mid-Sussex Water Company
  Portsmouth Water plc
  West Kent Water Company

South West Region
  South West Water Services Ltd

Thames Region
  Thames Water Utilities Ltd
  Colne Valley Water plc
  East Surrey Water plc
  Lee Valley Water plc
  Mid-Southern Water Company
  North Surrey Water Company
  Rickmansworth Water plc
  Sutton District Water plc

Wales Region
  Dŵr Cymru (Welsh Water)
  Chester Waterworks Company
  Wrexham and East Denbighshire Water
  Company

Wessex Region
  Wessex Water Services Ltd
  Bournemouth and District Water Company
  Bristol Waterworks Company
  West Hampshire Water Company

Yorkshire Region
  Yorkshire Water Services Ltd
  York Waterworks plc

companies not directly related with water and sewerage functions. The 29 statutory water companies are also subject to the same regulations as the new water service companies (Figures 1.3 and 1.4).

The situation elsewhere in the British Isles is rather different. In Scotland the functions of the regional water boards established under the Water (Scotland) Act 1967 were transferred in 1975 to 12 Regional or Island Councils. The seven River Purification Boards in Scotland have many of the functions of the NRA in England and Wales. They cover mainland Scotland, whereas the three island councils of Orkney, Shetland and the Western Isles act as the River Purification Board in their areas. The situation is to change as the Secretary of State for Scotland is to announce the setting up of a Scottish Environmental Protection Agency. The new agency will take over the functions of the River Purification Boards, as well as the responsibility for controlling all forms of pollution. In July 1993 the Secretary of State announced that responsibility for water supply in Scotland would be taken over by three new authorities. The structure has yet to be announced, although there will be an element of privatization involved. In Northern Ireland the responsibility for all water and sewerage services lies

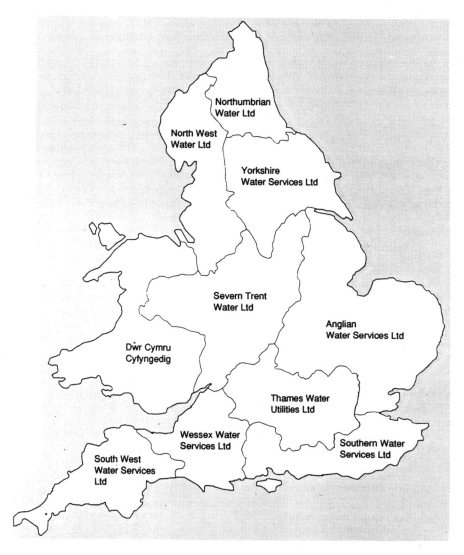

Figure 1.3. Areas covered by water service companies. These companies supply sewerage facilities for all of England and Wales, although in some areas water is supplied by the water-only companies (Figure 1.4). Reproduced by permission of the Water Services Association

with the Department of the Environment (NI). Known as Water Services, they are based at four regional centres throughout the province, with their headquarters in Belfast. There is also an Environmental Protection Division within the Department of the Environment (NI). In Ireland the local authorities, a mixture of county councils and urban district councils, are responsible for water services.

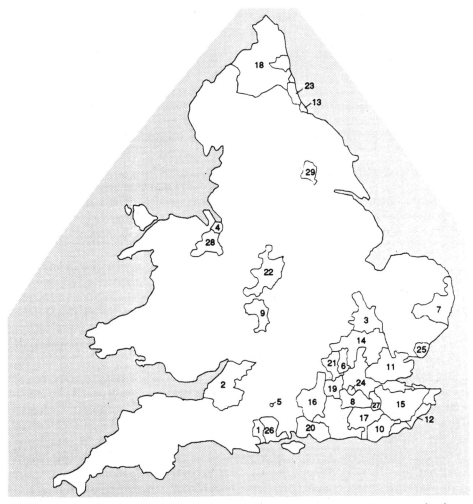

Figure 1.4. Areas covered by the 29 former statutory water companies. The companies do not provide sewerage facilities, these being provided by the water service companies. There has been a number of recent mergers of companies, reducing the number of water-only companies to 22. 1 = Bournemouth and District Water Co. (now Bournemouth and West Hampshire plc); 2 = Bristol Waterworks Co.; 3 = Cambridge Water Co.; 4 = Chester Waterworks Co.; 5 = Cholderton and District Water Ltd; 6 = Colne Valley Water Co. (now merged as Three Valleys Water Services plc); 7 = Suffolk Water plc; 8 = East Surrey Water plc; 9 = East Worcestershire Waterworks Co.; 10 = Eastbourne Water Co.; 11 = Essex Water Co.; 12 = Folkestone and District Water Co.; 13 = Hartlepools Water Co.; 14 = Lee Valley Water Co. (now merged as Three Valleys Water Services plc); 15 = Mid-Kent Water plc; 16 = Mid-Southern Water Co.; 17 = Mid-Sussex Water Co. (now South East Water plc); 18 = Newcastle and Gateshead Water plc (now North East Water plc); 19 = North Surrey Water Co.; 20 = Portsmouth Water plc; 21 = Ricksmansworth Water Co. (now merged as Three Valleys Water Services plc); 22 = South Staffordshire Waterworks Co.; 23 = Sunderland and South Shields Water plc (now North East Water plc); 24 = Sutton District Water plc; 25 = Tendring Hundred Waterworks Co.; 26 = West Hampshire Water Co. (now Bournemouth and West Hampshire Water plc); 27 = West Kent Water Co. (now South East Water plc); 28 = Wrexham and East Denbighshire Water Co.; and 29 = York Waterworks plc. Reproduced by permission of the Water Services Association

1.2.2 Regulation of the Water Industry

There are two forms of regulation on water undertakers. Economic regulation is very different in England and Wales to the rest of the UK or Europe. Environmental and quality regulation is imposed on all water undertakers by the implementation of EC Directives, although the rate of implementation and the exact nature of their implementation may differ slightly.

For England and Wales, the most important piece of legislation in recent years in the water industry is the Water Act 1989. This re-enacted legislation that previously applied to the regional water authorities. The act enabled privatization of the water industry, but at the same time redefined, extended and often made more stringent the regulatory arrangements which had been in force before 1989. It also provided a more comprehensive framework of enforcement. The Water Act 1989 makes the Secretary of State, the Director General of Water Services and the NRA the principal regulators of the industry (Macrory, 1989).

The Office of Water Services (Ofwat) is the Government's statutory watchdog. It is to this office that consumers should ultimately take their complaints and problems relating to pricing and standards of service when they have failed to obtain satisfaction from the water company itself. The cost of this is borne by the licence fee paid by the water service companies and statutory water companies (known as appointees) in England and Wales. It came into operation on the same day as the new water companies took over, 1 September 1989.

The Director General of Ofwat, in consultation with the Secretary of State, has the primary duty to (1) ensure that water and sewerage functions are properly carried out in England and Wales and (2) ensure that the water undertakers are able to finance the proper operation of these functions by securing reasonable returns on their capital. Subject to these primary duties the Director General must also endeavour to: (1) protect the interests of customers and potential customers in respect of charges (having particular regard to the interests of customers in rural areas and to ensuring that in fixing charges there is no undue preference towards, or undue discrimination against customers or potential customers); (2) protect the interests of customers and potential customers in respect of other terms of supply, the quality of services (taking into account in particular those who are disabled or of pensionable age) and the benefits that could be secured from the proceeds of the disposal of certain land; (3) promote economy and efficiency on the part of undertakers in the carrying out of the water and sewerage functions; and (4) facilitate effective competition between persons holding or seeking appointments under the Water Act 1989 as water or sewerage undertakers.

To act as a buffer between consumers with complaints and Ofwat itself, the Director General of Water Services has established and is responsible for the maintenance of 10 locally based Consumer Service Committees to represent the interests of consumers. The Consumer Service Committees are independent of the water companies and their main duty is to represent consumers' views on the operations of the companies and to investigate consumers' complaints.

Since 1 September 1989 all the water undertakers in England and Wales have

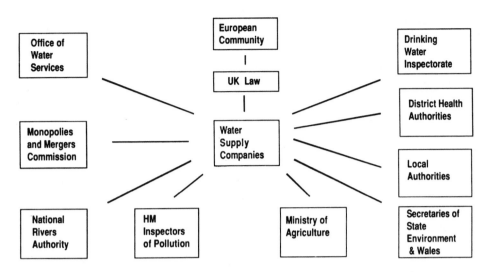

Figure 1.5. Regulation of the water supply and service companies in England and Wales. Reproduced by permission of the Water Services Association

operated under the terms of an individual appointment, which is in essence a kind of operator's licence. The terms of the appointment set out the maximum charge increase (*K*) for all regulated services, mainly water supply and sewerage charges. Also defined are the circumstances under which unforeseen or previously unquantifiable obligations placed on the companies may be eligible for costs to be passed onto customers, a process known as *cost pass through*. The Water Act 1989 enables the Director General to monitor standards of customer service, which includes the assessment of the state of underground assets such as mains and sewers to ensure that these are protected against progressive deterioration. One of the major financial problems facing the water service companies is the rehabilitation of thousands of kilometres of old and leaking water mains and sewers.

The Water Act 1989 requires each water undertaker to develop and maintain an efficient and economic system of water supply within its area. It must also ensure that all such arrangements have been made for providing supplies of water to premises in its area, making such supplies available to people who demand them and for maintaining, imposing and extending its mains and other pipes, as necessary to enable it to meet its water supply obligation

Environmental quality regulation in England and Wales is carried out by a number of government bodies (Figure 1.5). The local authorities have retained their public health responsibilities as to the wholesomeness of drinking water, and have special powers to deal with private supplies. The regulations referring to water abstractions, impoundments and discharges of waste waters is the responsibility of the NRA. The NRA is split into 10 regions based on the former water authority areas and corresponding more or less to the new areas covered by the water service companies. Their main regulatory functions cover (1) water resources (mostly abstractions, for which they issue licences as well as independently monitoring river quality), (2) pollution control (the NRA issues discharge licences, also known as consents, to both industries and sewage treatment works operated by the water service companies), (3) fisheries, (4) land drainage and flood protection, and finally (5) conservation, amenity, recreation and navigation. The discharge of substances dangerous to the aquatic environment, the so-called red list substances, which are usually by-products from industrial processes, are controlled by Her Majesty's Inspectorate of Pollution. The Ministry of Agriculture also has regulatory functions, but it is the activities of the Drinking Water Inspectorate (DWI) that are of most importance with respect to our drinking water.

Under Section 60 of the Water Act 1989 the Secretary of State for the Environment and the Secretary of State for Wales are empowered to appoint technical assessors to check that the water undertakers are conforming to their legal obligations. The DWI was formed at the beginning of January 1990 and has nine main tasks

1. *To carry out technical audits of water companies.* This is a system used by the DWI to check that water supply companies are complying with their statutory obligations and whether they are following good practice. There are three elements to this technical audit. First an annual assessment based on information provided by the companies of the quality of water in each supply zone, water treatment works and service reservoirs, compliance with sampling and other requirements and the progress made on improvement programmes. The second element is the inspection of individual companies covering all the above points at the time of the inspection, but also an assessment of the quality and accuracy of the information collected by the company. The third and final element are interim checks, which are made based on the information provided by the companies.

2. *To instigate action as necessary to secure or facilitate compliance with legal requirements.*

3. *To investigate incidents which adversely affect water quality.*

4. *To advise the Secretary of State* in the prosecution of water companies who have supplied water found to be unfit for human consumption.

5. *To provide technical and scientific advice to Ministers and Officials of the Department of the Environment and Welsh Office on drinking water issues.*

6. *To assess and respond to consumer complaints when local procedures have been exhausted.* Almost all of these problems are resolved by referring the complaint back to the water supply company concerned, requesting them to investigate the matter and report back to the DWI. The DWI may also ask the local environmental health officer to investigate the matter and to report back. The DWI also works closely with Ofwat and will investigate and liaise with them on complaints. Ofwat receives complaints from Customer Services Committees and these may be passed onto the DWI.

7. *To identify and assess new issues or hazards relating to drinking water quality and initiate research as required.*

8. *To assess chemicals and materials used in connection with water supplies.* The DWI operates a statutory scheme which assesses and approves (if appropriate) the use of

chemicals in treating drinking water. This scheme also covers the construction materials used to build water treatment plants and distribution systems. This scheme is to protect the public by ensuring that all chemicals added to water are safe, and that the chemicals which leach from construction materials are also safe. It is also important that such chemicals do not encourage microbial growth in distribution systems which would affect the taste or odour of the water.

9. *To provide authoritative guidance on analytical methods used in the monitoring of drinking water.*

All the water supply companies are inspected annually, although it appears that the level of inspection will vary from year to year. For example, in 1990, four of the larger companies were inspected using a small team of inspectors, 24 companies were inspected by teams of consultants acting on behalf of the DWI, whereas the remaining 11 companies were visited by inspectors when only a limited range of checks were made. In its first year of operation, the DWI found that an inspection of a single large water supply company would take on average 50 person-days, including all preparatory work and report writing. In the report that follows the inspection, areas of compliance and non-compliance are identified and recommendations made to ensure full compliance with statutory requirements, which includes prosecution where necessary. The DWI checks the sampling procedures and the location of sampling points to ensure that they are representative of water quality within each water supply zone. About 90 different laboratories analyse water supplies for the water companies. During inspection the procedures used and the training and competence of laboratory staff are examined. The actual results from sampling are also scrutinized, as is the data handling system used to analyse them, this being done to ensure that the integrity of the data is maintained. The DWI concentrates on those results which have not complied with the prescribed concentration value (PCV) or revised PCV limits, and ensures that the correct follow up action has been taken. Every day procedures are checked and all the data recorded on public registers. In England and Wales there are about 1100 programmes of work under way (in 1992) to improve drinking water quality. The DWI assesses the progress of each scheme to check that completion dates will be met.

The Water Act 1989 was almost entirely replaced by a number of Consolidation Acts. These came into force on 1 December 1991 and include the Water Industry Act 1991, the Water Resources Act 1991, the Statutory Water Companies Act 1991, the Land Drainage Act 1991 and the Water Consolidation (Consequential Provisions) Act 1991. The major provisions relating to drinking water quality are contained within the Water Industry Act 1991. The provisions of earlier Acts, especially the Water Act 1989, are now covered in the new Act. The new Act is examined at length in the Service Pipe Manual (Foundation for Water Research, 1993). Under Section 18 of the new Act the Secretary of State is required to take enforcement action to secure compliance by companies. This will include contraventions of the wholesomeness, monitoring or treatment requirements of the regulations. Also included are contraventions in respect of records and the provision of information. Enforcement action is taken on behalf of the Secretary of State for Wales or England by the DWI. There are three circumstances when action can be taken by the DWI, although it is their practice not to initiate action when the water supply company, after being informed of non-compliance, takes immediate remedial action to enforce full compliance in the future. Enforcement can follow when (1) a water quality standard set by the Water Supply (Water Quality) Regulations is breached and the breach is not considered trivial and is likely to occur again; (2) a breach of one of the other enforceable regulations such as those covering sampling, analysis or water treatment is identified; or (3) when existing Section 20 undertakings or time-limited relaxations expire before the required improvements have been completed (Department of the Environment, 1992).

1.2.3 Charges

One of the Director General of Water Services' most important tasks is reviewing increases in charges made by the water undertakers. Under a complex pricing formulae the 39 water companies in England and Wales can impose price increases in line with the current rate of inflation plus an individual sum known as the K factor. This extra charge K takes account of the need to finance the major improvements programme and the amount is decided by the Director General's office each year. The Director General uses the Producer Prices Construction Output Index when considering cost pass through. For capital expenditure repricing, an index specifically applied to the water industry, the public works non-roads index (PWNR) is used. The representative weighting in the retail price index (RPI) for gas, telecoms and water are 2.1, 1.6 and 0.7% respectively. The basic regulation of charge increases is by the formulae $RPI + K$, so most charges rise each year by K percentage points more than the annual rise of inflation. Those charges covered by this formula are the basket items, which are unmeasured water supply and sewerage services, measured (metered) water supply and sewerage services, and also trade effluent, which covers most water company charges. The value of K was fixed initially for each water service company, but from 1991 it can be varied to allow for cost pass through. This is where certain costs will qualify for being passed onto the customer. The value of K was set for the first time in January 1991 when Ofwat calculated the RPI as 9.7% and allowed varying K values for each water undertaker. Overall in 1991 water charges rose on average by 15%, taking the standard water charge fee from £133 to £154, representing a span of increases of 9.7% by the Suffolk Water Company to 26.9% for the West Kent and the Eastbourne Water Companies. So in some areas combined water and sewerage charges exceeded £200 for the first time. In the Anglian Water area the rise of 15.2% meant that the average charge was £204.

From 1985–6 to 1990–1, the average household water bill increased from £86 to £134, an increase of 56% (Table 1.7). According to Ofwat further increases are inevitable if the 10 water and sewerage and 29 water-only companies are to finance the planned improvement schemes. These include replacing leaky water mains, the introduction of new sewage treatment plants and bringing tap water up to EC standards. It is estimated that this will cost £28 billion.

The charge for drinking water ranges from £61.29 (Severn Trent Water) to £104.28 (Dŵr Cymru) for 1991–2 for unmeasured supplies (Table 1.8). Metered water costs

Table 1.7. Average household water charges for the 10 water service companies. Reproduced by permission of the Water Services Association

Year	Cost of water supply (£)	Cost of sewerage and environmental services (£)	Combined water service bill (£)
1985–6	39.92	45.93	85.85
1986–7	43.20	49.64	92.84
1987–8	45.82	52.99	98.81
1988–9	49.58	57.57	102.15
1989–90	55.12	63.65	118.77
1990–1	63.52	70.30	133.82
1991–2	73.51	81.00	154.51

between 37.33 pence/m^3 (Thames Water) to 54.30 pence/m^3 (Dŵr Cymru). Of course, water service bills also contain sewerage and environmental service charges so the average household bill is £154.51 for 1991–2, being lowest in the Thames Water area at £130.60 and highest in Wales at £204.44 (Table 1.8). To a large extent the cost of water reflects the dispersed nature of the population.

Metered costs have also risen constantly (Table 1.9), but have not kept up with the rate of inflation, although since privatization prices have soared.

The average charges given here are calculated from the 10 water service companies only; they do not include the statutory water company charges which tend to be lower due to lower operating costs. For example, in 1990–1 the average statutory water company charge was £57.23 compared with £63.52 for the water service companies. Metered water charges can be complicated. For example, the tariffs set by Southern Water on the Isle of Wight for the financial year 1988–9 included a £10 per year service or standing charge for reading the meter and billing twice a year. The actual charge for water was 28 pence/m^3 for the first 90 m^3 used each year with the excess usage charged at 53 pence/m^3. The 90 m^3 allowed at the lower rate is sufficient water for all basic needs for one year, so in an area where water is in short supply the higher tariff is used as an incentive to conserve water. Water meters are often also used to calculate sewerage charges, with a nominal figure of 95% taken as the proportion of water used actually discharged to the public sewer (Section 2.4).

In England and Wales in 1991–2 the average household bill for water and sewerage services was £2.95 per week or £155 per year. Charges for water services were on average £73.51 per year or 20p per day. This works out at 4.3 pence for a bath (that uses 90 l), or 1.0 pence for a shower (20 l), and a mere 0.4 pence to flush the lavatory (9 l) (Table 1.5). So in 1992 £1.00 bought 2120 l of tap water or just over 1 l of bottled water.

Table 1.8. Average household bills in 1990–1 and 1991–2 by water service company. Reproduced by permission of the Water Services Association

| | Average household bill (£) | | | | | | Cost of metered supply (pence/m³) | |
| | Water supply | | Sewerage | | Total | | | |
	1990–1	1991–2	1990–1	1991–2	1990–1	1991–2	1990–1	1991–2
Anglian	78.34	93.79	96.67	110.65	175.01	204.44	40.91	48.46
Northumbrian	56.59	66.05	65.50	75.98	122.09	142.03	33.21	38.74
North West	58.00	66.62	66.33	75.95	124.33	142.57	40.00	46.40
Severn Trent	53.39	61.29	66.78	76.50	120.12	137.79	41.80	48.15
Southern	57.69	66.38	82.33	96.17	140.12	162.55	35.60	40.70
South West	71.17	82.56	93.84	109.32	165.01	191.88	44.40	51.10
Thames	58.17	66.31	56.15	64.29	114.32	130.60	31.48	37.33
Dŵr Cymru	89.71	104.28	78.25	90.93	167.96	195.21	45.62	54.30
Wessex	68.65	80.09	85.19	96.07	153.84	176.16	44.90	53.81
Yorkshire	70.07	80.50	65.15	73.99	135.22	154.49	41.20	47.40
Weighted average	63.52	73.51	70.30	81.00	133.82	154.51	40.44	47.20

Table 1.9. National average metered cost for water compared with the retail price index (RPI) for October each year where January 1974 = 100. The year 1974 is taken from which to calculate the RPI value as this is when the water authorities were formed. Reproduced by permission of the Water Services Association

Year	Charge (pence/m^3)	RPI
1978–9	14.0	210
1979–80	14.9	236
1980–1	18.1	272
1981–2	20.1	304
1982–3	22.1	324
1983–4	22.8	341
1984–5	23.5	358
1985–6	25.8	377
1986–7	27.6	388
1987–8	29.3	406
1988–9	31.9	432
1989–90	35.2	464
1990–1	40.4	—
1991–2	47.2	—

1.3 DRINKING WATER LEGISLATION

1.3.1 EC Regulations

The EC Directive 'relating to the quality of water intended for human consumption' was approved by the Council of Ministers on 15 July 1980 (EC, 1980). The Directive applies to all water supplied for consumption by humans and that used in the production of food or in processing or marketing products intended for human consumption. Natural mineral waters and medicinal waters do not fall under the provisions of this Directive (Section 8.1). Directives have to be enacted in Member States by the introduction of national legislation, usually in the form of Statutory Instruments. The Drinking Water Directive, as it is more widely known, was enacted in Ireland in 1988 by the publication of the European Communities (Quality of Water Intended for Human Consumption) Regulations 1988. In England and Wales the Directive was finally implemented by the enactment of the Water Supply (Water Quality) Regulations 1989 and as amended by the Water Supply (Water Quality)(Amendment) Regulations 1989 and 1991. In Scotland the Directive was implemented through amendments of the Water (Scotland) Act 1980 and the Water Supply (Water Quality) (Scotland) Regulations 1990. The Scottish Regulations are essentially identical to those set for England and Wales. Neither the Water Act 1989 or the Water

Supply (Water Quality) Regulations apply to Northern Ireland. However, as it is the policy of the Department of the Environment (NI) to follow the standard practice used in the UK, this legislation will have a similar, albeit indirect, impact in Northern Ireland.

The Regulations covering the UK encompassed both the EC Drinking Water Directive (80/778/EEC) and the two earlier Directives relating to the abstraction of raw water for human consumption (75/440/EEC and 79/869/EEC). The new Regulations have a number of important elements, these are: (1) a formal definition of the term 'wholesomeness of water' combined with mandatory sampling frequencies and public reporting of compliance results; (2) facilities for non-compliance with required parameters in the form of relaxations, provided public health is not compromised and an approved programme of remedial works to achieve compliance is implemented at an early date; (3) a 'catch all' clause which requires (i) 'that the water does not contain any element, organism or substance (other than a parameter) at a concentration or value which would be detrimental to public health,' and (ii) 'that the water does not contain any element, organism or substance (whether or not a parameter) at a concentration or value which in conjunction with any other element, organism or substance it contains (whether or not a parameter) would be detrimental to public health'; and (4) requirements for the maintenance of registers and publication of information on compliance in prescribed forms including an annual report on water quality. So, in essence, it has became an offence for water supply companies to supply water which is unfit for human consumption.

1.3.2   Standards

The Directive itself lists 66 drinking water parameters and categorizes them into six groups: (A) organoleptic (four); (B) physicochemical (15); (C) substances undesirable in excessive amounts (24); (D) toxic substances (13); (E) microbiological (six); and (F) minimum concentration for softened water (four).

The EC normally sets two standards for parameters in the form of concentrations not to be exceeded. The G value is the guide value, which the Commission desires Member States to work towards in the long term. The I value is the mandatory value in Directives, and is the minimum standard that can be adopted. In the Drinking Water Directive, the I value is called the maximum admissible concentration or MAC. National standards must conform to the MAC values, although it is permissible for Member States to set stricter standards, which does happen occasionally. The Directive sets out G or MAC values for 60 compounds, with MAC values only set for 44 of these parameters, although guidance is given for the others. Therefore there is a considerable amount of leeway in setting standards for certain of the parameters listed. Each parameter listed has a number. This refers to the paragraph in the Directive which relates to this parameter. The G value and the MAC given in the Directive are given in Table 1.10, along with the PCV given in the Water Supply (Water Quality) Regulations 1989. The PCV is the national standard for England, Scotland and Wales. In the Republic of Ireland it is called the national limit value (NV), although the term MAC is generally used. In Denmark, local and district councils are responsible for water supply

Table 1.10. Comparison of the water quality standards set for the 66 parameters in the Drinking Water Directive by the EC [guide (G) and maximum admissable concentration (MAC)]. England and Wales [prescribed concentration value (PCV)]. Ireland [national limit value (NV)] and Denmark [guideline (GL) and MAC]

Parameter	Unit of measurement	EC G	EC MAC	UK PCV	IRL NV	Denmark GL	Denmark MAC
A. Organoleptic parameters							
1 Colour	mg/l Pt-Co	1	20	20	20	5	15
2 Turbidity[a]	Formazin turbidity units	0.4	4	4	4	0.3	0.5
3 Odour[b]	Dilution number at 25°C	—	3	3	3	—	3
4 Taste	Dilution number at 25°C	—	3	3	3	—	3
B. Physicochemical parameters[c]							
5 Temperature	°C	12	25	25	25	—	12
6 Hydrogen ion concentration	pH unit: Min.	6.5	[d]	5.5	6.0	7	—
	Max.	8.5		9.5	9.0	8	8.5
7 Conductivity	uS/cm at 20°C	400	[e]	—	1500	>300	—
8 Chloride	mg Cl/l	25	—	250	250	50	300
9 Sulphate	mg SO_4/l	25	250	250	250	50	250
10 Silica	mg SiO_2/l	—	—	—	[f]	—	—
11 Calcium	mg Ca/l	100	—	—	200	—	—
12 Magnesium	mg Mg/l	30	50	50	50	30	50
13 Sodium[k]	mg Na/l	20	150[g]	150[g]	150[g]	20	175[M]
14 Potassium	mg K/l	10	12	12	12	—	10
15 Aluminium	mg Al/l	0.05	0.2	0.2[h]	0.2	0.05	0.2
16 Total hardness[i]							
17 Dry residuals[j]	mg/l[j]	1500	1500	1000	—	1500	—
18 Dissolved oxygen		—	[k]	[k]	[k]	[N]	—
19 Free carbon dioxide			[l]	[l]	[l]	[P]	—
C. Parameters concerning substances undesirable in excessive amounts							
20 Nitrates	mg NO_3/l	25	50	50	50	—	50
21 Nitrites	mg NO_2/l	—	0.1	0.1	0.1	[P]	0.1
22 Ammonium[m]	mg NH_4/l	0.05	0.5	0.5	0.3	0.05	0.5
23 Kjeldahl nitrogen[n]	mg N/l	—	1	1	1	—	1

continued overleaf

Table 1.10. (Continued)

Parameter	Unit of measurement	EC		UK PCV	IRL NV	Denmark	
		G	MAC			GL	MAC
24 Oxidizability (permanganate value)	mg O_2/l	2	5	5	5	1.5	3
25 Total organic carbon (TOC)[o]	mg C/l	—	—	p	—	—	—
26 Hydrogen sulphide	μg S/l	—	UO	—	UO[q]	—	UO[q]
27 Substances extractable in chloroform	mg/l dry residual	0.1	—	—	NSI[r]	0.1	—
28 Dissolved or emulsified hydrocarbons (after extraction by petroleum ether); mineral oils	μg/l	—	10	10	10	P	10
29 Phenols[s]	μg C_6H_5OH/l	—	0.5	0.5	0.5	P	0.5
30 Boron	μg B/l	1000	—	—	2000	—	1000
31 Surfactants	μg/l as lauryl sulphate	—	200	200	200	P	100
32 Other organochlorine compounds not covered by parameter 55[t]	μg/l	1	—	—	100	1	—
33 Iron	μg Fe/l	50	200	200	200	50	200
34 Manganese	μg Mn/l	20	50	50	50	20	50
35 Copper	μg Cu/l	100[u] 3000[v]	—	3000	500[u] 3000[v]	—	100
36 Zinc	μg Zn/l	100[u] 5000[v]	—	5000	1000[u] 5000[v]	—	3000
37 Phosphorus	μg P_2O_5/l	400	5000	2200	5000	P	100
38 Fluoride	μg F/l	—	1500	1500	1000[w]	—	3000
39 Cobalt	μg Co/l	—	—	—	—	—	687
40 Suspended solids	mg/l	0	—	x	NPV[y]	P	1500
41 Residual chlorine	μg Cl/l	—	—	—	z	100	—
42 Barium	μg Ba/l	100	10	10	10	P	z
43 Silver	μg Ag/l	—	80[A]	80[A]	80[A]	P	10

D. Parameters concerning toxic substances

No.	Parameter	Unit						
44	Arsenic	μg As/l	—	50	50	50	P	50
45	Beryllium	μg Be/l	—	—	—	—	—	—
46	Cadmium	μg Cd/l	—	5	5	5	P	5
47	Cyanides	μg CN/l	—	50	50	50	P	50
48	Chromium	μg Cr/l	—	50	50	50	P	50
49	Mercury	μg Hg/l	—	1	1	1	—	1
50	Nickel	μg Ni/l	—	50[B]	50	50[B]	P	50[O]
51	Lead	μg Pb/l	—	50	50	50	P	50
52	Antimony	μg Sb/l	—	10	10	10	P	10
53	Selenium	μg Se/l	—	10	10	10	P	10
54	Vanadium	μg V/l	—	—	—	—	—	—
55	Pesticides and related products[C]							
	(i) individual substances	μg/l	—	0.1	0.1	0.1	P	0.1
	(ii) total[D]	μg/l	—	0.5	0.5	0.5	P	0.5
56	Polycyclic aromatic hydrocarbons[E]	μg/l	—	0.2	0.2	0.2	—	0.2

E. Microbiological parameters

No.	Parameter	Unit						
57	Total coliforms[F]	No./100 ml[G]	—	0	0	0	—	P
58	Faecal coliforms	No./100 ml[G]	—	0	0	0	—	P
59	Faecal streptococci	No./100 ml[G]	—	0	0	0	—	P
60	Sulphite-reducing clostridia	No./20 ml[H]	—	<1	<1	<1	—	P
61	Total bacteria colony counts	No./ml at 22°C[I]		10	NSI	NSI	NSI[r]	5
		No./ml at 37°C[I]		100	NSI	NSI	NSI[r]	50
62	Total bacteria counts for water in enclosed containers	No./ml at 37°C	5	20	—	20	—	20
		No./ml at 22°C	20	100	—	100	—	200

continued overleaf

Table 1.10. (*Continued*)

Parameter	Unit of measurement	EC G	EC MAC	UK PCV	IRL NV	Denmark GL	Denmark MAC
F. Minimum required concentration for softened water intended for human consumption[J,K]							
1 Total hardness	mg CaCO$_3$/l	60	60	60	—	—	—
2 Hydrogen ion concentration[L]	pH[L]	—	—	—	—	—	—
3 Alkalinity[L]	mg HCO$_3$/l	30	30	30	>100	—	—
4 Dissolved oxygen[L]	mg O$_2$/l	—	—	—	—	—	—

[a]Including suspended solids.
[b]Including hydrogen sulphide.
[c]In relation to the natural structure of the water.
[d]Maximum admissible value is 9.5.
[e]Approximate concentration above which effects might occur is 200 mg/l.
[f]See Article 8 of Directive.
[g]With a centile of 80, calculated over a reference period of three years.
[h]Expressed in μg/l in the Water Supply Regulations 1989, so multiply mg/l value by 1000 to convert. Therefore 0.2 mg/l is 200 μg/l.
[i]See part F of table.
[j]After drying at 180°C.
[k]Saturation should be above 75% except for groundwaters.
[l]The water should not be aggressive.
[m]Ammonia and ammonium ions.
[n]Kjeldhal nitrogen is total nitrogen (ammonium + organic nitrogen) excluding oxidized forms (nitrate + nitrite).
[o]The reason for any increase in the usual concentration must be investigated.
[p]No significant increase over that normally observed.
[q]UO is undetectable organoleptically, i.e. not detected by either taste or smell.
[r]NSI is no significant increase above background.
[s]Excluding natural phenols which do not react with chlorine.
[t]Haloform concentrations must be as low as possible.
[u]At outlets of pumping and/or treatment works and their substations.
[v]After water has been standing for 12 hours in the piping and at the point where the water is made available to the consumer.
[w]At 8–12°C.
[x]Included in turbidity (parameter 2).
[y]NPV is no persistently visible suspended solids.

^ZSee Article 8 of Directive.

^AIf silver is used in a water treatment process (unusual).

^BIn running water. Where lead pipes are present the lead content should not exceed 50 µg/l in a sample taken either directly or after flushing. If a sample is taken after flushing. If a sample is taken after flushing and the lead content either frequently or to an appreciable extent exceeds 100 µg/l, suitable measures must be taken to reduce the exposure to lead on the part of the consumer.

^CIncludes insecticides (persistent organochlorine compounds, organophosphorus compounds, carbamates), herbicides, fungicides, polychlorinated biphenyls and polychlorinated terphenyls.

^DSum of the detected concentrations of individual substances.

^ESum of the detected concentrations of fluoranthene, benzo-3,4-fluoranthene, benzo-11,12-fluoranthene, benzo-3,4-pyrene, benzo-1,12-perlyene and indeno(1,2,3-cd) pyrene.

^FProvided a sufficient number of samples is examined (95% consistent results).

^GAnalysis by membrane method.

^HAnalysis by multiple tube method.

^IEither/or.

^JAlso applies to desalinated water.

^KIf excessively hard natural water is softened, then MAC for sodium may be exceeded. However, level of sodium must be kept as low as possible and the essential requirements for the protection of public health may not be disregarded.

^LThe water should not be aggressive.

^MHigher value allowed with special approval.

^NMinimum of 5 mg/l if water treatment includes aeration.

^O20 µg Ni/l ex works.

^PNot detectable.

— No value given.

and are required to report to the county council about quality. The Directive was implemented in September 1983 (Statutory Order 468:1983) and the Ministry of the Environment updated the regulations in August 1988 (Statutory Order 515:1988). Unlike the UK, the new Danish regulations specify maximum permissible values and guide values. These values are compared in Table 1.10 with the EC, UK and Irish standards, although generally there is very close agreement between the PCV, NV and MAC values.

As is the case with other Directives, derogations (exemptions) from the values specified are permitted because of climatological or geographical factors which make the standards impossible to attain. Such derogations must not relate to toxic or microbiological factors, nor contribute a public health hazard. In 1990 the Spanish Government updated its 1982 legislation which implemented the Directive (Royal Decree 1423/1982) with new drinking water regulations (Royal Decree 1138/1990). The new Spanish legislation brings their standards more in line with the Directive, and also includes guide values for α and β radioactivity of 0.1 and 1.0 Bq/l, respectively.

The EC realizes that improvements to water supplies cannot be achieved overnight and that it requires a structured improvement programme over a number of years. The time-scale is also important so that the costs of such a programme can be spread over a reasonable period. Therefore relaxations in compliance have been granted for a number of parameters. Full compliance is required for all supplies with respect to temperature, hydrogen ion concentrations, dry residuals, Kjeldahl nitrogen, oxidizability, phenols, surfactants, silver, arsenic, cyanide, chromium, mercury, nickel, antimony, selenium, faecal streptococci and sulphate-reducing clostridia. Under Article 9 of the Directive, derogations have been granted in the UK subject to review on 31 December 1994, or earlier, for the parameters colour, sulphate, magnesium, sodium, potassium, aluminium, dissolved or emulsified hydrocarbons and manganese. These derogations affect about 3% of the population. Time-limited derogations have been granted with respect to the following parameters, subject to the requirement that improvement works are undertaken by the date that the derogation expires: colour, turbidity, odour, taste, aluminium, nitrite, ammonium, iron and manganese.

These improvement programmes, carried out by the water supply companies, should bring almost all drinking water supplies up to standard by 1995. This will include the standards for aluminium, nitrate and lead. This will cost the water industry about £1.4 billion at 1987–8 prices to achieve this by the target date. Pesticide values, however, will not effectively be reduced until full-scale treatment techniques have been developed. Until then alternative water supplies free of pesticides, or diluting the contaminants with clean water obtained from outside the catchment, will be the only way to effectively reduce concentrations (Section 4.3).

The Water Industry Act 1991 requires water supply companies to supply consumers with water that is wholesome at the time of supply. The time of supply is the moment when it leaves the water company's communication pipe and passes into the consumer's supply pipe leading to the house. This is at the point where the boundary stoptap (valve) is located (Section 3.2). The consumer is responsible for the maintenance and repair of his or her own pipework and plumbing, and so any subsequent deterioration which occurs to drinking water quality is the consumer's responsibility.

There is an exception, and this relates to lead, copper and zinc. If there is a danger that the water will dissolve these metals or corrode them from the consumer's plumbing systems so that the concentrations of these metals exceed the PCV limits, then the water must be modified at the treatment plant to prevent this. This was primarily introduced to control the exposure of consumers (especially children) to lead.

Under the Water Industry Act (Section 65) the term wholesome is defined with reference to the Water Supply (Water Quality) Regulations 1989. Here drinking water is regarded as wholesome if: (1) it meets the standards prescribed in the regulations; (2) the hardness or alkalinity of water which has been softened or desalinated is not below the prescribed standards; and (3) it does not contain any element, organism or substance, whether alone or in combination, at a concentration or value which would be detrimental to public health.

The Regulations incorporate all the standards set out in the Drinking Water Directive, including national standards for 11 parameters. In total, numerical standards have been set for 55 parameters, with descriptive standards for another two. The Regulations also include a number of standards which apply directly to water from treatment plants and to water held in service reservoirs.

1.3.3 Sampling Water Quality

There is a statutory requirement to monitor the quality of water supplies. In the UK this is carried out by the water companies themselves, although it is checked by the local authorities and the DWI. Each water supply company area is split into a number of separate water supply zones and these are the basic unit for monitoring water quality. Supply zones are designated by the water supply companies and must not supply more than 50 000 people each. A discrete area served by a single source will always be delineated as a single supply zone unless it serves a population in excess of 50 000, in which case it is divided into two or more supply zones.

Water supply companies are required to take a specified number of samples from each supply zone each year. The exact number of samples depends on the population served, the parameters being measured and the source of the water (Table 1.11). The number of samples taken is given in the column marked 'standard'. If the PCVs are not exceeded for a period of three years, then the annual number of samples taken can be reduced. The reduced sampling frequency is expressed for a number of parameters at two different rates, one for groundwaters and another for surface waters, the former frequency being slightly less. If, however, the supply contains a mixture of groundwater and surface water, it will then be treated as a surface water and the higher sampling frequency will apply. If any of the listed parameters exceed the PCVs set, then the sampling frequency must increase to the higher rate until the Secretary of State is satisfied that the problem has been resolved. Samples are taken from strategic points which are representative of the water in that zone. Sampling points are usually the taps of consumers. These can be selected randomly, taken from fixed points or a mixture of the two, except where copper, lead and zinc are monitored, in which case the houses used as sampling points must be chosen at random. For microbiological parameters at

Table 1.11. Sampling frequencies at consumers' taps in the UK, as required by the Water Supply (Water Quality) Regulations for specific parameters, depend on the population supplied within a water supply zone, as well as other factors (see text). Reproduced with the permission of the Controller of Her Majesty's Stationery Office from HMSO (1989a)

Parameters	Water supply zone		Sampling frequency (number per year)			
	Volume distributed for domestic purposes (m³/d)	Population supplied	Reduced		Standard	Increased
			Ground-water	Surface water		
Conductivity or hydrogen ion	≤100	≤500			4	12
	101– 1000	501– 5000	4	4	6	12
Odour (quantitative)	1001– 2000	5001–10 000	4	6	12	24
	2001– 4000	10 001–20 000	6	12	24	
Taste (qualitative)	4001– 7000	20 001–35 000	11	21	42	
	7001–10 000	35 001–50 000	15	30	60	
Odour (quantitative) Taste (quantitative) Turbidity	≤100	≤500			4	12
Temperature	101– 1000	501– 5000			4	12
Hydrogen ion	1001– 2000	5001–10 000			4	24
Nitrate	2001– 4000	10 001–20 000	4	4	6	24
Nitrite	4001– 7000	20 001–35 000	4	5	10	36
Ammonium Iron Aluminium Manganese	7001–10 000	35 001–50 000	4	5	10	48
Colour Trihalomethanes Tetrachloromethane Trichloroethene Tetrachloroethene Copper	≤7000	≤35 000	1		4	12
Lead Zinc Pesticides* Polycyclic aromatic hydrocarbons†	7001–10 000	35 001–50 000	1		4	24
Total coliforms	≤100	≤500			12	
Faecal coliforms	101–1000	501– 5000			12	
Residual disinfectant	1001–2000	5001–10 000			24	
Colony counts	2001–4000	10 001–20 000			48	
		20 001–50 000			‡	

Table 1.11. (*continued*)

Parameters	Water supply zone		Sampling frequency (number per year)			
	Volume distributed for domestic purposes (m^3/d)	Population supplied	Reduced		Standard	Increased
			Ground-water	Surface water		
Chloride						
Sulphate						
Calcium						
Magnesium						
Sodium						
Potassium						
Dry residues						
Permanganate value						
Total organic carbon						
Boron				1	12	
Surfactants						
Phosphorus						
Fluoride						
Barium						
Silver						
Arsenic						
Cadmium						
Cyanide						
Chromium						
Mercury						
Nickel						
Antimony						
Selenium						
Total hardness						
Alkalinity						

*Any fungicide, herbicide or insecticide; increased or reduced sampling, where required or permitted, may be confined to the substance in question.
†Sum of the detected concentrations of fluoranthene, benzo-3,4-fluoranthene, benzo-11, 12-fluoranthene, benzo-3,4-pyrene, benzo-1,12-perylene and indeno (1,2,3-cd) pyrene.
‡At the rate of 12 per 5000 population except in the case of colony counts where the standard number is 52.

least 50% of the sampling points (taps) must be selected randomly. The DWI is required to check that the sampling locations selected are the most representative of the water supply zone. Water is also monitored as it leaves the treatment plant and in service reservoirs (which include water towers) for total and faecal coliforms, residual disinfectant and colony counts. This analysis is carried out at a much higher frequency than is measured at the consumers' taps. Sampling frequency depends on the volume of water supplied, so if the flow exceeds 12 000 m^3/d the analysis is daily, except for colony counts.

All the water quality information collected from routine monitoring must be made openly available in public registers. Such registers are not uncommon in local authorities, but the Regulations require that a copy of the information in the register is provided to consumers on demand. Each water supply company is also required to publish an annual report on the water they have supplied, including a discussion on its quality.

1.3.4 Enforcement

The Water Industry Act 1991 made the supplying of water that is unfit for human consumption a criminal offence for the first time. This provision was designed particularly to deal with gross contamination arising mainly from accidents or operational problems. Any water supplied which does not comply with the definition of wholesomeness given in the Regulations should be considered as unfit for human consumption. Penalties are very severe and can include imprisonment of those directly responsible, which could include a director, manager or other employee of a water company. The Act also makes it a duty of the Secretary of State to take enforcement action against a water supply company that has failed to supply wholesome water. 'Trivial' breaches of the Regulations are not normally followed by enforcement action, neither are more serious breaches if the water company gives an undertaking to take steps which are considered adequate over an acceptable time-scale to rectify the problem. However, if companies fail to give such an undertaking or fail to comply with it once given, then enforcement action follows.

The Secretary of State has two enforcement powers. First, under Section 18 of the Water Industry Act 1991 for contraventions of the wholesomeness of water, of monitoring, or treatment requirements as laid down in the Regulations. Second, he may institute prosecutions under Section 70 of the Act for supplying water unfit for human consumption. There is a set procedure of enforcement. After notification, the DWI on behalf of the Secretary of State serves a 'notice of intention to enforce' on the company. The company will then usually give an undertaking to carry out a programme of work to put right the problem to ensure full compliance within a given time-scale. The DWI will initiate enforcement action only if the company subsequently fails to comply (Section 1.2).

The Water Industry Act 1991 puts into place a new legal framework to deal with private supplies. The local authorities have the duty to be informed about the wholesomeness of all drinking water supplies within their area and this includes private supplies. If private supplies are found to fail the Drinking Water Regulations' definition of wholesomeness, then they are able to serve notices to (1) secure an improvement in the quality of the private supply, or (2) force the householder to be connected to a mains supply. Such notices may be appealed and can result in local inquiries. New regulations dealing with private supplies came into force on 1 January 1992. The Private Water Supplies Regulations 1991 (HMSO, 1991) specify how private supplies should be classified, the parameters to be monitored, the frequency and methods of sampling to determine whether they meet the Water Quality Regulations and the maximum charges that local authorities can make for sampling and analysis. The DWI has the role of ensuring that local authorities in England and Wales are complying with these new regulations.

It is in everyone's interest that private water supplies should also conform to EC standards, so the efforts of local authorities in England and Wales are to be welcomed in this respect. This is sadly not the case in many other EC Member States, such as the Republic of Ireland, where private supplies are not monitored, and with a dispersed rural population predominately not receiving a mains supply, the quality of individual private supplies in particular is often very poor.

1.3.5   European Implementation

Although there has been much criticism of the level of compliance in the UK, it is difficult to compare it with the situation in the rest of Europe. Many European countries, such as Ireland, have decentralized monitoring systems, which means that it is far more difficult to obtain an accurate picture of water quality performance and compliance with the EC Directive. In this respect the Water Quality Regulations in England and Wales and the two-tier regulatory role of the local authority and the DWI have exerted strong pressure on the water industry. This should result in the UK having some of the best quality drinking water in Europe by 1995.

All EC Member States have now incorporated the Drinking Water Directive into law, with France the last to do so. In general, national standards have closely followed the MAC values, although there are some national differences. For example, the German MAC for lead is 40 μg/l and for arsenic 40 μg/l, although the latter is to be reduced to 10 μg/l from January 1996. Most countries, including the UK and Ireland, have failed to set guide values. Of the EC Member States, Denmark probably has the most stringent regulations in terms of both G and MAC values (Table 1.10). With the exception of The Netherlands and Denmark, where drinking water quality is regulated by the environment ministries, other countries treat drinking water as a foodstuff with the Ministry of Health the regulatory body for quality. The application of the Directive to private supplies has been weak except where such supplies have been used for food processing, hotels, camp sites and similar situations. In Italy and Germany private supplies are regulated, but not in France or The Netherlands. There is no requirement in the Directive to inform the public of quality compliance, so unlike the UK, other EC countries do not generally make the results of sampling readily available. However, all countries have some form of external auditing of potable water quality and mechanisms for checking water quality by separate sampling.

1.3.6   US Standards

Standards in the USA arise from the Safe Drinking Water Act 1974, which directed the US Environmental Protection Agency (USEPA) to establish minimum National Drinking Water Standards. The Act was to ensure uniformity and consistency of state drinking water quality regulation. The standards cover all public water supply systems serving 25 or more people, or from which at least 15 service connections are taken (Section 3.2). This means that the standards apply to 85% of all Americans, who are served by over 200 000 separate supply systems, with the remainder served largely by private wells (NEHA, 1989). The Act also covers private supplies serving public areas such as camp sites. In 1986 the US Congress passed a set of amendments significantly expanding the original Act. These included: (1) accelerating the USEPA regulation of drinking water parameters and increasing the total number of regulated parameters to nearly 200 by the end of the century; (2) prohibiting the future use of lead pipes and lead solder in plumbing used to supply water for human consumption; (3) increasing the protection of groundwater sources of drinking water—although groundwater was originally protected under the 1974 Act, the amendments establish special programmes to protect critical groundwater resources, including protecting areas around wells supplying public drinking water, and by regulating the underground injection of wastes above and below drinking water resources; (4) requiring water treatment plants to filter surface waters and to disinfect all surface waters and many groundwaters, primarily to control Giardia (Section 4.7); and (5) streamlining enforcement procedures, which includes raising penalties to $25 000 per day. Up to 1986 standards for 26 parameters had been set; however, the amendments required the USEPA to set standards for another 57 parameters in the short term.

States, and Indian tribes that meet the same criteria as states, can assume primary enforcement authority over their drinking water. To do this the state or tribe must adopt drinking water standards at least as stringent as the national standards and be able to carry out adequate monitoring and enforcement requirements. The USEPA will carry out this requirement where a state or tribe cannot, or does not, wish to do so. So far 54 states or territories have been granted primary enforcement authority. By 1990 only Wyoming, Indiana, the District of Columbia and the Indian tribes had not taken up primary enforcement authority, leaving the USEPA responsible for implementation and enforcement in these areas. Since 1977 federal law has required water undertakers to sample and test water supplied to consumers. Where standards are not complied with the appropriate state agency and the consumers must be informed.

The 1986 amendments to the Act required the Office of Drinking Water of the USEPA to issue a maximum contaminant level goal (MCLG) and a maximum contaminant level (MCL) for various parameters. The MCLs are enforceable standards that apply to specified parameters which the USEPA has determined to have an adverse effect on human health above certain levels. The MCLs are set as close as feasibly possible to MCLGs taking into account both technological and cost considerations. The MCLGs are non-enforceable health-based goals that have been established as levels at which no known or anticipated adverse effects on the health of humans occurs, and which will allow an adequate margin of safety. For example, for known carcinogens the USEPA sets the MCLG at zero as a matter of policy. This is based on the theory that there exists some cancer risk, albeit very small, at any exposure level.

The standards are split into primary and secondary drinking water standards (Table 1.12). The Primary National Drinking Water Standards cover 30 parameters at

Table 1.12. Primary and Secondary National Drinking Water Regulations set by the US Environmental Protection Agency (EPA). Reproduced by permission from USEPA (1993)

Contaminant	Level*
Primary National Drinking Water Regulations	
Biological contaminants	
Total coliform bacteria (per 100 ml)	1
Turbidity (turbidity units)	0.5–1.0
Inorganic contaminants (mg/l)	
Arsenic	0.05
Barium	2.0
Cadmium	0.005
Chromium	0.1
Fluoride	4.0
Lead	0.015**
Mercury (inorganic)	0.002
Nitrate (as nitrogen)	10.0
Nitrite (as nitrogen)	1.0
Selenium	0.05
Silver	None
Organic contaminants (mg/l)	
Benzene	0.005
Carbon tetrachloride	0.005
1,2-Dichloroethane	0.005
1,1-Dichloroethylene	0.007
p-Dichlorobenzene	0.075
Endrin	0.002
Lindane	0.0002
Methoxychlor	0.04
Total trihalomethanes	None
Toxaphene	0.003
Trichloroethylene	0.005
1,1,1-Trichloroethane	0.2
Vinyl chloride	0.002
2,4,5-TP (Silvex)	0.05
2,4-D	0.07
Radionuclides	
Gross α particle activity (pCi/l)	15
Gross β particle activity (mrem/y)	4
Radium-226 and 228 (pCi/l)	20
Secondary National Drinking Water Regulations	
pH	6.5–8.5
Chloride (mg/l)	250
Copper (mg/l)	1
Foaming agents (mg/l)	0.5
Sulphate (mg/l)	250
Total dissolved (solids hardness) (mg/l)	500
Zinc (mg/l)	5
Fluoride (mg/l)	2.0
Colour (colour units)	15
Corrosivity	Non-corrosive

continued overleaf

Table 1.12. (*continued*)

Contaminant	Level*
Iron (mg/l)	0.3
Manganese (mg/l)	0.05
Odour (odour number)	3 threshold

*Level is maximum contaminant level (MCL) for the Primary Regulations and suggested level for Secondary Regulations; **Action level see page 244.

Table 1.13. Recent additions and changes to the Primary National Drinking Water Regulations. Reproduced by permission from US Environmental Protection Agency (1993)

	January 1991: final MCLGs* (mg/l)	January 1991: final MCLs* (mg/l)
Inorganics		
Asbestos (MFL)*	7.0	7.0
Barium	—	—
Cadmium	0.005	0.005
Chromium	0.1	0.1
Mercury	0.002	0.002
Nitrate	10.0 (as N)	10.0 (as N)
Nitrite	1.0 (as N)	1.0 (as N)
Total Nitrate and Nitrite	10.0 (as N)	10.0 (as N)
Selenium	0.05	0.05
Organics		
o-Dichlorobenzene	0.6	0.6
cis-1,2-Dichloroethylene	0.07	0.07
trans-1,2-Dichloroethylene	0.1	0.1
1,2-Dichloropropane	0	0.005
Ethylbenzene	0.7	0.7
Monochlorobenzene	0.1	0.1
Styrene	0.1	0.1
Tetrachloroethylene	0	0.005
Toluene	1.0	1.0
Xylenes (total)	10.0	10.0
Pesticides/PCBs*		
Alachlor	0	0.002
Aldicarb	0.007	0.007
Aldicarb sulphoxide	0.007	0.007
Aldicarb sulphone	0.007	0.007
Atrazine	0.003	0.003
Carbofuran	0.04	0.04
Chlordane	0	0.002
Dibromochloropropane	0	0.0002
2,4-D	0.07	0.07
Ethylene dibromide	0	0.00005
Heptachlor	0	0.0004

Table 1.13. (continued)

	January 1991: final MCLGs* (mg/l)	January 1991: final MCLs* (mg/l)
Heptachlor epoxide	0	0.0002
Lindane	0.0002	0.0002
Methoxychlor	0.04	0.04
PCBs (as decachlorobiphenyl)	0	0.0005
Pentachlorophenol	0	0.001
Toxaphene	0	0.003
2,4,5-TP (Silvex)	0.05	0.05

*Abbreviations: MCLG = maximum contaminant level goal; MCL = maximum contaminant level; PCB = polychlorinated biphenyl; MFL = million fibres per litre.

Table 1.14. Proposed additions to the Primary National Drinking Water Regulations. Reproduced by permission from US Environmental Protection Agency (1993)

	MCLGs* (mg/l)	MCLs* (mg/l)
Inorganics		
Antinomy	0.006	0.006
Beryllium	0.004	0.004
Cyanide	0.2	0.2
Nickel	0.1	0.1
Sulphate	400–500	400–500
Thallium	0.0005	0.002–0.001
Organics		
Adipates [Di(ethylhexyl)adipate]	0.04	0.4
Dalapon	0.2	0.2
Dichloromethane (methylene chloride)	0	0.005
Dinoseb	0.007	0.007
Diguat	0.02	0.02
Endothall	0.1	0.1
Endrin	0.002	0.002
Glyphosphate	0.7	0.7
Hexachlorobenzene	0	0.001
Hexachlorocyclopentadine (HEX)	0.05	0.05
Oxamyl (Vydate)	0.2	0.2
PAHs [Benzo(a)pyrene]*	0	0.0002
Phthalates [Di(ethylhexyl)phthalate]	0	0.004
Picloram	0.5	0.5
Simazine	0.004	0.004
1,2,4-Trichlorobenzene	0.07	0.07
1,1 + 2Trichloroethane	0.003	0.005
2,3,7,8-TCDD (Dioxin)	0	3×10^{-8}

*Abbreviations: MCLG = maximum contaminant goal; MCL = maximum contaminant level; PAH = polycyclic aromatic hydrocarbon.

Table 1.15. Revised World Health Organization drinking water guide values for bacteriological quality of drinking water. Reproduced by permission from World Health Organization (1993)

Organisms	Guideline
All water intended for drinking	
E. coli or thermotolerant coliform bacteria*†	Must not be detectable in any 100 ml sample
Treated water entering the distribution system	
E. coli or thermotolerant coliform bacteria*	Must not be detectable in any 100 ml sample
Total coliform bacteria	Must not be detectable in any 100 ml sample
Treated water in the distribution system	
E. coli or thermotolerant coliform bacteria*	Must not be detectable in any 100 ml sample
Total coliform bacteria	Must not be detectable in any 100 ml sample. In the case of large supplies where sufficient samples are examined, must not be present in 95% of samples taken throughout any 12 month period

Immediate investigative action must be taken if either *E. coli* or total coliform bacteria are detected. The minimal action in the case of total coliform bacteria is repeat sampling; if these bacteria are detected in the repeat sample, the cause must be determined by immediate further investigation.

*Although *E. coli* is the more precise indicator of faecal pollution, the count of thermotolerant coliform bacteria is an acceptable alternative. If necessary, proper confirmatory tests must be carried out. Total coliform bacteria are not acceptable indicators of the sanitary quality of rural water supplies, particularly in tropical areas where many bacteria of no sanitary significance occur in almost all untreated supplies.

†It is recognized that in most rural water supplies in developing countries faecal contamination is widespread. Under these conditions, the national surveillance agency should set medium-term targets for the progressive improvement of water supplies, as recommended in Volume 3—*Surveillance and Control of Community Supplies* (in press).

present which are considered to be potentially harmful to health. The Secondary National Drinking Water Standards are intended to protect public welfare as opposed to public health. These parameters are not considered to pose a threat to public health but are more concerned with aesthetic factors such as taste, odour and colour. There are currently 13 parameters listed (Table 1.12). The USEPA recommends these standards to the states and tribes as reasonable goals, but federal law does not require water undertakers to comply with them. However, states tend to adopt their own enforceable regulations governing these parameters.

Tables 1.13 and 1.14 show recent changes (December 1993) proposed to the Primary National Drinking Water Standards (USEPA, 1993).

1.3.7 WHO Revised Standards

The WHO guidelines for drinking water are perhaps the most important standards relating to water quality. They are used universally and are the basis for EC and USEPA legislation. The original guidelines were published in two volume in 1984.

Table 1.16. Revised World Health Organization drinking water guide values for chemicals of significance to health in drinking water. Health-based guide value for asbestos, tin and silver have not been set as they are considered not hazardous to health in the concentrations usually found in drinking water. Reproduced by permission from World Health Organization (1993)

Chemical	Guide value	Remarks
Inorganics (mg/l)		
Antimony	0.005(P)	
Arsenic	0.01(P)	For 6×10^{-4} excess skin cancer risk*
Barium	0.7	
Beryllium	—	NAD
Boron	0.3	
Cadmium	0.003	
Chromium	0.05(P)	
Copper	2(P)	ATO
Cyanide	0.07	
Fluoride	1.5	Climatic conditions, volume of water consumed and intake from other sources should be considered when setting national standards
Lead	0.01	It is recognized that not all water will meet the guideline value immediately; meanwhile, all other recommended measures to reduce the total exposure to lead should be implemented
Manganese	0.5(P)	ATO
Mercury (total)	0.001	
Molybdenum	0.07	
Nickel	0.02	
Nitrate	50 ⎫	The sum of the ratio of the concentration of
Nitrite	3 ⎭	each to their respective GV should not exceed unity
Selenium	0.01	
Uranium	—	NAD
Organics (μg/l)		
Chlorinated alkanes		
Carbon tetrachloride	2	
Dichloromethane	20	
1,1-Dichloroethane	—	NAD
1,2-Dichloroethane	30	For 10^{-5} excess risk*
1,1,1-Trichloroethane	2000(P)	
Chlorinated ethenes		
Vinyl chloride	5	For 10^{-5} excess risk*
1,1-Dichloroethene	30	
1,2-Dichloroethene	50	
Trichloroethene	70(P)	
Tetrachloroethene	40	
Aromatic hydrocarbons		
Benzene	10	For 10^{-5} excess risk*
Toluene	700	ATO

continued overleaf

Table 1.16. (*continued*)

Chemical	Guide value	Remarks
Xylenes	500	ATO
Ethylbenzene	300	ATO
Styrene	20	ATO
Benzo(*a*)pyrene	0.7	For 10^{-5} excess risk*
Chlorinated benzenes		
Monochlorobenzene	300	ATO
1,2-Dichlorobenzene	1000	ATO
1,3-Dichlorobenzene	—	NAD
1,4-Dichlorobenzene	300	ATO
Trichlorobenzenes (total)	20	ATO
Miscellaneous organics		
Di(2-ethylhexyl)adipate	80	
Di(2-ethylhexyl)phthalate	8	
Acrylamide	0.5	For 10^{-5} excess risk*
Epichlorohydrin	0.4(P)	
Hexachlorobutadiene	0.6	
EDTA	200(P)	
Nitrilotriacetic acid	200	
Dialkyltins	—	NAD
Tributyltin oxide	2	
Pesticides (μg/l)		
Alachlor	20	For 10^{-5} excess risk*
Aldicarb	10	
Aldrin/dieldrin	0.03	
Atrazine	2	
Bentazon	30	
Carbofuran	5	
Chlordane	0.2	
Chlortoluron	30	
DDT	2	
1,2-Dibromo-3-chloropropane	1	For 10^{-5} excess risk*
2,4-D	30	
1,2-Dichloropropane	20(P)	
1,3-Dichloropropane		NAD
1,3-Dichloropropene	20	For 10^{-5} excess risk*
Ethylene dibromide		NAD
Heptachlor and heptachlor expoxide	0.03	
Hexachlorobenzene	1	For 10^{-5} excess risk*
Isoproturon	9	
Lindane	2	
MCPA	2	
Methoxychlor	20	
Metolachlor	10	
Molinate	6	
Pendimethalin	20	
Pentachlorophenol	9(P)	

Table 1.16. (*continued*)

Chemical	Guide value	Remarks
Permethrin	20	
Propanil	20	
Pyridate	100	
Simazine	2	
Trifluralin	20	
Chlorophenoxy herbicides other than 2,4-D and MCPA		
Dichlorprop	100	
2,4-DB	90	
2,4,5-T	9	
Silvex	9	
Mecoprop	10	
MCPB	—	NAD
Disinfectants and disinfectant by-products		
Disinfectants		
Monochloramine	3	
Di- and trichloramines	—	NAD
Chlorine	5	ATO. For effective disinfection-free chlorine residual $\geqslant 0.5$ mg/l after at least 30 minutes contact time at pH <8.0
Chlorine dioxide		A guideline value has not been established because of chlorine dioxide's rapid breakdown and because the chlorite guideline value is adequately protective for potential toxicity from chlorine dioxide
Iodine	M	NAD
Disinfectant by-products		
Bromate	25(P)	For 7×10^{-5} excess risk*
Chlorite	200(P)	
Chlorate	—	NAD
Chlorophenols		
2-Chlorophenol	—	NAD
2,4-Dichlorophenol	—	NAD
2,4,6-Trichlorophenol	200	For 10^{-5} excess risk* ATO
Formaldehyde	900	
MX	—	NAD
Trihalomethanes		The sum of the ratio of the concentration of each to their respective GV should not exceed unity
Bromoform	100	
Dibromochloromethane	100	
Bromodichloromethane	60	For 10^{-5} excess risk*
Chloroform	200	For 10^{-5} excess risk*
Chlorinated acetic acids		
Monochloroacetic acid	—	NAD
Dichloroacetic acid	50(P)	
Trichloroacetic acid	100(P)	
Trichloroacetaldehyde/ chloral hydrate	10(P)	

Table 1.16. (*continued*)

Chemical	Guide value	Remarks
Chloropropanones	—	NAD
Haloacetonitriles		
Dichloroacetonitrile	90(P)	
Dibromoacetonitrile	100(P)	
Bromochloroacetonitrile		NAD
Trichloroacetonitrile	1(P)	
Cyanogen chloride (as CN⁻)	70	
Chloropicrin		NAD

(P) Provisional guideline value. This term is used for constituents for which there is some evidence of a potential hazard but where the available health effects information is limited and/or where an uncertainty factor greater than 1000 is used in the derivation of the tolerable daily intake (TDI). Provisional guideline values are also recommended (1) for those substances for which the calculated guideline value would be (i) below the practical quantification level or (ii) below the level that can be achieved through practical treatment methods, or (2) where disinfection is likely to result in the GV being exceeded.
NAD = No adequate data to recommend a health-based GV.
ATO = Concentrations of the substance at or below the health-based GV may affect the appearance, taste or odour of the water.
*For substances that are considered to be carcinogenic, the guideline value is the concentration in drinking water associated with an excess lifetime cancer risk of 10^{-5} (one aditional cancer per 100 000 of the population ingesting drinking water containing the substance at the GV for 70 years). Concentrations associated with estimated excess lifetime cancer risks of 10^{-4} and 10^{-6} can be calculated by multiplying and dividing, respectively, the GV by 10. In cases in which the concentration associated with a 10^{-5} lifetime excess cancer risk is not feasible as a result of inadequate analytical or treatment technology, a provisional GV is recommended at a practicable level and the estimated associated cancer risk presented. It should be emphasized that the guideline values for carcinogenic substances have been computed from hypothetical mathematical models that cannot be experimentally verified and that the values should be interpreted differently than TDI-based values because of the lack of precision of the models. At best, these values must be regarded as rough estimates of cancer risk. However, the models used are conservative and probably err on the side of caution. Because a linear relationship between dose and effect is assumed, the model overestimates cancer risks, which may be as low as zero. Moderate short-term exposure to levels exceeding the GV for carcinogens does not significantly affect the risk.

Volume 1 is the guidelines while Volume 2 contains the scientific evidence on which the recommendations in Volume 1 are based. The existing guidelines were based on the available toxicological evidence up to 1981, and so are very much out of date. The revision began in 1987, with new guidelines finally agreed in Geneva in September 1992. The new guidelines include microbiological, chemical and radiological parameters. The chemical parameters include 17 inorganic chemicals, 27 organic chemicals, 33 pesticides and 17 disinfectants and associated by-products. The revised guidelines are shown in Tables 1.15–1.18.

The guidelines are generally based on lifetime exposures. For lead the guideline is based on the intake of a bottle-fed infant (i.e. 50% of the weekly intake of an infant drinking 0.75 l of water daily). It is this new guide value for lead of 10 μg/l which is going to be the most contentious and costly to implement in most developed countries. Some guide values are designated as provisional due either to a lack of toxicity information (e.g. trichloracetic acid) or because of the absence of sufficiently reliable or sensitive analytical methods (e.g. bromate). Microbial guidelines are still based on

Table 1.17. Revised World Health Organization drinking water guide values for radioactive components. Reproduced by permission from World Health Organization (1993)

	Screening value (Bq/l)	Remarks
Gross-α activity	0.1 ⎫	If a screening value is exceeded, more detailed radionuclide analysis is necessary. Higher values do not necessarily imply that the water is unsuitable for human consumption
Gross-β activity	1 ⎭	

Table 1.18. Revised World Health Organization drinking water guide values for substances and parameters that may give rise to complaints from consumers. Reproduced by permission from World Health Organization (1993)

Parameter	Levels likely to give rise to consumer complaints*	Reasons for consumer complaints
Inorganics (mg/l)		
Aluminium	0.2	Depositions, discoloration
Ammonia	1.5	Odour and taste
Chloride	250	Taste, corrosion
Colour	15 TCU	Appearance
Copper	1	Staining of laundry and sanitary ware (health-based provisional GV 2 mg/l)
Hardness	—	High hardness: scale deposition, scum formation Low hardness: possible corrosion
Hydrogen sulphide	0.05	Odour and taste
Iron	0.3	Staining of laundry and sanitary ware
Manganese	0.10	Staining of laundry and sanitary ware (health-based provisional GV 0.5 mg/l)
Dissolved oxygen	—	Indirect effects
pH	—	Low pH: corrosion High pH: taste, soapy feel Preferably <8.0 for effective disinfection with chlorine
Sodium	200	Taste
Sulphate	250	Taste, corrosion
Taste and odour	—	Should be acceptable
Temperature	—	Should be acceptable
Total dissolved solids	1000	Taste
Turbidity	5 NTU	Appearance. For effective terminal disinfection median ⩽1 NTU, single sample ⩽5 NTU
Zinc	3	appearance, taste
Organics (μg/l)		
Toluene	24–170	Odour, taste (health-based GV 700 μg/l)
Xylenes	20–1800	Odour, taste (health-based GV 500 μg/l)
Ethylbenzene	2.4–200	Odour, taste (health-based GV 300 μg/l)

Table 1.18. (*continued*)

Parameter	Levels likely to give rise to consumer complaints*	Reasons for consumer complaints
Styrene	4–2600	Odour, taste (health-based GV 20 μg/l)
Monochlorobenzene	10–120	Odour, taste (health-based GV 300 μg/l)
1,2-Dichlorobenzene	1–10	Odour, taste (health-based GV 1000 μg/l)
1,4-Dichlorobenzene	0.3–30	Odour, taste (health-based GV 300 μg/l)
Trichlorobenzenes (total)	5–50	Odour, taste (health-based GV 20 μg/l)
Synthetic detergents	—	Foaming, taste, odour
Disinfectants and disinfectant by-products (μg/l)		
Chlorine	600–1000	Taste and odour (health-based GV 5 mg/l)
Chlorophenols		
2-Chlorophenol	0.1–10	Taste, odour
2,4-Dichlorophenol	0.3–40	Taste, odour
2,4,6-Trichlorophenol	2–300	Taste, odour (health-based GV 200 μg/l)

*These are not precise numbers. Problems may occur at lower or higher values according to local circumstances. Range of taste and odour threshold concentrations given for organics.

Escherichia coli or thermotolerant coliforms as indicators of faecal pollution. Total coliforms are recommended only as indicators of treatment efficiency and the integrity of the distribution system, not as indicators of the presence of pathogens. No guide values have been set for viruses, protozoans or specific bacterial pathogens due to the absence of suitable routine analytical methods. Although the guide values are all health related, the guidelines recognize that for many compounds the health-based guide value may be very much higher than the concentration affecting the aesthetic quality of the water, in particular taste and odour. So guide values based on aesthetic quality are also given on the premise that although the water may be safe to drink, consumers may turn to other less safe sources of water if their mains supply is unaesthetic. The WHO felt it unnecessary to recommend a health-based guide value for asbestos, tin or silver as they considered them not to be hazardous to health at the concentrations usually found in drinking water (Table 1.16). The revised guidelines will have a significant effect on existing standards world-wide. The current EC Drinking Water Directive is expected to be revised in the near future.

Chapter 2
Sources of Water

2.1 INTRODUCTION

Water supplies are not pure in the sense that they are devoid of all dissolved chemical compounds like distilled, deionized water. In the early days of chemistry, water was known as the universal solvent as a result of its ability to slowly dissolve into solution anything it comes into contact with, from gases to rocks. So as rain falls through the atmosphere, flows over and through the earth's surface, it is constantly dissolving material, forming a chemical record of its passage from the clouds. Therefore, water supplies have a natural variety in quality, which depends largely on the source of the supply. All our water comes from the water cycle and it is this process which controls our water resources.

2.2 THE WATER CYCLE

The total volume of water in the world remains constant. What changes is its quality and availability. Water is constantly being recycled, a system known as the water or hydrological cycle. Hydrologists study the chemical and physical nature of water and its movement on and below the ground. In terms of total volume, 97.5% of the world's water is saline with 99.99% of this found in the oceans, the remainder making up the salt lakes. This means that only 2.5% of the volume of water in the world is actually non-saline. However, not all of this freshwater is readily available for use by humans. About 75% of this freshwater is locked up as ice-caps and glaciers, with a further 24% located underground as groundwater, which means that less that 1% of the total freshwater is found in lakes, rivers and the soil. Therefore, only 0.01% of the world's water budget is present in lakes and rivers, with another 0.01% present as soil moisture but unavailable to humans for supply. So although there appears to be a lot of water about, there is in reality very little which is readily available for use by humans (Table 2.1). Within the hydrological cycle, water is constantly moving, driven by solar energy. The sun causes evaporation from the oceans, which forms clouds and precipitation (rainfall). Evaporation also occurs from lakes, rivers and the soil, with plants contributing significant amounts of water by evapotranspiration. Although about 80% of precipitation falls back into the oceans, the remainder falls onto land. It is this water that replenishes the soil and groundwater, feeds the streams and lakes, and provides all

Table 2.1. Total volumes of water in the global hydrological cycle. Reproduced from Charley 1969) by permission of Methuen Publishers Ltd

Type of water	Area (km^2 × 10^3)	Volume (km^3 × 10^3)	Percentage of total water
Atmospheric vapour (water equivalent)	510 000 (at sea level)	13	0.0001
World ocean	362 033	1 350 400	97.6
Water in land areas	148 067	—	—
Rivers (average channel storage)	—	1.7	0.0001
Freshwater lakes	825	125	0.0094
Saline lakes; inland seas	700	105	0.0076
Soil moisture; vadose water	131 000	150	0.0108
Biological water	131 000	(Negligible)	—
Groundwater	131 000	7 000	0.5060
Ice-caps and glaciers	17 000	26 000	1.9250
Total in land areas (rounded)		33 900	2.4590
Total water, all realms (rounded)		1 384 000	100
Cyclic water			
Annual evaporation			
From world ocean		445	0.0320
From land areas		71	0.0050
Total		516	0.0370
Annual precipitation			
On world ocean		412	0.0291
On land areas		104	0.0075
Total		516	0.0370
Annual outflow from land to sea			
River outflow		29.5	0.0021
Calving, melting, and deflation from ice-caps		2.5	0.0002
Groundwater outflow		1.5	0.0001
Total		33.5	0.0024

the water needed by plants, animals and, of course, humans (Figure 2.1). The cycle is continuous and so water is a renewable resource (Franks, 1987). In essence, the more it rains the greater the flow in the rivers and the higher the water-table rises as the underground storage areas (i.e. the aquifers) fill with water as it percolates downwards into the earth. Water supplies depend on the rainfall so when the amount of rain decreases then the volume of water available for supply will decrease, and in cases of severe drought it will fall to nothing. To provide sufficient water for supply all year round, careful management of resources is required.

Nearly all our supplies of freshwater come from precipitation which falls onto a catchment area. Also known as a watershed or river basin, the catchment is the area of land, often bounded by mountains, from which any water that falls into it will drain into a particular river system. A major river catchment will be made up of many smaller subcatchments, each draining into a tributary of the major river. Each subcatchment will have different rock and soil types, and each will have different land-use activities

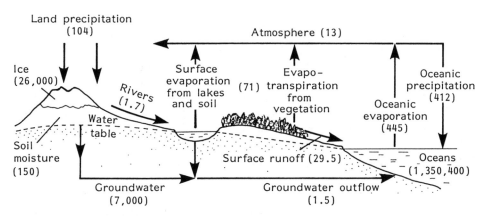

Figure 2.1. Hydrological cycle showing the volume of water stored and the amount cycled annually. All volumes are expressed as 10^3 km^3

which also affect water quality. The water draining from each subcatchment may therefore be different in terms of chemical quality. As the tributaries enter the main river they mix with water from other subcatchments upstream, constantly altering the chemical composition of the water. Water from different areas of the country are therefore chemically unique.

When precipitation falls into a catchment one of three major fates befalls it. (1) It may remain on the ground as surface moisture and eventually be returned to the atmosphere by evaporation. Alternatively, it may be stored as snow on the surface until the temperature rises sufficiently to melt it. Storage as snow is an important source of drinking water in some regions. For example, throughout Scandinavia lagoons are constructed to collect the runoff from snow as it melts, and this provides the bulk of their drinking water during the summer. (2) Precipitation flows over the surface into small channels to become surface runoff into streams and lakes. This is the basis of all surface water supplies and will eventually evaporate into the atmosphere, percolate into the soil to become groundwater, or continue as surface flow in rivers back to the sea. (3) The third route is for precipitation to infiltrate the soil and slowly percolate into the ground to become groundwater, which is stored in porous sediments and rocks (Section 2.4). Groundwater may remain in these porous layers for periods ranging from just a few days to possibly millions of years. Eventually groundwater is removed by natural upward capillary movement to the soil surface, plant uptake, groundwater seepage into surface rivers, lakes or directly to the sea, or artificially by pumping from wells and boreholes.

Water supplies therefore come from two principal resources within the water cycle: surface water and groundwater. Each of these resources is interrelated and each has its own advantages and disadvantages as a source of drinking water. Clearly, as water moves through this system of surface and underground pathways its quality is altered, often dramatically, so that the quality of water leaving the catchment will be different from the water which entered it as precipitation.

Our almost exclusive dependence on rainfall to provide our drinking water requires careful and long-term management. Although in theory the amount of rain falling on the British Isles is currently more than adequate to meet all our foreseeable needs, there are two practical problems. The first is that more water than is required for our immediate needs must be collected and stored during periods of heavy rainfall, usually during the winter, so that this excess can be used to supplement supplies during periods of low rainfall. Secondly, the areas where rainfall is highest are the areas of least population, with most of the population centred in the areas of lowest rainfall. This means transferring water from areas of high rainfall to areas where the demand is greatest, and finding and exploiting as many alternative supplies as possible.

Water for drinking is abstracted from rivers, reservoirs, lakes or underground aquifers (groundwater). The National Rivers Authority (NRA) in England and Wales licenses all abstractions of water by water supply companies, industry, agriculture and private supplies. Some industries require large volumes of water regardless of quality, whereas others treat their own water to the standard required by their process, which is often much higher than drinking water standards. It is therefore common for industry to abstract water directly from the water source. Domestic supplies are treated and are at present generally not metered. Currently 99.2% of the population in England and Wales are connected to a public water supply. This is comparable with the best in Europe, with the exception of The Netherlands, where nearly 100% of the population is connected. In contrast, nearly all industrial and many commercial supplies are metered. There are currently about 100 000 private supplies in England and Wales, of which only 200 or so supply more than 500 people.

All the major water resources are considered below, with some newer resources and management techniques considered in Sections 2.5 and 2.6. The effect of climate change and details of water resources in the USA have been reviewed by Waggoner (1990).

2.3 SURFACE WATERS: LAKES, RESERVOIRS AND RIVERS

2.3.1 Surface Water

Surface water is a general term describing any water body which is found flowing or standing on the surface, such as streams, rivers, ponds, lakes and reservoirs. Surface waters originate from a combination of sources: (1) surface runoff—rainfall which has fallen onto the surrounding land and that flows directly over the surface into the water body; (2) direct precipitation—rainfall which falls directly into the water body; (3) interflow—excess soil moisture which is constantly draining into the water body; and (4) water-table discharge—where there is an aquifer below the water body and the water-table is high enough, the water will discharge directly from the aquifer into the water body (Bowen, 1982).

The quality and quantity of surface water depends on a combination of climatic and geological factors. The recent pattern of rainfall, for example, is less important in enclosed water bodies such as lakes and reservoirs where water is collected over a long period and stored, whereas in rivers and streams where the water is in a dynamic state of constant movement, then the volume of water is dependent on the preceding weather conditions.

In rivers the flow is generally greater in winter than in summer due to a greater amount and longer duration of rainfall. Short fluctuations in flow, however, are more dependent on the geology of the catchment. Some catchments yield much higher percentages of the rainfall as stream flow than others. Known as the runoff ratio, the rivers of Wales and Scotland can achieve values of up to 80% compared with only 30% in lowland areas in southern England. So although the Thames, for example, has a vast catchment area of about 9869 km^2, it has only half the annual discharge of a river such as the Tay, which has a catchment of only 4584 km^2. Of course, in Scotland there is a higher rainfall than in south-east England, and also lower evaporation rates due to lower temperatures.

Even a small reduction in the average rainfall in a catchment area (e.g. 20%), may halve the annual discharge from a river. This is why when conditions are only marginally drier than normal a drought situation can readily develop. As we have seen in drought areas of the UK, it is not always the case that the more it rains the more water there will be in the rivers. Groundwater is also an important factor in droughts (Beran and Rodier, 1985). In some areas during the dry summer of 1975, despite the rainfall figures which were far below average, the stream flow in rivers which received a significant groundwater input were higher than normal due to the excessive storage built up in the aquifer over the previous wet winter. The drought beginning in 1989, with three successive dry winters, resulted in a significant reduction in the amount of water stored in aquifers with a subsequent fall in the height of the water-table (Section 2.4). This resulted in some of the lowest flows on record in a number of south-eastern rivers in England, with sections completely dried up for the first time in living memory. Before these rivers return to normal discharge levels, the aquifers which feed them must be fully replenished and this may take several years. Groundwater contributes substantially to the base flow of many lowland rivers, so any steps taken to protect the quality of groundwaters will also indirectly protect surface waters.

Precipitation carries appreciable amounts of solid material to earth such as dust, pollen, ash from volcanoes, bacteria, fungal spores and even, on occasions, larger organisms. The sea is the major source of many salts found dissolved in rain such as chloride, sodium, sulphate, magnesium, calcium and potassium ions. Atmospheric discharges from the home and industry also contribute material to clouds which are then brought back to earth in precipitation. These include a wide range of chemicals such as organic solvents and the oxides of nitrogen and sulphur which cause acid rain. The amount and type of impurities in precipitation varies with the location and time of year, and can affect both lakes and rivers. Land use, including urbanization and industrialization, significantly affects water quality, with agriculture having the most profound effect on supplies due to its dispersed and extensive nature (Eriksson, 1985).

The quality and quantity of water in surface waters is also dependent on the geology of the catchment. In general, chalk and limestone catchments result in clear, hard waters, whereas impervious rocks such as granite result in turbid, soft waters. Turbidity is caused by fine particles, both inorganic and organic in origin, which are too small to readily settle out of suspension and so the water appears cloudy. The reasons for these differences is that rivers in chalk and limestone areas rise as springs or are fed from aquifers through the river bed. Because appreciable amounts of the water come from groundwater resources, the river retains a constant clarity, constant flow and indeed a constant temperature throughout the year, except after periods of the most prolonged rainfall. The chemical nature of these rivers is also very stable and rarely alter from year to year. The water has spent a very long time in the aquifer before entering the river and during this time dissolves the calcium and magnesium salts comprising the rock, resulting in a hard water with a neutral to alkaline pH. In comparison, soft water rivers usually rise as runoff from mountains, so the flow is linked closely to rainfall. Such rivers suffer from wide fluctuations in flow-rate, with sudden floods and droughts. Chemically, these rivers are turbid due to the silt washed into the river with the surface runoff, and because there is little contact with the bedrock they contain low concentrations of cations such as calcium and magnesium, which makes the water soft with a neutral to acid pH. Such rivers often drain upland peaty soils and the water therefore contains a high concentration of humus, giving the water a clear brown–yellow colour, similar to beer in appearance.

2.3.2 Lakes and Reservoirs

As large cities expanded during the 19th century they relied on local water resources, but as demand grew they were forced to invest in reservoir schemes, often remote from the point of use. Examples include reservoirs built in Wales, the Pennines and the Lake District to supply major cities such as Birmingham, Manchester and Liverpool, with water in some instances being pumped over 50 miles to consumers. Most are storage reservoirs where all the water collected is used for supply purposes. Such reservoirs are sited in upland areas at the headwaters (source) of rivers. Suitable valleys are flooded by damming the main streams. They can take many years to fill and once brought into use for supply purposes must be carefully managed. A balance must be maintained between the water taken out for supply and that being replaced by surface runoff. Usually the surface runoff during the winter far exceeds demand for supply so that the excess water can be stored and used to supplement periods when surface runoff is less than the demand from consumers. There is, of course, a finite amount of water in a reservoir and water rationing is often required to prevent storage reservoirs drying up altogether during dry summers. A major problem is when there is a dry winter and so the expected excess of water does not occur, resulting in the reservoirs not being adequately filled at the beginning of the summer. Under these circumstances water shortages may occur even though the summer is not excessively dry.

Most lakes have an input and an output, and so in some ways they can be considered as slowly flowing rivers. The long period of time that water remains in the lake or reservoir ensures that the water becomes cleaner due to bacterial activity removing any organic matter present, and physical flocculation and settlement processes, which remove small particulate material. Storage of water therefore improves the quality, which then reduces the treatment before supply to a minimum (Section 3.2). However,

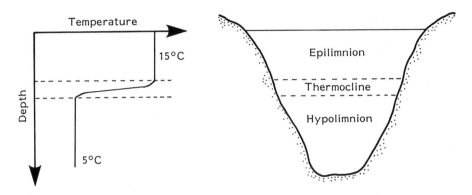

Figure 2.2. Thermal stratification in deep lakes and reservoirs

this situation is complicated by two factors. Firstly, in standing waters much larger populations of algae can be supported than in rivers and, secondly, deep lakes and reservoirs may become thermally stratified, particularly during the summer months. These two factors can seriously affect water quality.

Thermal stratification is caused by variations in the density of water in lakes and reservoirs. Water is at its densest at 4°C, when it weighs exactly 1000 kg/m^3. However, either side of this temperature water is less dense (999.87 kg/m^3 at 0°C and 999.73 kg/m^3 at 10°C). During the summer the sun heats the surface of the water reducing its density, so that the colder denser water remains at the bottom of the lake. As the water continues to heat up, then two distinct layers develop. The top layer or epilimnion is much warmer than the lower layer, the hypolimnion. Owing to the differences in density, the two layers, separated by a static boundary layer known as the thermocline, do not mix but remain separate (Figure 2.2).

The epilimnion of lakes and reservoirs is constantly being mixed by the wind and so the whole layer is a uniform temperature. As this water is both warm and exposed to sunlight it provides a favourable environment for algae. Usually the various nutrients required by algae for growth, in particular phosphorus and nitrogen, are not present in large amounts (i.e. limiting concentrations). When excess nutrients are present, due to agricultural runoff, for example, then massive algal growth may occur. These so-called algal blooms result in vast increases in the amount of algae in the water, a phenomenon known as eutrophication. The algae are completely mixed throughout the depth of the epilimnion and in severe cases the water can become highly coloured. This top layer of water is usually clear and full of oxygen, but if eutrophication occurs then the algae must be removed by treatment. The algae can result in unpleasant tastes in the water even after treatment (Section 4.9). Like all plants algae release oxygen during the day by photosynthesis, but at night they remove oxygen from the water during respiration. When eutrophication occurs the high numbers of algae will severely deplete the oxygen concentration in the water during the hours of darkness, possibly resulting in fish kills, and certainly causing further problems at the water treatment plant. In contrast, there is little mixing or movement in the hypolimnion which rapidly becomes deoxygenated and stagnant. Dead algae and organic matter settling from the upper layers are degraded in this lower layer of the lake. As the hypolimnion has no source of oxygen to replace that already used, its water may become completely devoid of oxygen. Under anaerobic conditions iron, manganese, ammonia, sulphides, phosphates and silica are all released from sediments in the reservoir into the water, whereas nitrate is reduced to nitrogen gas. This makes the water unfit for supply purposes. For example, iron and manganese will result in complaints about discoloured water and bad taste. Ammonia interferes with chlorination, depletes oxygen faster and acts as a nutrient to encourage eutrophication (as do phosphorus and silica). Sulphides also deplete oxygen and interfere with chlorination; they also have an awful smell and impart an obnoxious taste to the water.

The thermocline, the zone separating the two layers, has a tendency to move slowly to lower depths as the summer progresses. This summer stratification is usually broken up in autumn or early winter as the air temperature falls and the temperature of the epilimnion decreases. This increases the density of the water making up the epilimnion to a comparable level with the hypolimnion, making stratification unstable. Subsequently, high winds will suddenly cause the whole water body to turn over and so the stratification fails and the layers become mixed. Throughout the rest of the year the whole lake remains completely mixed, resulting in a significant improvement in water quality. Limited stratification can also develop during the winter as surface water temperatures approach $0°C$ while the temperature of the lower waters remains at $4°C$. This winter stratification is broken up in the spring as the temperature increases and the high winds return. Stratification is mainly a phenomenon of deep lakes.

During stratification each layer or zone has its own characteristic water quality and this can pose serious operational problems to the water supply company. Some are able to abstract water at various depths, thus ensuring that the best quality water or combination of waters is always used. Others try to prevent nutrients entering reservoirs by controlling agricultural activity and other possible sources of nutrients within the catchment area. Occasionally algal growth is controlled by chemical addition. Another method widely used is to pump colder water from the base of the lake to the surface, thus ensuring that the lake does not stratify. The management of reservoirs and lakes for supply is a complex business, dependent on a number of external factors such as air temperature, amount of sunshine, nutrient inputs (both natural and from human activity), and many others.

Most of the land around reservoirs, the catchment area, is usually owned by the water supply company. They impose strict restrictions on farming practice and general land use to ensure that the quality of the water is not threatened by indirect pollution. Restricted access to catchment areas and reservoirs has been relaxed over the past few years, although it is still strictly controlled. This controlled access and restrictions on land use has caused much resentment, especially in Wales where as much as 70% of all the water in upland reservoirs is stored for use in the English Midlands. The problem is not only found in Wales, but in England as well. For example, as much as 30% of the Peak District is occupied by reservoir catchment areas. Water supply companies want to ensure that the water is kept as clean as possible, because water collected in upland reservoirs is of a very high quality. Storage also significantly improves its purity further. The cleaner the raw water, the cheaper it is to treat. Restricted access to catchment areas therefore means that this quality is less likely to be reduced. Conflict is inevitable between those who want access to the land for recreation or other purposes and the water companies, who want to supply water to their consumers at the lowest price possible.

2.3.3 Water Abstraction

As a result of the very large capital investment in land and building reservoirs, and the enormous opposition from the public to such schemes on environmental, social and aesthetic grounds, there has been a swing away from the construction of storage reservoirs in the latter half of this century. During this period river abstraction has been exploited along with groundwater resources.

Water is abstracted from rivers by constructing weirs to ensure a minimum depth of water behind the weir, or by using floating pontoons. Abstraction must not interfere with other river uses such as navigation, but must ensure that water can be removed at all times of the year. The amount of water that can be abstracted is limited by the minimum flow required to: (1) protect the biological quality of the river including fisheries; (2) dilute industrial and domestic wastes—remember that rivers are vital for removing wastes and to a certain extent treating wastes through natural self-purification processes; (3) to ensure other river uses are not affected by abstraction (such as sufficient depth for navigation); and (4) to allow adequate flow to prevent the tide encroaching further upstream and turning freshwater sections brackish.

To maintain the integrity of rivers the minimum flow under dry weather conditions must be calculated and maintained at all times. Once calculated, any water in excess of this minimum dry weather flow can in theory be used for abstraction. Conversely, if the flow in the river falls close to or even below this minimum flow, then abstraction must be reduced or stopped altogether.

The quality of river water is also an important factor. River water requires complex and expensive treatment before being supplied to the consumer. The complexity and cost of treatment increases as the quality of the raw water deteriorates. Also, as rivers drain large areas of land, pollution is inevitable. All the waste disposed of, or chemicals used, within a catchment area will eventually find their way into the river, and so extreme care must be taken to ensure that the water quality is protected and monitored continuously. Most intakes have a storage capability so that raw water can be stored for up to seven days before being treated and supplied. This has a dual function. Firstly, it protects the consumer from the effects of pollution in the river, or accidental spills of toxic materials, allowing sufficient time for the pollution to disperse in the river before abstraction is resumed without cutting off supplies to consumers. Secondly, storing water in this way improves the quality of the water before treatment (Section 3.2).

After supply the water is returned to the river as treated sewage effluent and may well be abstracted again further downstream. This is certainly the case in many major lowland rivers such as the Thames, Severn and Trent (Section 2.6). However, as demands have continued to grow the natural flows of many rivers have proved inadequate to meet the volumes currently needed for abstraction. Also, the quality of many rivers has deteriorated through our exploitation of rivers as effluent carriers.

To maximize water availability for supply, hydrologists examine the hydrological cycle within the catchment, measuring rainfall, stream flow and surface runoff, and where applicable groundwater supplies. Often they can supplement water abstracted from rivers at periods of very low flow by taking water from other resources such as groundwater or small storage reservoirs, using these limited resources to top up the primary source of supply at the most critical times. More common is the construction of reservoirs at the headwaters which can then be used to control the flow in the river itself, a process known as compensation.

These compensation reservoirs are designed as an integral part of the river system. Water is collected as surface runoff from upland areas and stored during wet periods. The water is then released when needed to ensure that the minimum dry weather flow is maintained downstream to allow abstraction to continue. In winter, when most precipitation falls and flows are generated in the river, all this excess water is lost. Storing the excess water by constructing a reservoir and using it to regulate the flow in the river maximizes the output from the catchment area. A bonus is that such reservoirs can also play an important function in flood prevention. The natural river channel itself is used as the distribution system for the water, unlike supply reservoirs where expensive pipelines or aqueducts are required to transport the water to the point of use. River management is also easier because most of the water abstracted is returned to the same river. Among the more important UK rivers which are compensated are the Dee, Severn and Tees.

Reservoirs are not a new idea and were widely built to control the depth in canals and navigable rivers. Smaller reservoirs, often called header pools, were built to feed mill races to drive water wheels. Without a reservoir, there are times after dry spells where there is negligible discharge from the soil so that the only flow in the river will be that coming from groundwater seeping out of the underlying aquifer. Some rivers, rising in areas of permeable rock, may even dry up completely in severe droughts. In many natural rivers the natural minimum flow is about 10% of the average stream flow. Where river regulation is used this minimum dry weather flow is often doubled, and although this could in theory be increased even further, it would require an enormous reservoir capacity.

Reservoirs are very expensive to construct. However, there are significant advantages to regulating rivers using compensation reservoirs rather than supplying water directly from a reservoir via an aqueduct. With river regulation much more water is available to meet different demands than from the stored volume only, as the reservoir is only fed by the upland section of the river contained by the dam. Downstream all the water draining into the system is also available. Compensation reservoirs are therefore generally much smaller in size and so cheaper to construct.

Reservoirs can only yield a limited supply of water and so the management of the river system to ensure adequate supplies every year is difficult. Owing to the expense of construction, reservoirs are designed to provide adequate supplies for most dry summers. However, it is not cost-effective to build a reservoir large enough to cope with the severest droughts which may only occur once or twice a century. Water is released from the reservoir to ensure the predicted minimum dry weather flow. Where more than one compensation reservoir is available within the catchment area, water will first be released from those which refill quickly. Water regulation is a difficult task requiring operators to make intuitive guesses as to what the weather may do over the next few months. For example, many water companies have been severely criticized for maintaining water restrictions throughout the winter of 1990–1 to replenish reservoirs which failed to completely fill during the previous dry winter. Imposing bans may ensure sufficient supplies for essential uses throughout a dry summer and autumn; however, if it turns out to be a wet summer after all, such restrictions will be deemed by the consumers to have been unnecessary. Although water planners have complex computer models to help them predict patterns in water use and so to plan the best use of available resources, it is all too often impossible to match supplies with demand. This has mainly been a problem in south-east England, where the demand is greatest due to a high population density and also a high industrial and agricultural demand, but where the least rainfall is recorded.

Dry winters are often more of a problem than dry summers as reservoirs are not fully replenished. This is, of course, a major problem with storage reservoirs, where most of the water for supply in the summer and autumn is collected during the winter. If the reservoir is not full by the beginning of spring, restrictions in supply are almost inevitable. Similar problems occur with compensation reservoirs, for if the winter is dry the reservoir will be required to augment low flows early in the summer and so will be dangerously depleted if required to continue augmenting the flow throughout the rest of the summer. Where reservoirs are used to prevent flooding there needs to be room to store the winter flood water. This may mean allowing the level of the reservoir to fall deliberately in the autumn and early winter to allow sufficient capacity to contain any flood water. This is the practice at the Clywedog reservoir on the upper Severn. However, if the winter happens to be drier than expected, then the reservoir may be only partially full at the beginning of the summer. Operating reservoirs and regulating rivers is a delicate art, and as the weather is so unpredictable, decisions made on the best available information many months previously may prove to have been incorrect (Parr *et al.*, 1992).

Many of the world's larger river systems flow through more than one country, with the countries downstream dependent on the behaviour of those upstream to ensure adequate water volume and quality for their needs. Increasingly, overabstraction and the construction of dams is dramatically reducing flows in some major international rivers, resulting in severe water shortages for downstream countries. This is leading to increasing hardship and in many cases conflict (Pearse, 1992).

2.4 GROUNDWATER SOURCES

2.4.1 Groundwater Supplies

About a quarter of potable water supplies in Britain come from groundwater resources, although in other countries the dependence on groundwater is much greater (Tables 2.2 and 2.3). Economically groundwater is much cheaper than surface water, as it is available at the point of demand at relatively little cost and it does not require the

Table 2.2. Percentage of drinking water derived from groundwater resources in various European countries. Reproduced by permission of the World Health Organization

Country	Percentage from groundwater	Country	Percentage from groundwater
Denmark	98	Luxemburg	66
Austria	96	Finland	49
Portugal	94	Sweden	49
Italy	91	Greece	40
Germany (former West)	89	UK	25
Switzerland	75	Ireland	25
France	70	Spain	20
Belgium	67	Hungary	10

Table 2.3. Amount of drinking water supplied from surface and groundwater in the UK. Reproduced by permission of the Water Services Association

Country	Amount (Ml/d)	Source(%)	
		Surface water	Groundwater
England and Wales	19 500	72	28
Scotland	2210	97	3
Northern Ireland	680	92	8
Total	22 390	75	25

construction of reservoirs or long pipelines. It is usually of good quality, usually free from suspended solids and, except in limited areas where it has been affected by pollution, free from bacteria and other pathogens. Therefore it does not require extensive treatment before use (Section 3.2).

British groundwater is held in three major aquifer systems, with most of the important aquifers lying south-east of a line joining Newcastle upon Tyne and Torquay (Figure 2.3). An aquifer is an underground water-bearing layer of porous rock through which water can flow after it has passed downwards (infiltration) through the upper layers of soil. On average 7000 Ml of water is abstracted from these aquifers each day. Approximately 50% of this vast amount of water comes from Cretaceous chalk aquifers, 35% from Triassic sandstones and the remainder from smaller aquifers, the most important of these being Jurassic limestones. Of course, groundwater is not only abstracted directly for supply purposes; it often makes a significant contribution to rivers also used for supply by discharging into the river as either base flow or springs (Figure 2.4). The discharge of groundwater into rivers may be permanent or seasonal, depending on the height of the water-table within the aquifer. The water-table separates the unsaturated zone of the porous rock comprising the aquifer from the saturated zone; in essence it is the height of the water in the aquifer (Figure 2.5). The water-table is measured by determining the level of the water in boreholes and wells. If

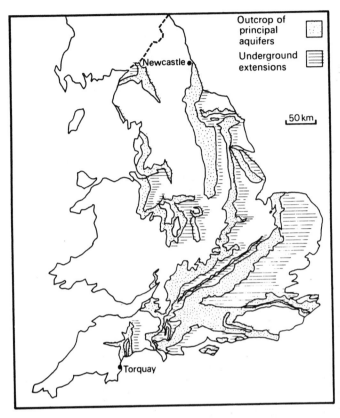

Figure 2.3. Location of principal aquifers in England and Wales. Reproduced by permission from the Open University (1974)

numerous measurements are taken from wells over a wide area, then the water-table can be seen to fluctuate in height depending on the topography and climatic conditions. Rainfall replenishes or recharges the water lost or taken from the aquifer and so raises the level of the water-table. If the level falls during periods of drought or due to overabstraction for water supply, then this source of water feeding the river may cease. In periods of severe drought groundwater may be the only source of water feeding some rivers and so if the water-table falls below the critical level, the river itself could dry up completely (Owen, 1993).

Groundwater abstraction in the southern regions of England has increased substantially over the past 20 years. In some areas water levels in wells have been reported to be falling year by year. Many southern chalk streams are drying up due to overabstraction, not only for drinking water supply, but for crop irrigation and industrial use. Licences are currently being refused for new abstractions. Some of the best salmonid fishing streams are suffering and the headwaters of rivers such as the Test and Itchen have dried up in recent years. The flow in the River Allen in Dorset has been reduced to a very small stream where 25 years ago fishermen required thigh waders.

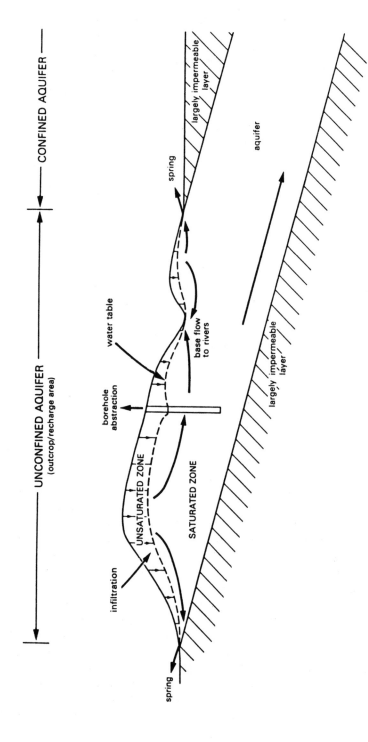

Figure 2.4. Schematic diagram of groundwater systems. Reproduced from Lawrence and Foster (1987) by permission of the British Geological Survey

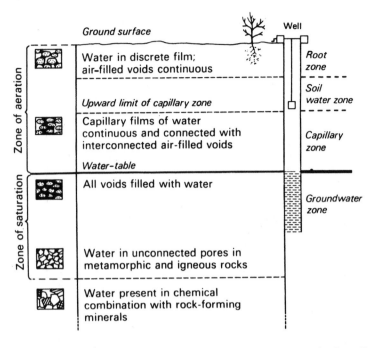

Figure 2.5. Cross-section through soil and aquifer showing various zones in the soil and rock layers and their water-bearing capacities. Reproduced by permission from the Open University (1974)

This has led to an outcry by fishermen and conservationists alike. The water companies have struggled through years of drought to maintain an ever-increasing demand for water by allowing more and more groundwater to be abstracted. Aquifers, especially chalk aquifers, refill with water (recharge) very slowly and so a rapid response by the NRA to prevent further reduction in flows is needed. The NRA has compiled a list of the 40 rivers most badly affected by overabstraction, and set in hand a programme of priority action for the top 20 rivers. Two priority action areas are tributaries of the River Test. The Wallop Brook is under threat from a nearby public supply borehole with a licence of right that is drawing river water through the stream bed. The river is literally flowing away down a plug-hole in its base. The flow in the River Bourne is also severely depleted by boreholes used to feed cressbeds. Such riverside groundwater abstraction will often induce inflow of river water into aquifers. Apart from reducing flows, if the river is polluted this will contaminate the groundwater causing quality problems in water supplies, and so the practice is undesirable.

As drinking water can come either directly or indirectly from groundwater sources, its quality is important to many more than simply those people who receive supplies directly via boreholes and wells. The principal aquifers lie beneath extensive areas of farmland in eastern, central and southern England, and in many of these areas groundwater may contribute more than 70% of the drinking water supplies (Table 2.4).

Table 2.4. Proportion of regional supplies abstracted from groundwater sources in England and Wales. Dependence on groundwater may be very much higher locally within these regions. Reproduced by permission of Water Services Association (1990)

Water supply region	Percentage of supplies from groundwater
Southern	73
Wessex	50
Anglian	44
Thames	42
Severn Trent	37
North West	13
Yorkshire	13
South West	10
Northumbria	9
Welsh	4

2.4.2 Aquifer Classification

Aquifers are classified as either confined or unconfined. An unconfined aquifer is one that is recharged where the porous rock is not covered by an impervious layer of soil or other rock. The unsaturated layer of porous rock is separated from the saturated water-bearing layer by an interface known as the water-table. The unsaturated layer is rich in oxygen. Where the aquifer is overlain by an impermeable layer, no water can penetrate into the porous rock from the surface; instead, water slowly migrates laterally from unconfined areas. This is a confined aquifer. There is no unsaturated zone because all the porous rock is saturated with water as it is below the water-table level, and of course there is no oxygen (Figure 2.4) (Brown *et al.*, 1983). Because confined aquifers are sandwiched between two impermeable layers the water is usually under considerable hydraulic pressure, so that the water will rise to the surface under its own pressure via boreholes and wells, which are known as artesian wells. Artesian wells are well known in parts of Africa and Australia, but are also found on a smaller scale in the British Isles. The most well known artesian basin is that on which London stands. This is a chalk aquifer which is fed by unconfined aquifers to the north (Chiltern Hills) and south (North Downs) (Figure 2.6). In the late 19th century the pressure in the aquifer was such that the fountains in Trafalgar Square were fed by natural artesian flow. However, if the artesian pressure is to be maintained then the water lost by abstraction must be replaced by recharge, in the case of London by infiltration at the exposed edges of the aquifer basin. Continued abstraction in excess of natural recharge has lowered the piezometric surface (i.e. the level that the water in an artesian well will naturally rise to) by about 140 m below its original level in the London basin. A detailed example of an aquifer is given later in this section.

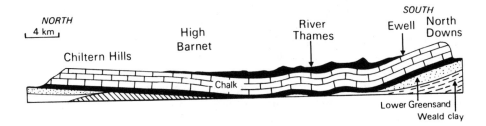

Figure 2.6. Cross-section through the London artesian basin. Reproduced by permission from the Open University (1974)

It is from unconfined aquifers that the bulk of the groundwater supplies are abstracted. It is also from these aquifers, in the form of baseflow or springs, that a major portion of the flow of some lowland rivers in eastern England arises. These rivers are widely used for supply purposes and so this source of drinking water is largely dependent on their aquifers, so good management is vital.

Groundwater in unconfined aquifers originates mainly as rainfall and so is particularly vulnerable to diffuse sources of pollution, especially agricultural practices and the fallout of atmospheric pollution arising mainly from industry. Increased chemical and bacterial concentrations in excess of the EC limits set in the Drinking Water Directive have been recorded in isolated wells for many years. This was a local phenomenon with the source of the pollution usually easily identified to a local point source such as a septic tank, a leaking sewer or farmyard drainage. In the 1970s there was concern over increasing nitrate levels in particular, which often exceeded the EC maximum admissible concentration. This was not an isolated phenomenon, but increased levels were found throughout all the principal unconfined aquifers in the UK. The areas affected were so large that clearly only a diffuse source could be to blame, rather than point sources (Royal Society, 1984). However, it was not until the mid-1970s that the widespread increase in nitrate concentrations in groundwater was linked to the major changes in agricultural practices that had occurred in Britain since the Second World War. The major practice implicated at that time was regular cereal cropping, which has to be sustained by the increasing use of inorganic fertilizers. The nitrate problem is fully explored in Section 4.2. Trace amounts of organic compounds are another major pollutant in groundwaters. As a result of the small volumes of contaminant involved, once dispersed within the aquifer it may persist for decades. Many of these compounds originate from spillages or leaking storage tanks. A leak of 1,1,1-trichloroethane (also used as a thinner for typing correction fluids) from an underground storage tank at a factory manufacturing microchips for computers in San Jose, California in the USA, caused extensive groundwater contamination resulting in serious birth defects, including miscarriages and stillbirths, in the community receiving the contaminated drinking water. Agriculture is also a major source of organic chemicals. Unlike spillages or leaks which are point sources, agricultural-based contamination is a dispersed source. In the UK, pesticides which were banned in the early 1970s such as DDT have only recently appeared in groundwater supplies. Many different pesticides are also being reported in drinking water and this is examined in more depth in Section 4.3. Other important sources of pollution are landfill and dump sites, impoundment lagoons including slurry pits, the disposal of sewage sludge onto land, runoff from roads and mining (OECD, 1989).

Unlike confined aquifers, unconfined aquifers have both an unsaturated and a saturated zone. The unsaturated zone is situated between the land surface and the water-table of the aquifer. Although it can eliminate some pollutants, the unsaturated zone has the major effect of retarding the movement of most pollutants, thereby concealing their occurrence in groundwater supplies for long periods. This is particularly important with major pollution incidents, where it may be many years before the effects of a spill or leakage from a storage tank, for example, will be detected in the groundwater due to this prolonged migration period (Eriksson, 1985).

Most of the aquifers in Britain have relatively thick unsaturated zones. In chalk they vary from 10 to 50 m in thickness, which means that surface-derived pollutants can remain in this zone for decades. Another problem is that the soils generally found above aquifer outcrops are thin and highly permeable, and so allow the rapid infiltration of water to the unsaturated zone taking the pollutants with them. Soil bacteria and other soil processes therefore have little opportunity to utilize or remove pollutants. This unsaturated zone is not dry; it does in fact hold large volumes of water under tension in a process matrix, along with varying proportions of air. However, below the root zone layer, the movement of this water is predominantly downwards, albeit extremely slowly (Figure 2.5).

It is in the saturated zone of unconfined aquifers that the water available for abstraction is stored. The volume of water in unconfined aquifers is many times the annual recharge from rainfall. It varies according to rock type and depth, but, for example, in a thin Jurassic limestone the ratio may be up to three, whereas in a thick porous Triassic sandstone it may exceed 100. The saturated zone also contains a large volume of water which is immobile, locked up in the microporous matrices of the rock, especially in chalk Jurassic limestone aquifers.

Where aquifers have become fissured (cracked), movement is much more rapid. However, the movement of pollutants through unfissured rock, by diffusion through the largely immobile water which fills the pores, will take considerably longer. This, combined with the time lag in the unsaturated zone, results in only a small percentage of the pollutants in natural circulation within the water-bearing rock being discharged within a few years of their originally infiltrating through agricultural soils. Typical residence times vary but will generally exceed 10–20 years. The deeper the aquifer, the longer this period. Other factors, such as enhanced dilution effects where large volumes of water are stored, and the nature of the porous rock all affect the retention time of pollutants. An excellent introduction to aquifers and groundwater has been prepared by Price (1991), whereas Raghunath (1987) has published a more technical and advanced monograph. The management and assessment of the risk of contamination in groundwaters is reviewed by Reichard et al. (1990).

2.4.3   Quality

The quality of groundwater depends on a number of factors: (1) the nature of the rain-water, which can vary considerably, especially in terms of acidity due to pollution and the effects of wind-blown spray from the sea which affects coastal areas in particular; (2) the nature of the existing groundwater which may be tens of thousands of years old; (3) the nature of the soil through which water must percolate; and (4) the nature of the rock comprising the aquifer.

In general terms groundwater consists of a number of major ions which form compounds. These are calcium, magnesium, sodium, potassium and to a lesser extent iron and manganese. These are all cations (they have positive charges), which are found in water combined with an anion (which have negative charges), to form compounds referred to as salts. The major anions are carbonate, hydrogencarbonate, sulphate and chloride.

Most aquifers in the country have hard water. Total hardness is made up of carbonate (or temporary) hardness caused by the presence of calcium hydrogencarbonate ($CaHCO_3$) and magnesium hydrogencarbonate ($MgHCO_3$), whereas non-carbonate (or permanent) hardness is caused by other salts of calcium and magnesium (Section 4.6). It is difficult to generalize, but limestone and chalk aquifers contain high concentrations of calcium hydrogencarbonate, whereas dolomite aquifers contain magnesium hydrogencarbonate. Sandstone aquifers are often rich in sodium chloride ($NaCl$), whereas granite aquifers have elevated iron concentrations. The total concentration of ions in groundwater, the conductivity, is often an order of magnitude higher than in surface waters. The Department of the Environment (1988a) has published an excellent review of groundwater quality for England and Wales.

The conductivity (the total amount of anions and cations present) also increases with depth due to less fresh recharge to dilute existing groundwater and the longer period for ions to be dissolved into the groundwater. In very old, deep waters the concentrations are so high that they are extremely salty. Such high concentrations of salts may result in problems due to overabstraction or in drought conditions when old saline groundwaters may enter boreholes through upward replacement, or due to saline intrusion into the aquifer from the sea.

In terms of the volume of potable water supplied, confined aquifers are a less important source of groundwater than unconfined aquifers. However, they do contribute substantial volumes of water for supply purposes and can locally be the major source of drinking water. Groundwater in confined aquifers is much older than in unconfined aquifers and so is characterized by a low level of pollutants, especially nitrates and micro-organic pollutants, including pesticides. This source is currently of great interest for use in diluting water from sources with high pollutant concentrations, a process known as blending (Section 4.2).

Confined aquifers are generally not used if there is an alternative source of water because of low yields from boreholes and quality problems, especially high salinity in some deep aquifers, excessive iron and/or manganese, problem gases such as hydrogen sulphide and carbon dioxide, and the absence of dissolved oxygen. These problems can

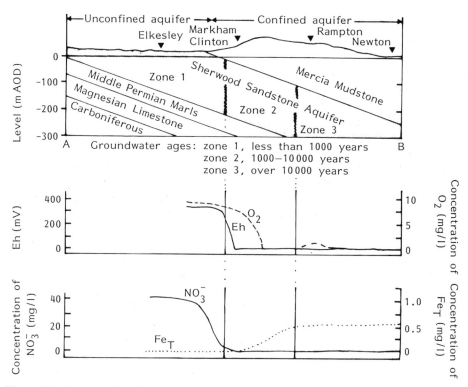

Figure 2.7. Hydrogeochemistry of the Sherwood Sandstone aquifer. Reproduced from Kendrick *et al.* (1985) by permission of the Water Research Centre

be overcome by water treatment, although this, combined with the costs of pumping, makes water from confined aquifers comparatively expensive.

An example of a major British aquifer is the Sherwood Sandstone aquifer in the East Midlands, which is composed of thick red sandstone. It is exposed in the western part of Nottinghamshire, which forms the unconfined aquifer, and dips uniformly to the east at a slope of approximately 1 in 50. It is overlain in the east by Mercia Mudstone to form a confined aquifer (Figure 2.7). Three groundwater zones have been identified in this aquifer, each of increasing age. In zone 1 the groundwater is predominantly modern and does not exceed a few tens or hundreds of years in age, whereas in zone 2 the groundwater ranges from 1000 to 10 000 years old. In zone 3 the groundwater was recharged 10 000–30 000 years ago. In this aquifer the average groundwater velocity is very slow, just 0.7 m/year (Edmunds *et al.*, 1982). The variations in chemical quality across these age zones is typical of unconfined and confined aquifers. This is seen in the other major aquifers, especially the Lincolnshire limestone aquifer in eastern England where the high nitrate concentrations in the youngest groundwater zone are due to the use of artificial fertilizers in this intensively cultivated area. There are large groundwater resources below some major cities; in the UK these include London, Liverpool, Manchester, Birmingham and Coventry. However, these resources are particularly at risk from point sources of pollution from industry, especially solvents and other organic chemicals (Section 4.3), as well as more general contaminants from damaged sewers and urban runoff. Such resources are potentially extremely important, although due to contamination they are often under-exploited (Lerner and Tellam, 1993).

2.4.4 Demand and Groundwater

Water demand is continuing to increase rapidly in southern England. However, there is already a danger that current abstraction rates for some major lowland rivers are approaching the total usable flow under dry weather conditions. In some localized areas this has already been exceeded with serious environmental consequences. There is now considerable opposition to flooding valleys for new reservoirs, mainly on landscape quality and aesthetic grounds, so the development of groundwaters has been proposed as the obvious way forward. Supporters cite a number of facts to support this approach, namely: (1) groundwater abstraction does not affect the amenity value or current land use and appears to have a low environmental impact; (2) the major aquifers in Britain coincide with centres of maximum water demand, where alternative supplies are scarce and where land prices are highest; (3) reservoir construction and operation is far more expensive than groundwater abstraction; and (4) the large losses of water from reservoirs due to evaporation in the summer do not occur with groundwaters, and so it is more efficient in terms of resource utilization. However, in practice it seems unlikely that existing groundwater resources can be exploited much further, with chalk aquifers in particular, such as the London Basin, becoming seriously depleted. Future increases in supplies must come from the conservation of existing supplies, reuse and the more prudent management of water supplies generally.

Throughout Europe groundwater supplies are being taken out of service as the levels of contamination exceed legal limits or become uneconomic to treat using conventional water treatment methods (OECD, 1989). Treating pollution within aquifers is currently not possible although there are some interesting developments. For example, nitrate concentrations tend to decrease with depth due to dilution. Deepening wells, or encasing the upper sections of boreholes to prevent contamination from the upper sections of the aquifer ensures that more of the supply water is abstracted from the less contaminated water lower in the aquifer, thus ensuring a reduction in nitrate concentrations. This is only a short-term measure, but recent research has used shallow scavenging boreholes, which are located around the major supply borehole, to draw off the high nitrate water preventing it from contaminating the supply. The high nitrate water is then used for irrigation where the nitrate is used by plants. This can reduce nitrate levels in water supplies in the long-term, especially where linked with improvements in agricultural practice (Section 4.2). Other ideas include seeding aquifers with bacteria and organic matter to encourage anaerobic denitrification (where the nitrate is converted to gaseous nitrogen). This process occurs naturally in some parts of confined aquifers due to the absence of oxygen, so the principle is to speed up this natural activity. There are a number of other options still very much in the experimental phase which may offer some hope for the future, especially on a local scale.

For those requiring further information on hydrology there are many excellent books available. Two particularly good texts are those by Wanielista (1990) and Ward and Robinson (1990).

2.5 CONSERVATION OF SUPPLIES AND METERING

2.5.1 Metering

Industry, as the major water user in Britain, has had its water supply metered for many years and so pays for the volume of water it uses. As the price of water has increased, and more importantly as shortages have become more frequent, industry has implemented strict conservation measures to minimize wastage and water use such as recirculation and reuse before discharge. The idea is that no process should use water of a higher quality than it needs, so that water can be reused over and over again with the processes requiring the purest water using it first, and those processes where quality is least important using it last.

Unlike industry, the private consumer has always seen water as a right, and one that should essentially be free. Since the formation of the regional water authorities in 1974, the idea of metering water supplied to homes and charging for what had been used rather than having a fixed charge has gradually become established. The prime motivation for the introduction of meters was really to conserve supplies and make consumers more responsible in their daily use of water. Certainly in earlier years when charges were low and water was costing on average just 2 pence/m^3 (1 m^3 = 1000 l), the argument that householders would be unresponsive to price changes and so unlikely to make economies was valid. The anti-metering lobby predicted widespread disease as people lowered standards of public hygiene to save a few litres of water. In response, the water authorities claimed at the time that water was absurdly cheap compared with other liquid commodities such as petrol at £100/m^3 or whisky at £3500/m^3, or the only other liquid commodity delivered to your door as is water, milk at £100/m^3. Of course, such comparisons themselves were, and remain, absurd. The important basis to metering was that there was, and still is, only a limited amount of water available for supply and that demand from consumers was, and remains, dangerously close to exceeding what was, and is, available. Also, people's attitude to water is generally one of disinterest and many customers carelessly squander water on an often staggering scale. Metering is a method of making people more aware of their use of water and rewarding them for using water wisely. The fact that those who use more water pay more if on a meter is perhaps of more significance to other consumers than to the water undertakers.

Since the Department of the Environment permitted water authorities to introduce metering in 1984, the installation of meters has been surprisingly slow and often restricted to new properties only. Although the percentage of consumers metered varies from area to area, overall only 2% of all consumers are on metered supplies. Although metering of domestic supplies is rare in Ireland, it is widespread in other European countries.

The Public Utility Transfers and Water Charges Act (1988), which was passed in April 1988, allowed water undertakers in the UK to set up metering experiments. This was to examine how the demand for water would alter using different charging methods and to investigate the practical problems which would arise from a large-scale implementation of water metering. This was carried out at 12 locations, including the Isle of Wight where this involved installing a water meter in all 53 000 properties on the island. This particular metering experiment is especially interesting as the island has limited water resources. In fact, in 1978 a water main was laid across the Solent to top up supplies during periods of peak use at the height of the tourist season, although this is now in continuous use throughout the year. Southern Water were looking at several future options to expand their supply, which included a second cross-Solent link or another reservoir. Both options are very expensive, so if they could be deferred by implementing conservation measures, considerable capital saving would ensue. The results so far have been very encouraging. The general response from customers is that water metering is a fairer method of payment. Customers have been acutely aware of the slightest leakage or wastage and many have reduced their personal consumption by careful use. This has led to a significant reduction in water usage. However, as with other trials, caution is needed in these early stages as consumers tend to be extra careful in their use during the first three or four years, and this is then followed by a gradual relaxation in water use as the system becomes more accepted and familiar (Smith and Rogers, 1990).

The average household bill for water supply has increased by 84% since 1984–5 from £39.92 to £73.51 in 1991–2. Metered charges have also increased significantly by 100.9% over the same period from 23.5 to 47.2 pence/m^3. However, in addition to the volume charge there are fixed charges for different sizes of meter. Most water supply companies charge for the meter to be installed, so that the saving made by consumers over a five year period may be very marginal indeed. Different companies have different charges, and indeed very different attitudes, to metering. Families of different sizes use different amounts of water, and this will be reflected in the metered charge (Table 2.5). This table does not include any standing charge or extra tariffs for exceeding an annual volume allowance, so in practice the actual bill will be larger. Water charging methods are currently under review by Ofwat. They examined the three main methods. Metering is thought to cause social hardship on heavy water users who are often those

Table 2.5. Estimated annual use of water by households of different sizes and average charge for metered water (1991/2). The average annual non-metered charge is £74 in all instances

Size of household	Water consumption (m^3/year)	Annual metered charge (£)*
Single person	40–60	19–28
Couple sharing	70–110	33–52
Small family	110–150	52–71
Large family	150–210	71–99

*Does not include any standing or rental charge, or other tariffs based on volume used.

on poor incomes, those doing dirty jobs requiring more washing and cleaning, or those who are disabled who may also be incontinent. The flat rate system does not take into account a person's ability to pay. The system based on the ratable value of the home, which was based on the potential usage as assessed by the number of bathrooms, does appear to be fairer, but penalizes those, generally the elderly, who live in larger and often older properties. Ofwat distributed 18 million leaflets to consumers to identify which method of payment customers would prefer. Results from a preliminary survey of 3700 adults in England and Wales showed that 54% of people favoured metering in preference to either a flat rate or a rateable charge, which were preferred by 9 and 25%, respectively (Ofwat, 1991). From 1 April 2000, water supply companies will need to develop a replacement system for charges to those customers not on meters. This is because after this date ratable values will not be available for calculating water charges. Currently all new properties are having water meters installed. The results of the National Metering Trials, the second report of which was published by the Water Services Association in July 1990, are of vital importance for the development of new methods of levying charges. Not all water supply companies are happy about metering. Welsh Water, for example, would prefer a flat rate due to the cost of supplying meters and subsequently reading them, as their customers are so dispersed. Metering is clearly most cost-effective in built-up areas. To overcome the cost and logistical problems of reading meters, research is underway to find ways of reading meters automatically from a central location using telemetry.

2.5.2 Conservation in the Home

Each consumer uses on average each day 50 l for flushing the toilet, 45 l for washing and bathing, 14 l for washing clothes, 14 l for washing up the dishes, 8 l for outside uses and 4 l for drinking and cooking. Of course, these figures are a gross generalization, with actual usage depending on the type of house, behaviour and many other factors. For example, if a household owns a swimming pool, has young children, animals or livestock, or a very large garden, then their usage will be very much greater than the average. Consumers may be out all day, or send their clothes to a laundry, but either way they will still be using that volume of water either at work or at the launderette. One study comparing male- and female-only student hall of residences indicated that women used the toilet, as well as washed and bathed, more often than men. So there are many factors affecting our water usage, and it appears that a household consisting mainly of women will use considerably more water than a household consisting of the same number of men. Saving water is not a valid excuse for poor hygiene, so how can consumers realistically save water?

2.5.2.1 *The WC*

Every time the toilet is flushed 9–16 l of water go down the drain and the water meter clocks up another 0.4–0.8 pence. A typical family of two adults and three children will flush the toilet between 25 and 40 times a day. Studies have shown that in fact the toilet can be flushed up to 60 times a day by a family of four when it is being used to flush away discarded non-sanitary material. So the first thing is to ensure that the toilet is used wisely. It should not be used as a general waste disposal system such as flushing it to get rid of a discarded paper tissue, cotton wool, or hair from a comb or brush. If you are replacing your existing lavatory there are a number of new and innovative designs available. The latest works on a vacuum system and only uses 1 l of water per flush. Dual-volume flushing cisterns are also available. A simpler and cheaper method is to place a brick or small plastic bag full of water in the cistern, so reducing the volume of water of each flush; certainly in times of drought urine only should not be flushed away every time. Urine itself is practically free from bacteria and no known disease can be passed via human urine, so although unpleasant (especially if it is someone elses!), it is quite safe. Careful use of the toilet is a priority for the conservation of water.

The higher volume of waste water produced in Scotland is due primarily to the widescale use of a larger flushing cistern, 13.6 l compared with 9.0 l in the rest of the UK, although other factors also contribute to this variation. New UK guidelines from the Department of the Environment stipulate that all new cisterns manufactured after 1993 will have to have a maximum flushing volume of 7.5 l. However, the reliance of water closets which function on a siphon rather than a valve to release water restricts the minimum volume to between 4 and 5 l. A study by the Building Research Establishment (1987) highlights the potential water saving from the adoption of new cistern designs and suggests the need for new British Standards.

2.5.2.2   *Washing and bathing*

We have never been so obsessed with personal hygiene as we are at present. The amount of money spent on advertising shampoo and soap, usually involving a foaming luxuriant bath speaks for itself. Baths, although luxurious, are wasteful of water, whereas showers use only 20–30% of the water and so are more economic both in terms of the energy used in heating the water and the volume of water used. Personal hygiene is obviously important and so the only way in which we can conserve water is to use less water doing it. We can do this by not filling the bath quite so full, or not filling the wash basin to the top. It is staggering how many people leave the tap running when brushing their teeth. Depending on how slow a person is at doing this chore, quite a few litres of water can be wasted in this way each day.

2.5.2.3   *Laundry*

Automatic washing machines with their double washes and treble rinses use a large amount of water. The old twin-tub was much more efficient, but of course much harder work, and also far less convenient. Washing machine manufacturers have responded fairly reasonably to criticism from all around the world by producing new machines which take bigger loads, use less water and less energy. They even allow you to wash half-loads. Of course, any washing machine will wash half a load, but only the more recent designs actually regulate the volume of water to the amount of clothes placed in the machine. Washing machines, like refrigerators and cookers, are now almost universally used in the developed world, so improvements in their efficiency will have a significant effect on the environment. When using a washing machine always make sure that it is used for a full load, use less washing powder to ensure that an extra rinse is not required and, if possible, change your clothes a little less often, thereby reducing the total volume of clothes which need to be washed each week. The water from the last couple of rinses from a washing machine are so clean that it seems a shame that they can't be put to some good use in washing the car or watering the garden. It only involves a simple modification to the outflow pipe, but do be careful not to restrict the flow from the outflow pipe or the kitchen will become flooded and the machine may be damaged.

2.5.2.4 Washing-up

There can be few of us that would not abandon washing-up to save water if at all possible. Yet we each use on average 14 l each day in cleaning our dishes, cutlery and pans. As with washing clothes, there is little we can do to conserve water except to be more economical and efficient in the way we use it. Only wash up when there is a reasonable amount; do not wash things under the tap and always fill a bowl rather than the sink; do not rinse articles separately under a running tap—instead add less detergent (which is better for the environment anyway) and allow the dishes to drain. Dishwashers use much higher volumes of water than washing-up manually. Garbage grinders also use large volumes of water. Both should be avoided during periods of water shortage. Dishwashers should never be used for a small number of items only, and if used they should always be full. As mentioned in Chapter 1, dishwashers use nearly 50 l per wash cycle—that is a lot of water, about 18 250 l/year assuming just a single wash cycle each day.

2.5.2.5 Cooking and drinking

Only a few per cent of the total volume of water we use each day is used for cooking and drinking. Good hygiene practice in food preparation is so important that no conservation measures should be taken, except for perhaps moderation in rinsing vegetables.

The water supply companies treat all their water to an exceptionally high standard so that it is safe to drink all of it, even the water used to flush the toilet.

2.5.2.6 Garden and other outside uses

Uses outside the house are those which traditionally have been concentrated on by water supply companies during periods of drought. They feel that when water is scarce it is totally wasteful, and in fact criminal, to water lawns and wash cars. Gardens are clearly a major problem during droughts. There were hosepipe bans in parts of the UK in 1989 and 1990 affecting over 20 million people. In 1991, although the spring was wet the summer became increasingly dry so that by mid-summer hosepipe bans were affecting over 7 million people. No-one likes to see plants die due to a lack of water.

Lawns will recover even after fairly long periods of drought, whereas vegetables and shrubs may not. So water only those plants that need it, or which are going to be eaten. Water them around the root area only, as watering the bare earth is just a waste. Always water in the late evening so that the water is not lost by evaporation. As our pattern of weather changes, and it does seem to be changing, then the way in which we garden must alter to accommodate these changes. Keen gardeners should be recommended to make provision for water shortages by installing water butts and reusing water whenever possible.

There are many things consumers can do. In Ireland there are several new systems available for saving water. For example, one pioneered by Biocycle Ltd and now available throughout Europe is a novel water-saving system to ensure the safe disposal of domestic waste water while recirculating as much of the water as possible. The system consists of two large tanks sunk into the ground. The first is simply a water storage tank into which all the water from the roof flows. Inside the tank there is a submersible pump so that the water can be used for the garden or any other outside job. The second tank receives all the household's waste water. It is a combined septic tank and biological treatment system which cleans the used water to the same quality as a good sewage treatment plant. The difference is that the treated effluent is disinfected to destroy any pathogens that may be present. The main advantage of this system is that the water can be sprayed onto the surface of the ground where subsurface disposal using traditional percolation systems is not practicable. However, it can also be used to water lawns, herbaceous borders and hedges. The tank from the waste water treatment system pumps out several times a day depending on the water usage, and discharges the water through small spray nozzles close to the ground. The spray nozzles can even be controlled to deliver a certain amount of treated effluent each. It is perhaps not advisable to use this water on garden produce such as fruit and vegetables, but for other plants, especially lawns and herbaceous borders, it is excellent.

2.5.2.7 *Leaks*

Leaks are a major problem in the home and at work. In the home dripping taps, overflowing water tanks in the attic, or broken connections to the mains are all major sources of waste. Meters are placed at the consumer's connection to the mains and all the pipework from the stoptap onwards is their responsibility. So if a consumer has a leak and is metered, then he or she is paying for it. Dripping taps and badly adjusted ballcocks should be repaired as soon as possible, before they result in a more serious plumbing problem. However, it is the water supply companies themselves that waste the most water through worn out distribution mains and leaking reservoirs. It is estimated that a fracture occurs each year for every 5 km of pipe, caused by a multitude of factors such as careless excavation when laying or serving gas, electricity or telecommunication lines, from soil compaction and from traffic vibration, but in iron pipes leaks are mainly due to weakness as a result of corrosion. The amount of loss due to damaged pipes varies from area to area, but on average it is thought to be about 25% of all water supplied, although this can rise to an excess of 35% in some areas. In some areas pressure in the mains is reduced at night to minimize water loss via cracks.

Therefore, rather than investing in more reservoirs and exploiting new resources it may be better to repair leaky water mains and make the maximum use of existing resources. Some water supply companies have instigated leak detection programmes which have been exceptionally successful. Wessex Water reduced losses from 27 to 20% by implementing an intensive detection and repair programme, whereas North West Water have reduced losses from 38 to 29%. It is estimated that 15% is a realistic figure for all the water supply companies to aim for, and so there is a lot of water still to be conserved.

2.6 OTHER RESOURCES

2.6.1 Forecasting Water Demand

Although groundwater and surface water currently supply most of our water, increasing demand and increasing pollution of existing resources in some areas is intensifying the search for new resources. In many rivers abstraction already exceeds the total usable dry weather flow, the danger being that rivers could simply be sucked dry, destroying their flora and fauna as well as eliminating their potential for other uses. In the late 1970s and 1980s there was a general move away from reservoir construction, not only on a cost basis, but also due to intense opposition from the public on environmental and social grounds, as explained in the earlier sections. However, by the end of the 1980s it was clear that groundwater sources in southern England especially, but also throughout the UK, could not be significantly exploited any further to meet increased demands. Indeed, there is considerable evidence that many aquifers have been overexploited and are seriously depleted.

Perhaps the major reason for moving away from constructing new reservoirs is the difficulty of accurately predicting or forecasting water demand. Water usage soared from the end of the Second World War up to the end of the 1960s, by as much as 35% each decade. Water planners began to extrapolate from this rapid increase in water usage that demand for water would double again by the end of the century. New schemes were examined such as barrages and even a national water grid, and worries about massive water shortages were mooted. However, demand for water slackened and during the 1970s it increased by just 12%, and by only 10% in the 1980s. The current increase in demand is set to rise by only 6% in the 1990s. North West Water has seen no increase in demand this decade, due in part to a decrease in industry, and are planning for a gradual decrease over the next 10 years.

One of the most serious manifestations of the incorrect forecasting of the late 1960s was the construction of Kielder Water to satisfy totally erroneous water demands. Kielder Water was built to supply water to the great companies in the north-east such as British Steel. It was approved in October 1973 and its construction started in 1976, spurred on by the drought. However, by the time it was constructed the demand for water in the area had decreased dramatically due to a reduction in heavy industry and a fall off in the rate of increase in domestic usage. Figure 2.8 shows the type of forecast on

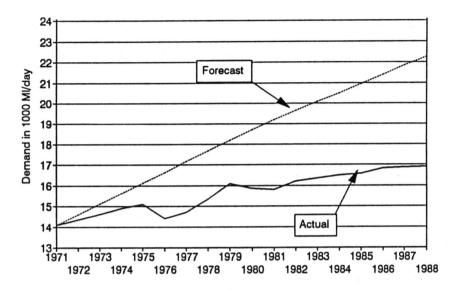

Figure 2.8. Type of forecasting used to predict future water consumption in the early 1970s compared with actual demand in England and Wales over the period 1971–88. Reproduced from Clayton and Hall (1990) by permission of the Foundation for Water Research

which the decision to construct Kielder Water was based. One of the largest reservoirs in Europe, Kielder Water has a storage capacity of 200 000 Ml, more than enough to satisfy the water demands in south-east England in full, including the current predicted increase for the next 20 years, if only it could be transported. Kielder Water, which required one and a half million trees to be felled during its construction, is just about as far north as you can get in England, just 5 km from the Scottish border. It captures water from the River Tyne and doubled the availability of water in the north-east. Since its completion in 1980 (it took two years to fill), demand has risen by just 8% and so far neither the reservoir nor the expensive tunnels and pipes built to supply the cities on the lower Tyne and Tees have ever supplied customers. The mistake of the construction of Kielder Water is perhaps a major contributing factor to the decrease in reservoir construction generally over the past 15 years.

However, although demand has decreased in the north, it has soared in the south and south-east. To meet this demand the water supply companies must build new reservoirs. There are a number of major new reservoirs being proposed at Broad Oak and Darwell (Southern Water), Didcot (Thames Water), Great Bradley (Anglian Water) and the Axe Valley (South West Water).

Alternative sources of water will have to be developed by using poor quality resources and reusing and recycling water. Aquifers are now managed, often with groundwater levels maintained by artificial recharge. Where alternative supplies are not available attention has focused on conservation, with metering becoming increasingly common. Transferring water from one catchment where water is plentiful to another where supplies are low has led to the idea of a national grid for water, the water being transferred from the wet north-west to the dry south-east. Other newer techniques are also being examined such as desalination, ion-exchange reactions and even importing water. All these are considered below with further details on resource management given by Blackburn (1978), McDonald and Kay (1988) and OECD (1989).

2.6.2 Use of Poor Quality Resources

Where new surface or groundwater supplies cannot be exploited, the only option may often be the utilization of water resources previously rejected for supply purposes due to poor quality or pollution. New advanced methods of water treatment such as granular activated carbon or ion exchange can in theory produce pure water from raw water of any original quality, no matter how poor. Although technically feasible, the quality of raw water permitted to be abstracted from surface waters, including lakes and reservoirs, is controlled by legislation. The quality of surface water abstracted for public supply must conform to EEC Directive 75/440/EEC. This lays down the required quality of water that can be abstracted for public supply and classifies all suitable waters into three broad categories: A1, A2 and A3. It also specifies the level of water treatment necessary to transform each category of water to drinking water quality. Water treatment processes are described in detail in Section 3.2.

A1 water resources are the cleanest and require the minimum of treatment such as simple physical treatment (e.g. rapid filtration) followed by disinfection. Waters in category A2 require normal physical treatment, chemical treatment and disinfection. A typical A2 treatment plant may include pre-chlorination, coagulation, flocculation, decantation, filtration and finally chlorination for disinfection. Most surface waters currently used in the UK fall into either the A1 or A2 category. However, many of the lowland rivers that flow through major conurbations where water demand is greatest, so that many will have to be used in the future as sources of drinking water, fall into the poorest category A3. To transform A3 surface waters to drinking water standard advanced treatment techniques are required which involve intensive physical and chemical treatment, extended treatment and disinfection. An example of the steps involved in A3 treatment is chlorination, coagulation, flocculation, decantation, filtration, absorption using activated carbon, and finally disinfection using chlorination. Clearly, the better the quality of the surface water (A1), the less treatment is required and the cost is at a minimum. Conversely, the poorest quality waters may often be prohibitively expensive to treat.

Where the quality of surface waters falls short of the mandatory limits for category A3, then they may not be used for abstraction. However, exceptions can be made for exceptional circumstances when special treatment must be used, such as blending, to bring the supply up to acceptable drinking water standards. The European Commission itself must be notified of the grounds for such exemption on the basis of a water resources management plan for the area concerned. This ensures that other more suitable resources are not being ignored and that there is no alternative to using the poorer quality water. In the list of exceptional circumstances are: (1) a reduction in quality due to floods or other natural disasters; (2) increased levels of specified parameters such as nitrate, temperature or colour, copper (for A1 waters), sulphate (for

A2 and A3 waters), and ammonia (A3 waters) due to exceptional meteorological or geographical conditions; (3) where surface waters undergo natural enrichment (eutrophication); and (4) nitrate, dissolved iron, manganese, phosphate, chemical and biochemical oxygen demand and dissolved oxygen in waters abstracted from lakes and stagnant waters which do not receive any discharge of waste, which are less than 20 m in depth, and which have a very slow exchange of water (less than once each year).

Exemptions cannot be made where public health is threatened, and of course the EC Drinking Water Directive is unaffected, so the water still has to be treated to an acceptable standard.

The EC Directive on surface water abstraction strictly covers surface waters for human consumption which are supplied to the consumer via a distribution network. It does not, for example, cover the quality of water used to recharge aquifers used for public supply. Groundwaters and brackish waters are also not covered.

About 46 chemical, physical and biological parameters are listed, although so far no values have been set for seven of the parameters listed, including nickel, total organic carbon and organic chlorine. The water must be regularly monitored to show that 95% of the samples of water taken conform with the mandatory I values. Where I values are not specified, as is the case with dissolved iron and manganese, then 90% of the samples must conform with the G values. There are exceptions to this that are rather complicated, so reference should be made directly to article 5 of the EC Directive if necessary (EC, 1975). The G values listed in this EC Directive are given in Table 2.6. The extent and frequency of sampling and analysis was not specified in this EC Directive; a separate Directive (79/869/EEC) known as *Methods of analysis and frequency of sampling* was adopted on 9 October 1979 and covers the technical and logistic aspects of the Surface Water Directive in considerable detail. Many Member States consider the Surface Water Directive to be obsolete since the introduction of the Drinking Water Directive and have proposed replacing it with an ecological Directive for surface waters.

2.6.3 Reuse and Recycling

In areas where water is scarce, treated sewage effluent may be reused after sufficient disinfection for uses not associated with human or animal consumption. At one time it was proposed to have a dual water supply system in UK towns, one with fully treated water and the other with a lower grade of water, usually untreated river water. The idea was to conserve water of the highest quality for consumption and household uses while also saving money by not having to treat all the water being supplied. Although this system is in use in some other countries it was never feasible in the UK because of the high cost of providing two separate mains, having two separate plumbing systems in buildings, plus the danger of people mistakenly using the lower grade of water for consumption. There is a growing fear that unless consumers demand drinking water on tap, then the few litres we consume daily may be delivered separately in bottles and cartons, rather like milk. In fact, several major dairies already supply bottled water along with their regular milk deliveries. In cost terms this could make a lot of sense to

Table 2.6. Standards set by the EC for surface waters intended for the abstraction of drinking water. Reproduced from EC (1975) by permission from the Commission of the European Union

Parameter (units)	A1 G	A1 I	A2 G	A2 I	A3 G	A3 I
1 pH	6.5–8.5		5.5–9		5.5–9	
2 Coloration (after simple filtration) (mg/l Pt scale)	10	20 (O)	50	100 (O)	50	200 (O)
3 Total suspended solids (mg/l SS)	25					
4 Temperature (°C)	22	25 (O)	22	25 (O)	22	25 (O)
5 Conductivity ($\mu s/cm^{-1}$ at 20°C)	1000		1000		1000	
6 Odour (dilution factor at 25°C)	3		10		20	
7* Nitrates (mg/l NO_3)	25	50 (O)		50 (O)		50 (O)
8† Fluorides (mg/l F)	0.7–1	1.5	0.7–1.7		0.7–1.7	
9 Total extractable organic chlorine (mg/l Cl)						
10* Dissolved iron (mg/l Fe)	0.1	0.3	1	2	1	
11* Manganese (mg/l Mn)	0.05		0.1		1	
12 Copper (mg/l Cu)	0.02	0.05 (O)	0.05		1	
13 Zinc (mg/l Zn)	0.5	3	1	5	1	5
14 Boron (mg/l B)	1		1		1	
15 Beryllium (mg/l Be)						
16 Cobalt (mg/l Co)						
17 Nickel (mg/l Ni)						
18 Vanadium (mg/l V)						
19 Arsenic (mg/l As)	0.01	0.05		0.05	0.05	0.1
20 Cadmium (mg/l Cd)	0.001	0.005	0.001	0.005	0.001	0.005
21 Total chromium (mg/l Cr)		0.05		0.05		0.05
22 Lead (mg/l Pb)		0.05		0.05		0.05
23 Selenium (mg/l Se)		0.01		0.01		0.01
24 Mercury (mg/l Hg)	0.0005	0.001	0.0005	0.001	0.0005	0.001
25 Barium (mg/l Ba)		0.1		1		1
26 Cyanide (mg/l Cn)		0.05		0.05		0.05
27 Sulphates (mg/l SO_4)	150	250	150	250 (O)	150	250 (O)
28 Chlorides (mg/l Cl)	200		200		200	
29 Surfactants (reacting with methyl blue) [mg/l (laurylsulphate)]	0.2		0.2		0.5	
30*† Phosphates (mg/l P_2O_5)	0.4		0.7		0.7	
31 Phenols (phenol index) paranitraniline 4-aminoantipyrine (mg/l C_6H_5OH)		0.001	0.001	0.005	0.01	0.1
32 Dissolved or emulsified hydrocarbons (after extraction by petroleum ether) (mg/l)		0.05		0.2	0.5	1

Table 2.6. (*continued*)

Parameter (units)	A1 G	A1 I	A2 G	A2 I	A3 G	A3 I
33 Polycyclic aromatic hydrocarbons (mg/l)		0.0002		0.0002		0.001
34 Total pesticides (parathion, BHC, dieldrin) (mg/l)		0.001		0.0025		0.005
35* Chemical oxygen demand (COD) (mg/l O_2)					30	
36* Dissolved oxygen saturation rate (%O_2)	>70		>50		>30	
37* Biochemical oxygen demand (BOD_5) (at 20°C without nitrification) (mg/l O_2)	<3		<5		<7	
38 Nitrogen by Kjeldahl method (except NO_3) (mg/l N)	1		2		3	
39 Ammonia (mg/l NH_4)	0.05		1	1.5	2	4 (O)
40 Substances extractable with chloroform (mg/l SEC)	0.1		0.2		0.5	
41 Total organic carbon (mg/l C)						
42 Residual organic carbon after flocculation and membrane filtration (5 μm) TOC (mg/l C)						
43 Total coliforms 37°C (/100 ml)	50		5000		50 000	
44 Faecal coliforms (/100 ml)	20		2000		20 000	
45 Faecal streptococci (/100 ml)	20		1000		10 000	
46 Salmonella	Not present in 500 ml		Not present in 1000 ml			

I = Mandatory; G = guide; O = exceptional climatic or geographical conditions.
* = see Article 8 (d).
†Values are upper limits set in relation to mean annual temperatures (high and low).
‡Included to satisfy ecological requirements of certain types of environment.

the water undertakers, especially if drinking water quality standards are increased further. However, in arid areas it is common to use sterilized treated effluents for non-consumable activities such as washing cars, flushing toilets and even washing clothes (Fewkes and Ferris, 1982).

In the UK at present about 30% of the raw water used for public supply is obtained from recycled effluent. This is a mean value for the whole of England and Wales, and in areas where supplies of upland water are very restricted, such as the south-east of England, this figure may be as high as 70%. Treated effluent is discharged from one consumer area into a lowland river and abstracted for reuse at the next urban area downstream. It is incredible, but true, that the rivers Thames and Lee consist of 95% effluent during dry summers. With the major areas of population centred in the Midlands and south-east England, it is unlikely that new upland supplies will be made available in the future and any subsequent increase in demand will have to be met by using groundwater or reclaimed water. The Thames River basin in England is an example of open cycle reuse, where the sewage from one community is converted to drinking water in another. Overall, the population of London is in excess of 10 million, with water being supplied from boreholes, the River Thames and its tributaries, including 1 200 000 m^3 of sewage daily (Figure 2.9). The recycle rate is, on average, 13%, but during the 1975–6 drought it exceeded 100%. The sewage receives full biological treatment followed by nitrification and denitrification when required, so that there is no ammonia or nitrate in the water. The water abstracted for supply from the Thames is stored for seven days before treatment, which normally involves slow sand filtration followed by chlorination. The reuse of the River Thames water is shown schematically in Figure 2.10. It is interesting that one community, Walton Bridge, actually discharges its effluent upstream of its own intake.

All municipal waste waters discharging to rivers used for public supply are fully treated biologically and the abstracted water is then subjected to full water treatment (which may also include treatment with activated carbon, membrane filtration or deionization if necessary) (Section 3.2). However, when water is recycled many times dissolved salts will accumulate in it, particularly the end products from biological sewage treatment. These include nitrate, sulphate, phosphate and chloride, all of which can cause unpleasant tastes and a decrease in quality, corrosion and scaling in pipes, and even toxicity. If there is no alternative source of supply then these inorganic salts may eventually have to be removed by advanced water treatment methods such as ultrafiltration, reverse osmosis or ion exchange, which makes the water expensive. Some pollutants are not easily removed and are very persistent. Although they are only normally present in trace amounts they are still toxic or in some instances carcinogenic.

Numerous surveys have indicated that there may be long-term effects in reusing water after it has been consumed, with a higher incidence of cancer among people drinking river water which has already been used for supply purposes compared with those drinking groundwater. However, a study carried out in London over the period 1968–74 showed that there was no significant difference between the mortality rate of those drinking reused River Thames water and groundwater (Figure 2.9) (Beresford, 1981). This work did, however, provide epidemiological evidence consistent with the earlier studies that a small risk to health exists from the reuse of drinking water. In this study, the percentage of domestic sewage effluent in the water was positively associated with the incidence of stomach and bladder cancers in women (Beresford et al., 1984). The health aspects of the reuse of waste water for human consumption is considered further in Section 4.7.

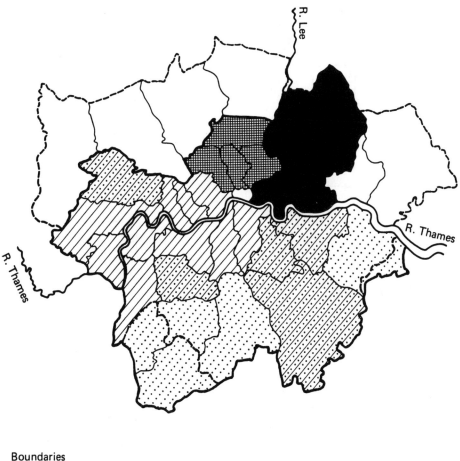

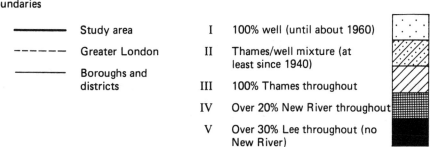

Figure 2.9. Sources of London's drinking water pre-1974. Reproduced from Beresford (1981) by permission of Oxford University Press. The effect of the new London ring main will be to mix water from all these sources

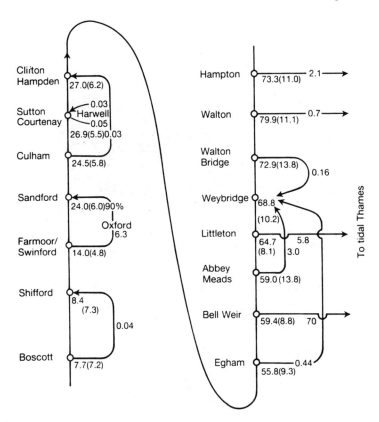

Figure 2.10. Example of the reuse of surface water for drinking water supply. The River Thames is shown schematically with the average river flows given in m^3/s; the percentage of this flow that consists of sewage effluent is given in parentheses. Reproduced from Dean and Lund (1981) by permission of Academic Press Ltd

2.6.4 Artificial Recharge and Aquifer Management

Water management schemes often involve two or more sources of supply. Usually one is used in normal circumstances, whereas the other is used only at times of peak demand or when there is a shortage of water from the normal source. Such schemes usually include a surface water supply, most probably a river, and a groundwater resource. The river is the primary source of water and the aquifer is used at times of low river flow.

Artificial recharge is now widely practised in the USA and the rest of Europe, although it is still uncommon in the British Isles. Unaided, aquifers recharge comparatively slowly, which means that the rate of water abstraction for supply is limited by the rate of natural recharge. Artificial recharge rapidly feeds water back into the aquifer during periods when it is plentiful on the surface. This ensures maximum utilization of the available storage capacity underground, thereby increasing the amount of water available to be abstracted during periods of shortages which would otherwise require reservoir construction for surface storage.

There are a number of methods of recharging aquifers. One of the simplest is to excavate a shallow lagoon and allow the water to percolate rapidly through the bottom into the aquifer. More commonly, recharge boreholes are used and the water is pumped directly into the aquifer. Its use is limited due to geological conditions in the UK, unlike The Netherlands where artificial recharge is widely practised. However, there are major concerns about this technique. River water is mainly used for artificial recharge and so the quality of this water will be very different from the groundwater and may alter the overall chemical nature of the water within the aquifer. More importantly surface water is of a lower quality and is also very susceptible to pollution. Therefore there is a very real risk of polluting all the groundwater within the aquifer by recharging with contaminated surface water. The significant amount of organic material and suspended solids in river waters in particular will increase biological activity within the aquifer and may even cause physical clogging. This is more of a problem where direct pumping is used via recharge boreholes, rather than infiltration lagoons where the water must percolate through the porous sediment in the unsaturated zone which improves the overall water quality. The high rate of infiltration can lead to clogging of the underlying sediment of lagoons if the water is too turbid, which seriously reduces the infiltration rate.

There has been much discussion of the possibility of using treated sewage effluent to recharge aquifers, thus conserving what water there is in areas of low rainfall. This is becoming an increasingly important issue as the winters have been comparatively dry, resulting in incomplete natural recharge of aquifers in some areas, with consequently less water for supply during the summer. In rural areas where the population is dispersed the effluent has traditionally been disposed of onto land either by surface irrigation or via percolation areas. In some areas, such as the chalk aquifer of southern England, there are so few rivers that it is sometimes difficult to find surface waters to discharge effluents into, so disposal to land in some instances has been going on for a long time. Water companies argue that in some areas the dependence on groundwater resources is so great (e.g. 70% in some parts of the Southern Water area, 50% in areas serviced by Thames and Anglian Water), that artificial recharge using treated effluent is the only practical option available if abstraction from aquifers is to be maintained at the same rate (Baxter and Clark, 1984).

Although it appears to be an efficient method of sewage disposal, it does cause concern about the long-term quality of groundwater resources. Although there is some removal of contaminants, including pathogenic micro-organisms, within the soil and the unsaturated zone, there is a tendency for most of these contaminants to slowly migrate downwards and eventually find their way into the saturated zone of the aquifer itself. Dilution is currently reducing the concentrations of contaminants to well below that required by the EC Drinking Water Directive, but with time concentrations will increase until an equilibrium is reached. There is a possibility of increased concentrations of organic chemicals and metals, in particular pesticides and the so-called sewage metals zinc, copper, chromium and lead, which are all found in slightly increased concentrations in treated sewage effluents. So far the most obvious increases have been in chloride and nitrate concentrations, but only time will tell exactly how safe effluent recharge is. Work carried out in other countries has shown that there may be a time lag of several years between the start of recharge and subsequent contamination of observation or supply boreholes, although sometimes rapid migration is seen, especially in fissured chalk and limestone aquifers. If problems are detected it may require a much longer period for the full recovery of water quality once recharge has stopped. As recharge begins a mixture of recharge effluent and natural groundwater will eventually be pumped from the supply borehole, so regular monitoring of water quality is essential to ensure potable standards are maintained. Owing to the potential long-term risks to important groundwater resources, where infiltration lagoons are used for recharge, it is advisable to use only waste water treated to as near drinking water quality as possible and to regard the lagoon system solely as a means of recharging the aquifer, much in the same way that injection wells are used. Any benefits from the potential treatment capacity of the soil and the unsaturated zone as the water percolates through to the aquifer should be regarded as incidental. The recharge method requires large areas of land, up to 7 m^2 per person, and so could never be feasible for large communities. The bulk of the water for artificial recharge will therefore most probably come from rivers.

However, it is possible to use shallow aquifers primarily as water treatment plants. For example, water from the River Rhine, which is of extremely poor quality, is pumped into settling basins near the town of Essen (Germany). After settlement for 12 hours the water flows into an enormous recharge lagoon (200 000 m^2 in area), which has been excavated in highly porous river gravels and sands. Under the sediments through which the water percolates, perforated pipes have been laid about 10 m deep via which the water is continuously abstracted. The water is then chlorinated and supplied directly to the 850 000 inhabitants of Essen at a very low unit cost.

The potential for the development of aquifers for water storage, treatment and supply is enormous, although at present many are being overabstracted. To achieve the maximum from these valuable resources water companies are appointing aquifer managers to control and develop specific groundwater resources. In certain areas of Britain, such as the south-east, aquifers are so important to the community in terms of household, industrial and agricultural water supplies that every effort must be made to preserve them and manage them wisely. A failure to do so could lead to severe social and economic hardship for many areas, simply because there is not enough water to go around!

2.6.5 Water Transfer

Rivers and long aqueducts are used to convey water collected in upland reservoirs to principal centres of demand. However, it has been suggested that larger regional transfer systems to transport water from areas of high water availability in the north of England, and especially Scotland, to the high demand areas where supplies are currently short in the south-east should be developed. Such transfer schemes could be co-ordinated to form a national network or grid, rather like the electricity power grid, allowing water to be transferred to wherever it was needed (Figure 2.11). Similar

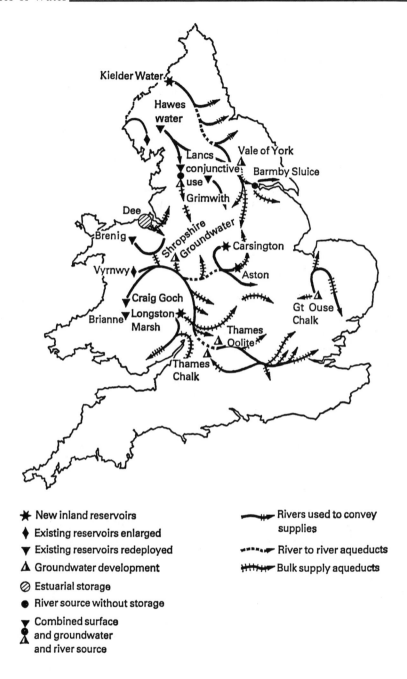

Figure 2.11. The idea of a water distribution network was first proposed in the early 1970s. Little progress has been made towards such a development, although the new water companies are once again looking at the possibility of a national grid for water. Reproduced from Kirby (1979) by permission of Penguin Books Ltd

schemes have been suggested throughout the world, especially in North America. There is considerable interest in the idea of importing water from France for use in East Anglia, via a water main laid in the Channel Tunnel. The construction of barrages across the mouths of major estuaries offers a solution to the provision of more surface water supplies. The estuarine barrage would form a huge freshwater lake behind it, providing vast amounts of surface water which would become available for use without the loss of further agricultural land for reservoir construction. It also offers the potential to provide water supplies to areas where the populations are greatest, releasing existing supplies for other users. Of course, there are disadvantages: the quality of such water is usually fairly poor, often contaminated by heavy industry, and there are many ecological, physical and social disadvantages. Certainly such a scheme in south-east England would make an enormous difference to the water supply problems facing that region.

2.6.6 Desalination

The process of evaporating freshwater from seawater using the sun's energy is old technology. However, modern distillation systems are able to produce continuous supplies of high purity water and are widely used on board ships at sea. To do this on a large enough scale to supply drinking water to even a small town requires enormous amounts of energy to remove the salt, so water produced in this way is very expensive. Currently the cost of desalination using the state of the art technology is between two and three and a half times the cost of A2 water, although it is very difficult to put a realistic cost on conventional treatment using existing water resources. Such comparisons may be meaningless in the long term. It is more realistic to compare desalination to new schemes such as water transfer, new reservoir construction, aquifer recharge and the use of poor quality resources. However, as poorer quality water is used and more advanced treatment technologies are employed to achieve the EC Drinking Water Directive required standards, especially for trace amounts of organic compounds and nitrates, then the cost of desalination becomes less prohibitive. Linked with the cost, indeed the feasibility, of providing new resources in certain areas such as the south-east of England, then desalination appears to be a strong possibility. The lack of water and the unpredictability of supplies in some areas is so restrictive, especially for industry and agriculture, that users in these areas may be prepared to pay marginally more for the water they use. In many arid countries where there is usually no surface water at all, and where groundwater supplies are inadequate, then desalination is the only method of water supply. It is these countries, in particular the Middle East, from which most of our current expertise in desalination technology derives. Desalination is also popular in arid countries such as Spain and Australia, and of course highly populated islands such as the Virgin Islands. Desalination is also used on Jersey to supplement the limited amount of surface water available. Worldwide there are about 7600 plants producing $14 \times 10^9 \, \text{m}^3$ of water annually. Sixty per cent of the capacity is in the Middle East, 13% in North America, 10% in Europe (including eastern Europe) and 7% in Africa. One of

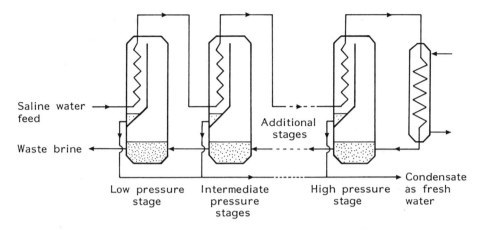

Figure 2.12. Principles of the multi-stage flash distillation process

the newest plants has been constructed by the City of Santa Barbara at a cost of $36 000 000. It is able to produce over 12×10^6 m^3 of drinking water each year.

Modern distillation plants such as that in Jersey are based on the multi-flash process (Figure 2.12). Cold salt water is fed into the system through a long pipe which passes through each chamber where it acts as a condenser. The salt water absorbs heat during condensation and is then heated in a steam-fed heat exchanger. The hot salt water enters the first chamber where part of the salt-free water flashes to salt-free vapour which condenses on the cold feed pipe as freshwater. The salt water passes to addition chambers which are each operated at slightly lower pressures than the preceding chamber, ensuring that further flashing occurs. The freshwater is collected for use as drinking water whereas the concentrated waste brine solution has to be carefully disposed of back to the sea (Wood, 1987).

Other processes of desalination, apart from distillation or evaporation, include freezing and reverse osmosis. Freezing a salt solution makes crystals of freshwater form and grow, leaving a concentrated brine solution behind. Although the technology exists, no large-scale facility has ever been developed. However, as freezing only needs 10–15% of the energy required by evaporation, it is attractive both economically and environmentally (Coughlan, 1991).

Osmosis is the movement of water (or any solvent) from a weak solution to a strong solution through a semipermeable membrane. Therefore if the membrane is placed between freshwater and salt water, the solvent (i.e. pure water) will move through the membrane until the salt concentration on either side is equal. Only water can pass through the membrane so the salts are retained The movement of water across the membrane is caused by a difference in pressure and continues until the pressure in both solutions is equal, limiting further passage. The pressure difference which causes osmosis to occur is known as the osmotic pressure. Reverse osmosis uses this principle to make the solvent (pure water) move from the concentrated solution (salt water) to the weak solution (freshwater) by exerting a pressure higher than the osmotic pressure on the concentrated solution, thus reversing the direction of flow across the membrane

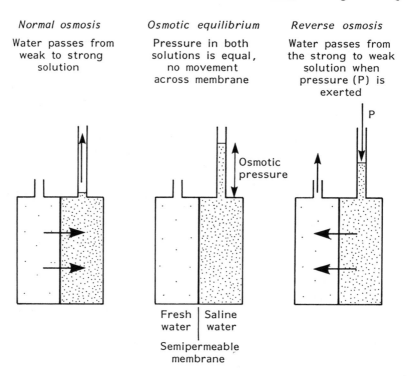

Normal osmosis

Water passes from
weak to strong
solution

Osmotic equilibrium

Pressure in both
solutions is equal,
no movement
across membrane

Reverse osmosis

Water passes from
the strong to weak
solution when
pressure (P) is
exerted

Osmotic
pressure

Fresh | Saline
water | water

Semipermeable
membrane

Figure 2.13. Principles of the reverse osmosis process

(Figure 2.13). By subjecting saline water to pressures greater than the osmotic pressure, pure water passes through the semipermeable membrane and can be collected for use as drinking water. Reverse osmosis is not membrane filtration, i.e. purification is not the result of passing the water through defined pores or holes. It is a far more complex action, where one molecule at a time diffuses through vacancies in the molecular structure of the membrane material. Filtration physically removes particles, although these particles can be very small. Commercial units are available, with membranes made from cellulose acetate, triacetate and polyamide polymers. Multi-stage units are generally used with the best results from the use of brackish water, rather than seawater. Reverse osmosis is also used for treating polluted river water and removing common contaminants such as nitrate (Parekh, 1988). Although expensive, reverse osmosis can also be used to soften water (Section 8.2).

2.6.7 Other Options

Transferring water from one region to another has already been examined. However, there are other transfer options that are currently under seriously consideration.

The importation of water by bulk tank to areas with water shortages is common during times of drought or when water mains burst. The importation of water by bulk sea tankers bringing water from other countries occurs in some arid countries, and has been seriously considered by some water companies during drought periods in coastal areas severely hit by water shortages. For example, during the drought of 1989 and 1990, Northumbria Water considered the idea of importing water from Gibraltar using ocean-going tankers. The disadvantage of importing water from another country is that in the long term you become very vulnerable. Everyone wants to be independent in terms of water supply.

Rain-making has become fairly successful, especially in south-western Australia. Once the province of the witch doctor or medicine man, it is now a high-tech scientific art. Silver iodide smoke is released from aircraft onto the tops of clouds where the temperature is less than $-10°C$. It is estimated that 1 g of silver iodide will release as much as $250\,000$ m^3 of rain-water, which makes it comparatively cheap. However, large-scale interference in natural rainfall patterns could prove ecologically and politically very serious.

The most widely studied transfer method in recent years is the use of natural ice, i.e. icebergs, as a source of drinking water as well as a cheap form of energy collected as they slowly heat up. It has been suggested that the recent increase in the breaking up of the ice sheets due to the greenhouse effect driving global warming will make this an easier and cheaper operation, although this remains to be seen. Towing icebergs the size of small islands is technically very difficult and extremely hazardous, but for arid countries, icebergs seem to be an ideal source of pure drinking water.

Chapter 3
_____ Supplying Water: How it Works

3.1 INTRODUCTION

Water rapidly absorbs both natural and man-made substances, generally making the water unsuitable for drinking without some form of treatment. Important categories of substances which can be considered undesirable in excess are:

1. *Colour*. This is due to the presence of dissolved organic matter from peaty soils, or the mineral salts of iron and manganese.
2. *Suspended matter*. This is fine mineral and plant material which is unable to settle out of solution under the prevailing conditions.
3. *Turbidity*. This is a measure of the clarity, or transparency, of the water. Cloudiness can be caused by numerous factors such as fine mineral particles in suspension, high bacteria concentrations, or even fine bubbles due to over-aeration of the water.
4. *Pathogens*. These can be viruses, bacteria, protozoa or other types of pathogenic organism which can adversely affect the health of the consumer. They can arise from animal or human wastes contaminating the water resource.
5. *Hardness*. Excessive and extremely low hardness are equally undesirable. Excessive hardness arises mainly from groundwater resources whereas very soft waters are characteristic of some upland catchments.
6. *Taste and odour*. Unpleasant tastes and odours are due to a variety of reasons such as contamination by waste waters, excessive concentration of certain chemicals such as iron, manganese or aluminium, decaying vegetation, stagnant conditions due to a lack of oxygen in the water, or the presence of certain algae.
7. *Harmful chemicals*. There is a wide range of toxic and harmful organic and inorganic compounds which can occur in water resources. These are absorbed from the soil or occur due to contamination from sewage or industrial waste waters.

Water treatment and distribution is the process by which water is taken from water resources, made suitable for use and then transported to the consumer. This is the first half of the human water cycle, before water is actually used by the consumer (Figure 3.1). The second half of the cycle is the collection, treatment and disposal of used water (sewage) (Gray, 1992).

The objective of water treatment is to produce an adequate and continuous supply of water that is chemically, bacteriologically and aesthetically pleasing. More specifically,

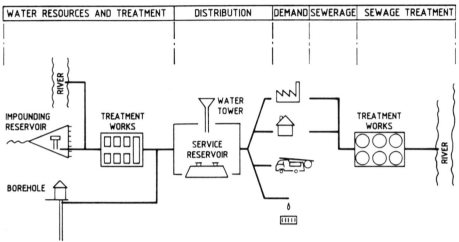

Figure 3.1. Water services cycle showing where water is used by humans during its movements within the hydrological cycle. Reproduced from Latham (1990) by permission of the Institution of Water and Environmental Management

water treatment must produce water which is:

1. Palatable—that is, has no unpleasant taste
2. Safe—it should not contain any pathogenic organism or chemical which could be harmful to the consumer
3. Clear—free from suspended matter and turbidity
4. Colourless and odourless—therefore aesthetic to drink
5. Reasonably soft—to allow consumers to wash clothes, dishes and themselves without excessive use of detergents or soaps
6. Non-corrosive—water should not be corrosive to pipework or encourage leaching of metals from pipes or tanks
7. Low organic content—a high organic content will encourage unwanted biological growth in pipes or storage tanks which can affect the quality of the water supplied

With the publication of the Water Supply (Water Quality) Regulations, water must conform with the standards laid down for a large number of diverse parameters. The Regulations also require that the water supply companies supply water to consumers that is wholesome and defines clearly what this term means.

Consumers expect clear, wholesome water from their taps 24 hours a day, every day. Although water which is unaesthetic, for example due to colour or turbidity, may be perfectly safe to drink, the consumer will regard it as unpalatable and probably dangerous to health. Problems not only originate from the resources themselves (Chapter 4), but during treatment, distribution and within the consumer's home (Chapters 5, 6 and 7, respectively).

Table 3.1. Main unit processes in water treatment in general order of use

Preliminary screening
Storage
Screening or microstraining
Aeration
Coagulation
Flocculation
Clarification
Filtration
pH adjustment
Disinfection
Softening and other tertiary processes
Sludge removal

3.2 WATER TREATMENT

Water treatment plants must be able to produce a finished product of consistently high quality regardless of how great the demand is. With the exception of particularly pure groundwaters, all water supplied for drinking requires purification. Although, in theory, the dirtiest water can be purified to drinking water quality, in practice treating even relatively pure water to produce a finished water of consistent quality, and of sufficient volume, is technically very difficult. Water treatment consists of a range of unit processes which are usually operated in series. These are listed in Table 3.1, although it is unusual for all of them to be used at any one particular plant. The cleaner the raw water, the fewer the number of steps or unit processes that are required, and hence the overall cost of the water is less. The most expensive operations are sedimentation and filtration in conventional treatment, whereas specialized processes for softening water or removing specific contaminants such as nitrates or pesticides can be very expensive.

Groundwater is generally much cleaner than surface waters and so does not need the same degree of treatment, apart from aeration and disinfection, before supply. Naturally occurring substances which may need to be reduced or eliminated in groundwaters include iron, hardness (if in excess of 300 mg/l) and carbon dioxide. Substances originating from humans which are becoming increasingly common in groundwaters and that need treatment include nitrates, pathogens (especially bacteria and viruses) and trace amounts of organic compounds such as pesticides. The water industry tries to obtain the cleanest water possible for supply, although the volume and consistency of supply are the major factors in the selection of a resource. The cleanest of the suitable resources available is usually selected; however, it may be necessary to blend several resources to dilute unwanted contaminants to below harmful concentrations.

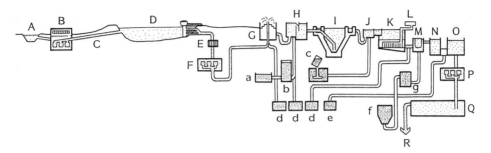

Figure 3.2. Schematic diagram of a large water treatment plant showing all the unit processes in series. A = river, B = intake and pumping station, C = rising main, D = reservoir, E = mechanical strainers, F = raw water pumping station, G = aeration tower, H = coagulant flash mixers, I = sedimentation tanks, J = flash mixer, K = rapid gravity filter, L = air blowers, M = flash mixer, N = chlorine retention chamber, O = clean water tank, P = out-going pump house, Q = storage reservoir, R = to supply distribution system, a = storage, b = coagulant solution, c = activated carbon slurry, d = chlorine supply, e = sulphonation, f = lime, g = lime slurry

3.2.1 Unit processes

The selection of the necessary unit processes depends on the quality of the raw water entering the treatment plant and the quality of the finished water required. A general layout of a typical large urban water treatment plant showing all the major processes is shown in Figure 3.2.

3.2.1.1 *Preliminary screening*

Large-scale treatment plants which serve large conurbations are rarely close to the sources of water, except where direct abstraction from rivers is practised. Most upland reservoirs are many miles from the point of consumption, so the raw water must be conveyed to the treatment plant either by pipe or open channel. The raw water is passed through a set of coarse screens to remove gross solids such as weeds, sticks and other large material before starting its journey to the plant. This is mainly carried out to protect pipes from becoming blocked or pumps being damaged.

3.2.1.2 *Storage*

Raw water is pumped from the intake to the storage reservoir where it is often stored to improve quality before treatment, as well as ensuring adequate supplies at periods of peak demand. There are a number of natural processes at work during storage which all significantly improve water quality. The filtration process can only deal effectively with suspended solids concentrations of less than 10 mg/l. For waters with suspended solids concentrations in excess of 50 mg/l, storage is necessary to allow this particulate matter to settle out of suspension. Ultraviolet radiation is another important natural treatment process. It destroys harmful bacteria and some other pathogenic organisms.

Other processes are also at work during storage. Colour is bleached by sunlight, while some organic impurities which are responsible for taste and odour problems are oxidized in the upper zones of the reservoir. Excessive hardness can also be reduced by the liberation of carbon dioxide by algae present in the reservoir during summer. This converts the hydrogencarbonates into carbonates which are then precipitated out of solution.

Storage can largely eliminate variations in water quality which can occur in surface waters especially due to floods or variations in the dilution of any pollutants present. Although the advantages largely outweigh the disadvantages, there are a number of important problems in the operation of storage reservoirs; for example, atmospheric pollution and fallout, pollution from birds (especially roosting gulls), algal development (if stored for longer than 10 days) and, of course, like all reservoirs they take up considerable areas of land so that their construction can be contentious. In deeper reservoirs the major problem is thermal stratification (Section 2.3).

3.2.1.3 *Screening and microstraining*

Before treatment the raw water is screened again, this time through fine screens. If considerable amounts of fine solids or algae are present, then microstraining may be used before the next stage. Microstrainers consist of fine stainless-steel mesh drums, with up to 25 000 apertures/cm^2. The microstrainer is a rotary drum which is partially immersed in the water. As it rotates the head difference drives the water through the micromesh which strains out the particles, especially the algae. The apertures normally used in water treatment are either 25 or 35 μm in diameter. When clean, no micro-organisms are retained but the particles and algae retained on the micromesh form a straining layer, allowing some retention of micro-organisms to occur, although this is incidental and cannot be either predicted or relied upon. Microstraining produces a wash-water in which all the strained particles, including the algae, are concentrated. It can represent as much as 3% of the total volume of water passing through the strainer and needs to be treated separately. There is evidence to show that such waters can be potentially rich in pathogens, especially cysts and oocysts of protozoans.

3.2.1.4 *Aeration*

Water from groundwater resources, from the bottom of a stratified lake or reservoir, or from a polluted river, will contain very little or no dissolved oxygen. If anaerobic water is allowed to pass through the treatment plant it will damage or affect other unit processes, in particular filtration and coagulation. Therefore the raw water needs to be aerated before it is treated further. This is achieved by bringing the water into contact with air. Although bubbling air through water is the way this is often done on a small scale, this is prohibitively expensive at larger plants because of the vast volumes of water that are treated each day. The simplest method is a cascade or fountain system. In a cascade, the water pours down a tower structure, which ensures reaeration through excessive turbulence. Alternatively, the pressure of the water as it is forced out of the reservoir by the weight of the water above it allows natural jets or fountains of water to spray up into the air. The jets do not rise high into the air as they are normally wide to cope with the large volumes of water being processed. Many such jets are partially enclosed to reduce evaporation losses. These two methods are visually spectacular and are often used to great effect in the overall design of treatment plants. There are many other types of aeration systems used, including packed towers and diffusers (Montgomery, 1975).

Apart from ensuring optimum treatment within the treatment plant, aeration also provides oxygen for purification and significantly improves the quality, especially the taste, of water. Aeration also reduces certain objectionable odours, and reduces the corrosiveness of water by driving off any carbon dioxide gas present, thus raising the pH. Aeration cannot, however, reduce the corrosive properties of acid waters alone and neutralization with lime may be needed. Iron and manganese can also be removed from solution by aeration. These metals are only soluble in water with a pH of less than 6.5 and in the absence of dissolved oxygen, and so are common in certain groundwaters. Aeration oxidizes the soluble metal salts into insoluble metal hydroxides which can then be removed by flocculation or filtration.

3.2.1.5 Coagulation

After fine screening most of the remaining suspended solids will be very small, usually < 10 μm (colloidal solids)—so small, in fact, that they may never settle out of suspension naturally. These solids are particles of clay, metal oxides, large protein molecules and micro-organisms. All small particles tend to be negatively charged, and as like charges repel, all the negatively charged colloidal particles in the water tend to repel one another, preventing aggregation into larger particles which could then settle out of suspension. A particle 100 μm in diameter will settle 200 000 times faster than a particle 0.1 μm in diameter (Table 3.2). The removal of colloidal matter is a two-step process: coagulation followed by flocculation.

A coagulant is added to the water to destabilize the particles and to induce them to aggregate into larger particles known as flocs. A variety of coagulants are used. The most common salts are aluminium sulphate (alum), aluminium hydroxide, polyaluminium chloride, iron(III) chloride, iron(III) sulphate and lime. Iron(II) sulphate ($FeSO_4 \cdot 7H_2O$),

Table 3.2. Settling velocity of particles as a function of size. Reproduced from Tebbutt (1979) by permission of Elsevier Science Ltd

Particle size (μm)	Settling velocity (m/h)
1000	600
100	2
10	0.3
1	0.003
0.1	0.00001
0.01	0.0000002

known as copperas, is also used although it is generally mixed with chlorine to give a mixture of iron(III) chloride ($FeCl_3$) and iron(III) sulphate [$Fe(SO_4)_3$], known as chlorinated copperas. There is a rapidly increasing range of synthetic organic polymers also available. These include polyacrylamides, polyethylene oxide and polyacrylic acid. The latter is also becoming widely used in 'green' ('environmentally friendly') washing powders as a substitute for polyphosphate builders in detergents. The use of coagulants and some of the problems arising from their use are considered in Chapter 5. The actual mechanisms of coagulation are complex and include adsorption, neutralization of charges and entrainment within the physical–chemical matrix formed (Montgomery, 1975).

The amount of coagulant added to the water is critical. Too little results in ineffective coagulation so that the filtration apparatus may become blocked too rapidly, while too much coagulant can lead to the excess chemical being discharged with the finished water. The coagulant is added to the process stream at a specific concentration (30–100 mg/l) using a mixing device. Either a mixing chamber (flash-mixer) is used with a high speed mixer, or the coagulant is added to the water in a mixing channel (using a hydraulic jump in a measuring flume) to induce mixing. Coagulation is complete within one minute of addition. Metal salts react with the alkalinity in the water to produce an insoluble metal hydroxide precipitate which enmeshes the colloidal particles. The precipitate formed appears fluffy when seen under a low power microscope, and apart from enmeshing any particles in the water it also adsorbs some of the dissolved organic matter present. As already stated, all natural particles in water have a small negative electrical potential on their surfaces. It is very small, usually less than -20 mV. As the hydroxide flocs carry a small positive electrical potential, there is a mutual attraction to the particles in the water. Coagulant aids, which are organic polymers with either positive or negative charges, are sometimes used to improve coagulation. They are a supplement to the normal coagulant and can cause problems in slow sand filtration.

3.2.1.6 *Flocculation*

When small particles collide in a liquid some naturally aggregate to form larger particles. As the larger particles settle they overtake smaller particles which are settling at much slower rates. If they collide, then the smaller particle will aggregate onto the larger particle.

The chance of particles colliding can be significantly increased by gently mixing the water; this process is known as flocculation. When there is a high concentration of colloidal particles, then flocculation can be effective on its own. However, at the lower concentrations usually found in water resources a coagulant must be used. In the water treatment process, flocculation therefore follows chemical addition (coagulation), which is required to destabilize the colloidal particles present. During this mixing larger flocs are produced that are easily removed during clarification. Flocculation occurs naturally by Brownian motion (perikinetic flocculation); however, for particles larger than 1 μm this is very slow and mechanical mixing devices are required (paddle or turbine mixers) to increase the rate of collisions (orthokinetic flocculation).

3.2.1.7 *Clarification*

Here the flocs formed by the addition of coagulant and flocculation are removed by settlement. The process is different from normal sedimentation processes found in waste water treatment and in industry where the water flows slowly along or across a tank allowing the particles to settle. In water treatment the water flows in an upward direction from the base of the tank. The flocs, which are heavier than water, settle towards the bottom, so the operator must balance the rate of settling against the upward flow of water to ensure that all the particles are held within the tank as a thick sludge blanket. There is a layer of clear, clarified water at the surface that overflows a simple weir to the next step of the treatment process. These tanks are referred to as floc blanket clarifiers and are extremely effective. As the flocs rise through the blanket further flocculation occurs, which increases the floc density. To maintain the sludge blanket at the required height, and the more sludge maintained within the tank the more efficient the separation will be, the excess sludge must be discharged from the tank. Sludge removal, or bleeding as it is commonly referred to, may be carried out continuously or intermittently. The sludge is a concentrated mixture of all the impurities found in the water, especially bacteria, viruses and protozoan cysts. It must therefore be handled carefully and disposed of safely. The volume of sludge is fairly large and is equivalent to between 1.5 and 3.0% of the flow through the clarifier. There are many different designs of sedimentation tanks, including the use of inclined plates to encourage settlement and the development of parallel plate and tube settlers. These are discussed by Montgomery (1975).

3.2.1.8 *Filtration*

After clarification, the water only contains fine solids (< 10 mg/l) and soluble material. Although some of these particles may have been in the natural raw water, many will have been formed during the coagulation process. Another process, filtration, is required to remove this residual material. The filters contain layers of sand (or anthracite) and gravel graded to ensure effective removal. In their simplest form, filters allow the downward passage of water through layers of fine sand, which is supported on layers of coarser gravels. Pipes at the base of the filter, underdrains, collect the filtered water. Particles which are removed by the sand clog the surface and reduce the rate of flow of water through the filter. Therefore the filter must be cleaned intermittently. This is done by either scraping the surface layer of sand containing the retained particles from the surface, or where possible by pumping water through the filter in the reverse direction under pressure. This washes all the small retained particles from the sand, a process known as back-washing. The efficiency of the filters depends on a number of factors such as the nature and quality of the material to be removed from the water, the size and shape of the filter media, and the flow rate of water through the filter.

There are two types of filter used in water treatment: rapid and slow sand filters. Rapid sand filters contain coarse grades of quartz sand (1 mm diameter) so that the gaps between the grains are comparatively large. This ensures that the water passes rapidly through at rates of between 5 and 10 $m^3/m^2/h$, operating at about 50 times the rate of a slow sand filter. Rapid sand filters are usually deeper, between 0.6 and 1.0 m in depth, and consist of sand, anthracite plus sand, or a similar material such as activated carbon and sand. These filters are used for water that has previously been treated by coagulation and sedimentation, and are less effective than slow sand filters in retaining very small solids. Therefore bacteria, taste and odours are less effectively removed than by slow sand filtration. Because of the high loading rates, rapid sand filters are much smaller and more compact than slow sand filters. In use, the amount of water passing through the filter each hour gradually decreases due to the gaps in the sand becoming blocked by the retained solids. When the filtration rate becomes too low the filter must be cleaned. Depending on the design and loading rates this is carried out several times a day or every few days by blowing air up through the layer of sand to scour material free from the grains of sand, and then washing away the solids by back-washing with clean water. Immediately after back-washing there is an increased risk of pathogenic micro-organisms penetrating the filter. It may take up to 20 minutes for rapid sand filters to reach their optimum performance in terms of water quality. The water used for back-washing should not be recycled through the plant (Section 4.7). Rapid sand filters are of two main designs: open-topped gravity-fed systems or more efficient closed systems where the filter is enclosed within a metal shell and the water is forced through the sand by pressure.

In contrast, slow sand filters have a layer of much finer quartz sand (0.5–2.0 m depth) overlying coarse sand or gravel (1.0–2.0 m depth) that physically removes fine solids (Visscher, 1988). However, these filters, apart from physical straining, also provide a degree of biological treatment. The top 2 mm of the sand is host to a mixture of algae and nitrifying bacteria (autotrophic layer). Here nitrogen and phosphorus are removed and oxygen is released. Below this autotrophic layer is a thicker layer of sand, up to 300 mm, which is colonized by bacteria (heterotrophic layer) and other micro-organisms that remove the residual colloidal and soluble organic material from the water (Duncan, 1988). Water treatment in sand filters is therefore a combination of physical and biological activity, with pathogenic bacteria, taste and odour (due to algae and organic compounds) largely removed. The quality of the water is excellent, unlike that from a rapid sand filter where further treatment may be required as there is little biological activity and only the larger solids are retained. The rate of filtration in slow sand filters is controlled by gravity alone and so combined with the small gaps between the particles of sand, water only passes through such filters slowly. On average the rate of filtration is between 0.1 and 0.3 $m^3/m^2/h$, so a very large area of filters is required. They are expensive to operate because the dirt layer that collects on the surface of the sand and impedes drainage must be mechanically skimmed off after the filter has been drained. Unlike rapid sand filters, slow sand filters cannot be back-washed. After 2–3 months the sand that has been removed during the skimming operations must be replaced to maintain the required depth of fine sand. This makes slow sand filtration labour intensive and operationally expensive.

3.2.1.9 *pH adjustment*

The pH of the finished water may require adjusting so that it is neither too acidic, which may corrode metal distribution pipes and household plumbing, or too alkaline, which will result in the deposition of salts within the distribution system causing a reduction in flow. The pH may be adjusted at a number of unit processes, such as coagulation, to ensure maximum efficiency. Alkalis such as lime, sodium carbonate or caustic soda are used to increase the pH, whereas acids are used to decrease it.

3.2.1.10 *Disinfection*

Although slow sand filters are extremely efficient at removing bacteria, and the coagulation process is good at removing viruses, the finished water still contains pathogenic viruses and bacteria which need to be removed or destroyed. In practice it is impossible to sterilize water, to kill off all the micro-organisms present, due to the very high concentration of chemicals required which would make the water very unpleasant and possibly dangerous to drink. Therefore the water is disinfected, rather than sterilized, by using one of the disinfection methods such as chlorination, ozone or ultraviolet radiation to ensure that pathogens are kept at safe levels. Of the three methods of disinfection chlorination is by far the most widely used.

Ozone has powerful oxidation properties and tends to be used where the natural water contains materials that would combine with chlorine to form unacceptable odours or tastes. Ozone, which is often used in combination with activated carbon, can eliminate all bacteria at a dose rate of 1 ppm within 10 minutes, and can also reduce colour, taste and odour. Apart from being more expensive than chlorination and that it has to be manufactured on site, the lack of residual disinfection action within the distribution mains is the major drawback (Masschelein, 1982). This allows biological growth to develop which causes taste and odour problems. Therefore low level chlorination is often used after ozonation to prevent such growth. When waters containing bromide are oxidized, especially using ozone or hydrogen peroxide, bromate is formed. Although not included in the EC Drinking Water Directive, bromate has been included in the revised WHO regulations (1993) as a disinfection by-product. Bromate is widely considered to be a genotoxic carcinogen and based on available toxicological data a guide value as low as 0.5–1.0 μg/l could have been derived, based on a lifetime risk level of 10^{-5} (Bull and Kopfler, 1991). Currently, analytical techniques for bromate are unable to detect levels below 1.0 μg/l. The WHO have set a provisional guide value of 25 μg/l as an interim measure, but a much lower value appears inevitable. The inclusion of bromate in the WHO guidelines poses a number of problems to the water industry. Bromide is common in surface and groundwaters, and although it can arise from industrial discharges, bromide is particularly common in rain-water and certain groundwaters as a residual from seawater. It is particularly common in Dutch surface waters. Bromate formation will therefore occur whenever ozone is used for disinfection, or during removal of pesticides with activated carbon. The industry is currently looking at the options available. Removal of bromide after ozonation would require membrane filtration and be extremely costly. Some control is possible by the optimization of current disinfection techniques.

Ultraviolet radiation is emitted from special lamps and is effective in killing all micro-organisms as long as the exposure time is adequate. Ultraviolet radiation is electromagnetic energy in the range 250–265 nm. To be effective this energy must reach the nucleic acid within the largest organism to induce structural changes that will prevent replication of the pathogen. The lamps are enclosed in stainless-steel reaction chambers (Kruithof *et al.*, 1991; Wolfe, 1990). They are used at small plants or for institutions where the chance of contamination after treatment is unlikely. Very effective household ultraviolet sterilization units are also available for single dwellings (Section 8.2).

Chlorine and its compounds are readily available in gas, liquid or solid forms. It is easy to add to water, has a high solubility (7000 mg/l) and is cheap. The residues it leaves in solution continue to destroy pathogens after the water has left the treatment plant and as it travels through the distribution network. Although it is toxic to micro-organisms, it is not generally thought to be harmful to humans at the concentrations used. Recent work in the USA, however, has identified a possible link between bladder cancer and chlorinated water. In the USA the concentration of chlorine used in drinking waters is generally significantly higher than in Europe, so at the concentrations used in the UK chlorine should be considered totally safe. Chlorine is, however, a dangerous chemical to handle in its concentrated form and produces a poisonous gas.

There is interest in the on-site generation of chlorine by the electrochemical oxidation of brine:

$$NaCl + H_2O \rightarrow NaOCl + H_2$$

Although this would eliminate the need for the bulk delivery of hazardous chemicals such as chlorine gas or sodium hypochlorite solution, there is concern about the possible formation of toxic by-products (e.g. chlorite, chlorate, perchlorate, bromate and chlorinated organic chemicals).

The chemistry of chlorine is complex. Essentially chlorine (Cl_2) reacts with water to form hypochlorous acid (HOCl) and hydrochloric acid (HCl)

$$Cl_2 + H_2O \rightarrow HCl + HOCl$$

In dilute solution this reaction is very rapid and is normally complete within one second. Hypochlorous acid is a weak acid which readily breaks down (dissociates) into the hypochlorite ion (OCl^-). This occurs almost instantaneously. Both hypochlorous acid and the hypochlorite ion act as disinfectants, although the hypochlorous acid is about 80 times more effective than the hypochlorite ion. A chemical equilibrium (i.e. balance) develops between the two forms, although dissociation is suppressed as the pH decreases (becomes more acid). In practice, at about pH 9, 100% of the chlorine is in the chlorite form, about 50% at pH 7.5 and at pH 5 or less it is all present as hypochlorous acid. Disinfection is therefore much more effective at acidic pH.

Chlorine is not as aggressive a disinfectant as ozone and there are a number of pathogenic micro-organisms which are resistant to chlorination. Effectively eliminating all the coliforms present does not necessarily indicate that all other pathogenic micro-organisms have also been destroyed (LeChevallier, 1990). Factors such as temperature and pH also affect chlorination, its efficiency decreasing at lower temperatures and in more alkaline waters. Many substances will readily combine with chlorine, especially reducing agents and unsaturated organic compounds. These compounds exert an immediate chlorine demand that must be satisfied before chlorine becomes available for disinfection. Excess chlorine must therefore be added to satisfy this demand as well as to leave a residual amount in the water long enough to penetrate and destroy all the micro-organisms present. Suspended organic and inorganic matter absorbs chlorine, whereas iron and manganese neutralize chlorine by forming insoluble chlorides. Thus it is better to remove these problematic substances by appropriate treatment before disinfection rather than increasing the dose of chlorine. Chlorine is very reactive and combines with almost everything in the water. Research has shown that chlorine reacts with organic compounds that occur naturally in water to form chlorinated hydrocarbons, many of which are toxic (Section 5.5). Chlorination is thus unsatisfactory when waters are rich in organic acids, such as those from peaty upland areas.

A major problem is the presence of ammonia. This reacts readily with chlorine to form a range of compounds known as chloramines, the exact nature of which depends on the relative concentrations of the two chemicals and the pH (White, 1972). Three chloramines are formed: these are monochloramine (NH_2Cl), dichloramine ($NHCl_2$) and trichloramine or nitrogen trichloride (NCl_3).

$$NH_4 + HOCl \rightarrow NH_2Cl + H_2O + H^+$$
Monochloramine
$$NH_2Cl + HOCl \rightarrow NHCl_2 + H_2O$$
Dichloramine
$$NHCl_2 + HOCl \rightarrow NCl_3 + H_2O$$
Trichloramine

When ammonia is present the dose of chlorine must be increased to ensure that sufficient excess chlorine is left in the water to destroy the pathogens. However, combined chlorine (combined residuals) such as mono- and dichloramines retain some of their disinfection potential and, although less effective than free chlorine present as hypochlorous acid and hypochlorite ions (free residuals), they have long-lasting disinfection properties. Therefore it has become the practice at some treatment plants to add ammonia at the chlorination stage to give a combined rather than a free residual effect. Combined residual chlorine requires a contact time of a hundred times longer than free residual chlorine to achieve the same degree of elimination of pathogens. When chlorine is added to water containing ammonia, which is either present naturally or added deliberately to produce a combined chlorine rather than a free chlorine residual, then a breakpoint curve is produced (Figure 3.3). If inorganic reducing agents such as iron and manganese are present then these will exert a chlorine demand that does not produce a residual and so is seen as a flat line at the start of the curve (a–b). Once satisfied the chlorine residual in the water is a function of the chlorine:ammonia weight ratio. When the ratio of chlorine:ammonia is <5:1 at pH 7–8, then the chlorine

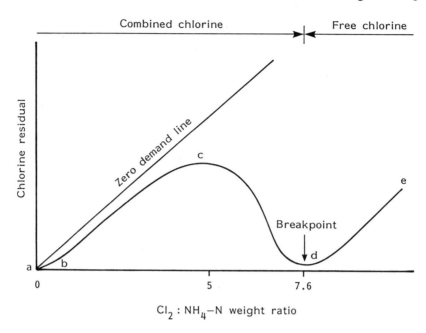

Figure 3.3. Breakpoint chlorination curve

residual is all monochloramine (b–c). Increasing the chlorine:ammonia ratio results in some of the monochloramine reacting to form small amounts of dichloramine. As the ratio approaches 7.6:1, the chloramines are oxidized by the excess chlorine to nitrogen gas, resulting in a rapid loss of residual chlorine in the water (c–d)

$$NH_2Cl + NHCl_2 \leftrightarrow N_2 + 3H^+ + 3Cl^-$$

After the breakpoint there is no ammonia left to react with the chlorine so the residual concentration increases in proportion to the amount of chlorine applied (d–e) (White, 1972). The disinfection potential before the breakpoint is therefore reached due to combined residuals (mono- and dichloramines) and after the breakpoint due to free chlorine, although trace amounts of dichloramine and trichloramine may remain at lower pH values (Figure 3.3). This is excellently reviewed by Bryant *et al.* (1992). Excess ammonia will find its way into the distribution main if the chlorine:ammonia ratio is too low, resulting in nitrification and the formation of nitrate, or more importantly, nitrite (Section 5.1).

Another reason for adding ammonia is to prevent the chlorine from reacting with trace amounts of organic compounds in the water, such as phenol, and forming unpleasant tastes. When ammonia is present in the raw water at concentrations in excess of 0.3 mg/l it can cause an unpleasant taste and odour if it is allowed to degrade anaerobically during treatment. Therefore it is not unusual for the raw water to be chlorinated as it enters the treatment plant to control the bacteria which cause the problem, although pre-chlorination will upset the biological activity in slow sand filters.

Chlorine is relatively easy to handle and cost effective, although chlorine dioxide is used under alkaline pH conditions, since under these conditions it will not form combined residuals with ammonia. As a general guide, the cleaner the water the larger the residual effect. The amount of chlorine used depends on the rate of flow and the residual chlorine concentration required, which is usually 0.2–0.5 mg/l after 30 minutes. The problem of inadequate residual chlorine allowing pathogens to survive in distribution systems, and the problems of the contamination of supplies after treatment is dealt with in Section 6.7

Super-chlorination is used to destroy problematic odours and tastes. Excess chlorine is added to the water to oxidize any remaining organic compounds. The excess chlorine is then removed after the required contact time by the addition of sulphur dioxide, a process known as sulphonation.

3.2.1.11 *Softening and other tertiary processes*

Conventional water treatment is unable to remove a number of soluble inorganic and non-biodegradable organic substances from water. Soluble inorganic material is removed by precipitation or ion exchange, whereas organic substances which are not biologically degraded can be removed by adsorption using activated carbon.

Chemical precipitation is more widely known as precipitation softening. It is used primarily to remove or reduce the hardness in water which is caused by excessive salts of calcium and magnesium. The cause of hardness and the problems caused by excessively hard waters are examined in detail in Section 4.6. Precipitation softening converts the soluble salts into insoluble ones, so that they can be removed by subsequent sedimentation. Lime or soda ash are normally used to remove the hardness, although the exact method of addition depends on the type of hardness present. Lime is most widely used but, like soda ash, produces a large volume of sludge which has to be disposed of (Pontius, 1990). Softening using ion-exchange reactions is becoming increasingly common.

Ion-exchange separation uses a resin, usually natural zeolites which are sodium aluminosilicates (Hill and Lorch, 1987). The zeolites exchange sodium ions for calcium and magnesium ions. The hardness is therefore removed and bound to the resin while sodium, which does not cause hardness, takes the place of calcium and magnesium in the water making it softer. Resins are housed in an enclosed metal tank similar to a pressurized rapid sand filter in design. Once the resin is exhausted and no more sodium ions are available for exchange, the exchange column is taken out of service and the resin regenerated. This is done by pumping concentrated salt solution through the resin, which removes all the calcium and magnesium held in the column, replacing them with sodium again ready for further softening. The regeneration process produces a concentrated solution of unwanted calcium and magnesium chlorides which must be disposed of.

The reactions are as follows:

Softening

$$Ca^{2+}/Mg^{2+} + Na_2(zeolite) \rightarrow Ca/Mg(zeolite) + 2Na^+$$

Hard water + Resin $\quad \rightarrow$ Resin $\quad\quad$ + Soft water

Regeneration

$$Ca/Mg(\text{zeolite}) + 2NaCl \rightarrow Na_2(\text{zeolite}) + CaCl_2/MgCl_2$$

Exhausted + Salt	Regenerated	Waste
resin solution	resin	solution

Synthetic ion-exchange materials have been developed to give much higher exchange capacities than natural compounds such as zeolites. Some organic ion-exchange resins can be used to remove anions such as nitrates, sulphates, chlorides, silicates and carbonates. Ion exchange is increasingly used to remove nitrates from drinking water, although denitrification is also widely practised.

Trace concentrations of synthetic organic compounds, especially pesticides and industrial solvents, are found in surface and groundwater resources. Activated carbon is used to remove these materials and other complex organic compounds responsible for taste and odour problems. Activated carbon works by adsorption of the organic molecule onto its porous structure. The material is continuously adsorbed until the activated carbon is saturated and its capability to adsorb more is exhausted. The carbon, which is wood ash or another lignin-based material which has been specially conditioned, comes in two forms: powdered or granular. If powder is used then it must be discarded when it is exhausted; granules, on the other hand, can be regenerated for reuse by heat treatment. Although more expensive, granules are used when a permanent activated carbon capacity is required due to significant concentrations of unwanted organic compounds in the water. The granules are used in a permanent bed similar to a rapid sand filter that incorporates a back-wash facility. The powdered form is only used for occasional or intermittent use. It is possible that the use of activated carbon may be curtailed in the USA due to concerns about the formation of dioxins during regeneration. The EC is preparing a new Directive on hazardous waste and has proposed that spent activated carbon is included in this category. When plants do not have on-site regeneration facilities, and only the largest water treatment plants do, then this will cause further problems relating to its use as the spent activated carbon will not be allowed to be transported to another site for regeneration (Montgomery, 1975; Mallevialle and Suffet, 1992). Other advanced processes have been reviewed by Lorch (1987) and Pontius (1990).

Apart from the routine monitoring of finished water quality, continuous monitoring equipment is often used at water treatment plants and fitted with alarm systems to warn operators of problems. Those most widely used are for pH, residual chlorine, fluoride, aluminium, iron, dissolved oxygen, colour, ammonia, turbidity, total organic carbon, nitrate and, of course, flow. Although these are excellent for specific parameters they do not measure the overall quality of the water, or warn of contamination of the raw water by trace contaminants such as pesticides. Fish monitors are now widely used to test both the raw water entering the plant and the finished water. They are generally based on the avoidance responses of fish (DeGraeve, 1982) or their ventilation frequency (Cairns and Garton, 1982) in relation to overall water quality. These responses can be monitored remotely and used to activate alarms to warn of a possible water quality problem.

3.2.1.12 *Sludge*

Water treatment produces considerable amounts of waste sludge in the form of a thin slurry. This is mainly gelatinous hydroxide sludge from coagulation and clarification, and the precipitation sludge from water softening. The water used to back-wash sand filters is rich in solids, up to 100 mg/l, whereas the wash-water from the microstrainers is rich in organic matter, especially algae. These solids are allowed to settle before the sludge is pumped to shallow lagoons to slowly solidify, or is dewatered using a filter press. The solid cake from the lagoons or filter press is then safely disposed of to a landfill site, spread on land, or incinerated. Care must be taken with the disposal of water treatment sludges and wash-waters due to the potential for the transfer of pathogens.

3.2.2 Process Selection

The cost of water treatment is dependent on three factors: (1) the quality of the raw water, with costs increasing as raw water quality deteriorates; (2) the degree of treatment required, so that the purer the finished water required, the more it will cost to produce it; and finally (3) the volume of water required and hence the size of the treatment plant, with the cost of water per unit volume decreasing as the capacity of the treatment plant increases.

The availability of water supplies was discussed in Chapter 2, but there are three broad categories of resources. These are, in decreasing order of raw water quality: groundwaters, impounded surface waters such as reservoirs and lakes, and finally upland and lowland rivers. As demand has increased, surface waters have been recycled, especially in the south of England where resources are short (Section 2.6).

Groundwaters are chemically and bacteriologically of good quality and so only minimum treatment is required. This makes groundwater supplies cheap compared with water from other resources. The general sequence of treatment for groundwater is given in Figure 3.4. The water contains carbon dioxide and iron which must be removed by aeration, followed by rapid sand filtration. If the water is excessively hard it may be necessary to soften it using either precipitation or ion exchange. All that is then required is to disinfect the water before distribution. Water from reservoirs may be exceptionally clean and as the raw water is aerobic, there is no carbon dioxide or iron to be removed. Such waters may only require microstraining and then disinfection (Figure 3.5). In contrast, river water supplies will require either storage followed by microstraining, then filtration or flocculation and clarification followed by rapid sand filtration before disinfection and subsequent supply (Figure 3.6). The quality of surface waters that can be abstracted for public supply now has to conform to an EC Directive. This specifies the required quality of water that can be abstracted and places the waters into three categories (A1, A2 and A3) for which specified treatment is required. For example, A1 resources are the cleanest and only require minimum treatment as shown in Figures 3.4 and 3.5, whereas A3 waters are the poorest quality supplies which require advanced treatment techniques such as chlorination, followed by coagulation, flocculation, decantation, filtration, adsorption using activated carbon and then full

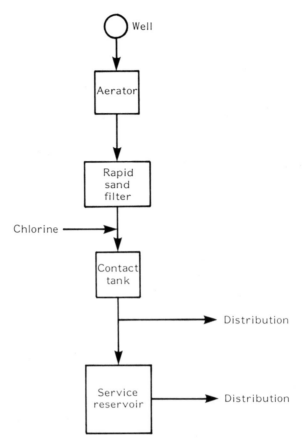

Figure 3.4. Normal sequence of treatment for a groundwater source to be used for supply

disinfection. This EC Directive was fully explored in relation to water quality and treatment in Section 2.6. Solt and Shirley (1991) and Pontius (1990) have produced two excellent reference texts on water treatment.

3.3 WATER DISTRIBUTION

After treatment the water has to be conveyed to the consumer. This is done by a network of pipes known as water mains that are laid underground, usually under roads and pavements. There is, however, much more to water distribution than just mains; there are service reservoirs and booster stations as well.

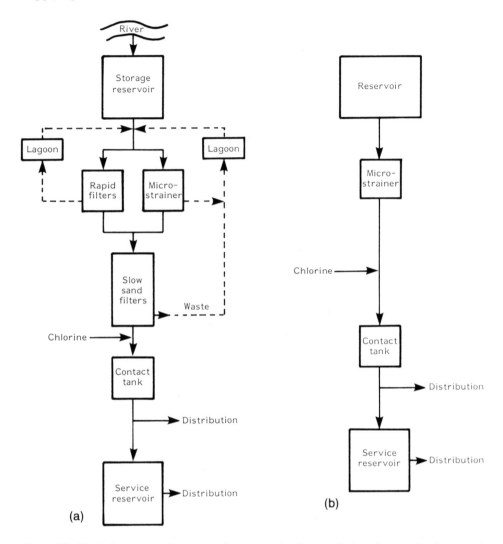

Figure 3.5. Typical sequence of treatment for water taken for supply from (a) an upland river and (b) an upland storage reservoir

3.3.1 Service Reservoirs

Service reservoirs are needed primarily because the water resource, and often the treatment plant, are usually considerable distances from the centre of population. Where groundwater resources are used, pumps need to be capable of pumping water at peak demand rates, rather than at the average daily demand rate. The service reservoir ensures that all peak demands for water are met, while smaller pumps and trunk mains

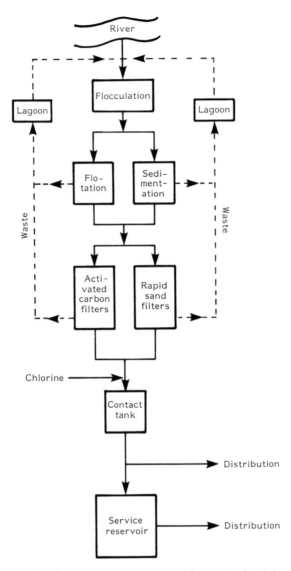

Figure 3.6. Sequence of unit processes to treat water from a lowland river for supply

can be used to cope with the average daily flow-rates rather than the peak demand rates which may be 50–80% greater.

Like electricity, the demand for water varies during the day (diurnal variation). Although the peaks (high) and troughs (low) in demand can be calculated fairly adequately from experience, service reservoirs and water towers are needed to ensure that these demands are fully met (Figure 3.7). They also have other important functions such as providing a reserve storage capacity in case of problems at the treatment plant

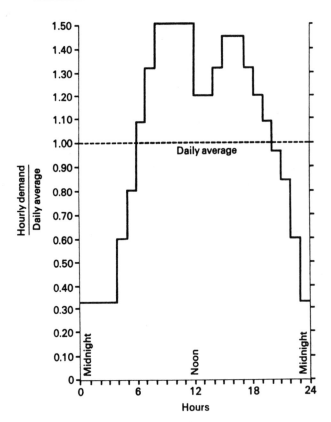

Figure 3.7. Variation in demand for water. Reproduced with permission from Open University (1975)

or with the trunk mains. They also compensate for any variation in water quality; for example, when water comes from more than one source.

Although the flow to the service reservoir remains constant, the level in the reservoir will rise and fall according to demand. The size of the reservoir varies according to the size of the area served, although enough storage for 24 or 36 hours is usually selected. Although generally constructed out of concrete, smaller reservoirs are often of steel or brick. An important design feature is that the tanks should be watertight, not only to prevent water loss, but to prevent contamination from outside the tank. The tanks are often split into a number of separate chambers to allow periodic cleaning, although more recent designs have been circular prestressed concrete tanks. Where this design is used at least two tanks are required if the supply is not to be interrupted during tank cleaning or maintenance. The reservoir must ensure an adequate hydraulic head to produce sufficient pressure within the main to push the water up into the storage tanks in household attics and even small blocks of flats. A minimum head of 30 m is required by the fire brigade for fire-fighting, whereas a head in excess of 70 m would result in an unacceptably high loss of water via leaks in the mains. High pressure also causes excess wear on household equipment such as taps, stoptaps (also widely called stopcocks or stop valves), washing machine valves and ball valves in WCs, all of which are designed to operate at moderate pressures. Noise from household plumbing also increases at high water pressures. Exceptionally tall buildings will probably require a system of pumps to raise the water to the storage tanks at the top. The installation, operation and maintenance of such pumps is the responsibility of the owner.

Where service reservoirs are required in flat areas, sufficient hydraulic head may be obtained by the construction of a water tower. Water towers serve the same purpose as a service reservoir except that they are generally much smaller, serving only small distribution zones. The water usually has to be pumped up to the tower from the main, thus increasing the cost of supply. Although service reservoirs are often below ground and rectangular, modern water towers have moved away from the traditional large brick or cast iron tanks that were once familiar to a variety of unusual shapes made from concrete. The most common design is the inverted mushroom shape supported on a remarkably slender tower.

3.3.2 Water Mains

There are two main categories of water main. The trunk mains are the largest and do not have any branch or service pipe connections. They are used for transporting large

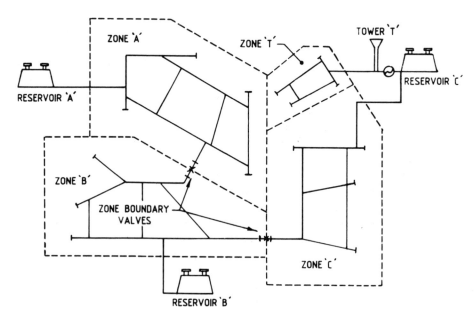

Figure 3.8. Schematic diagram showing that water distribution networks are broken down into operational zones each supplied by a service reservoir. Reproduced from Latham (1990) by permission of the Institution of Water and Environmental Management

volumes of water from the source to the treatment plant, from the plant to the service reservoir, and from one reservoir to another. The distribution mains consist of a pipe network of smaller, varying sized pipes, which is highly branched. It is the distribution main which supplies individual houses.

In large towns or cities the distribution main system is usually arranged into pressure zones controlled from specific service reservoirs (Figure 3.8). In practical terms this allows for better leakage control and metering of water usage in each district. Although these zones are independent in terms of supply, such zones are often interconnected by zone boundary valves to allow for the transfer of water between zones if the need arises. Within each zone any specific part of the distribution network can be isolated by valves, to isolate leaks, make new connections and to carry out maintenance or repairs. Within each pressure zone the distribution network consists mainly of loop (ring) circuits although spurs (dead-ends) are often necessary where the housing pattern does not allow a loop main to be used. In a long spur the water can stay in the main for a considerable time before being consumed, which can adversely affect its quality, especially towards the end of the pipework. Also, if repairs are required, then it is necessary to isolate the entire length of pipe. In contrast, it is possible to isolate a small section of pipework for repairs in a loop main without cutting off the supply to houses either side. This also ensures that the water is not retained for excessively long periods within the distribution system before it is used.

The distribution network consists of pipes of varying sizes ranging from 450 mm (18 in) and sometimes even bigger, down to 50 mm (2 in) with 75 mm (3 in) 100 mm (4 in) and 150 mm (6 in) all common. New mains normally use either 100 or 150 mm pipes as standard. At the end of each system, or in new housing estates or cul de sacs, you will often see the new flexible plastic pipes, which come in coils, being used. These generally come in two standard sizes, 63 and 90 mm.

The pipes come in a variety of materials. The most commonly used are iron (cast, spun or ductile), asbestos cement, uPVC (unplasticized polyvinyl chloride) and also MDPE (medium density polyethylene). There are new draft specifications for all pipe materials published by CEN, the European Committee for Standardization. Owing to the effects on water quality, asbestos cement pipes are no longer installed and are being replaced by plastic pipework whenever possible (Section 6.4). However, the cost of renovation or replacement can be horrendously expensive, so older or less favoured materials are not likely to be replaced very quickly. In fact, existing pipes are only replaced when the size is no longer adequate to supply the amounts of water now required, or if their performance in terms of leakage is no longer acceptable.

To overcome the logistic problem of replacing the inadequate and leaking trunk water mains system serving London, Thames Water have invested £250 million in the construction of a large ring main. The 2.5 m diameter concrete tunnel, which forms a 50 mile loop under the city, will supply 1320×10^6 l of water daily to six million consumers. Already a 28 mile section is in operation serving south London, with the entire system due to be completed by 1995. Five treatment plants feed water into the ring main, which is controlled centrally using sophisticated computer systems at Hampton in Middlesex. Water is fed from the ring main into local supply zones from 18 unmanned pump-out shafts (Figure 3.9). The system incorporates many new initiatives,

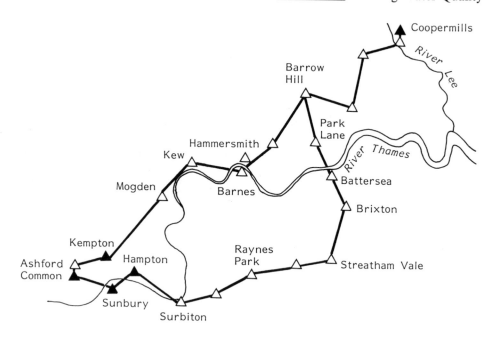

Figure 3.9. Map of London's new ring water main

including automatic monitoring along the main to test the water and add extra chlorine where necessary. The new main was tunnelled through Thames clay, which was considered to be cheaper and far less inconvenient to the public than replacing the existing trunk mains using more conventional surface based construction techniques.

Hydrants are important not only for fire-fighting, but for flushing out the main. The location of hydrants is decided by the needs of the fire brigade and the requirements of the water company. In fact, water companies have a legal requirement to locate hydrants where the fire brigade require them. In existing mains, however, if there appears to be insufficient water available for fire-fighting, then the fire brigade can demand a larger main, but they must bear the cost. Fire hydrants are tested regularly by the fire service, with the water company undertaking to rectify any defects identified, although the cost of repair is charged to the fire service. In the UK and Ireland, hydrants are located in an underground chamber to protect them against frost and general damage. The exact location of the hydrant is indicated by a yellow plaque marked with the letter H painted in black. The figures above and below the cross-piece of the letter H indicate the diameter of the main and the distance of the hydrant from the plaque, respectively. The optimum diameter for fire-fighting is 100 mm, although 75 mm is satisfactory for a ring main. The effects of distribution on water quality are considered in Chapter 6.

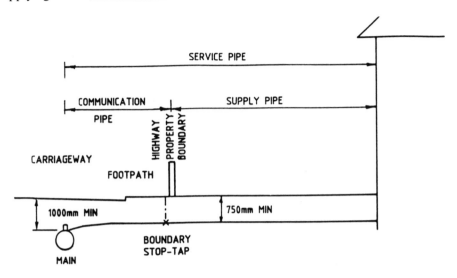

Figure 3.10. Arrangement of service pipes showing minimum depths to which pipes should be buried. Reproduced from Latham (1990) by permission of the Institution of Water and Environmental Management

3.3.3 Service Pipe

The pipe that conveys the water from the mains to the consumer's house is the service pipe. For single dwellings this is a small pipe of less than 25 mm in diameter, although it is usually only 13 mm (0.5 in). When more than one dwelling is served (blocks of flats, institutions such as schools or hospitals, or industrial premises) then the service pipe may be larger and possibly even the same diameter as the main itself when supplying large industrial complexes.

The service pipe is split into two sections, usually by the water company's stoptap. The communication pipe is owned by the water company, who are responsible for its maintenance and repair. It runs from the water mains to the boundary of the street, which is usually the outer wall or fence of the property to which the supply is going. The company installs a boundary stoptap as close to the boundary as possible, usually just outside the boundary fence (Figure 3.10). The company is also responsible for the maintenance and repair of the boundary stoptap, although the household it serves may operate it if necessary. The service pipe then runs from the boundary stoptap into the house. The householder usually has his or her own stoptap within the building, as close as possible to where the supply enters. This section of pipe between the two stoptaps is known as the supply pipe. The property owner is responsible for the maintenance and repair of the supply pipe, not the water company. In fact, all the company's responsibilities end at the boundary stoptap. This division of responsibility is perhaps the most common cause of bad feeling among customers, who are used to other utilities

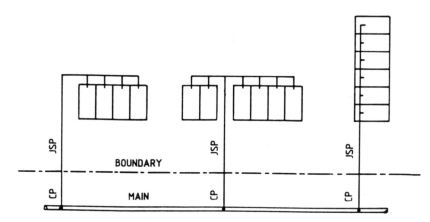

Figure 3.11. Examples of typical joint supply arrangements. CP = Communication pipe; JSP = joint supply pipe. Reproduced from Latham (1990) by permission of the Institution of Water and Environmental Management

such as gas, electricity and telephone companies taking responsibility for the service to within the building, or at least to the meter. This is not the case with water. In fact, water meters may be fitted anywhere on the service pipe, including within the property, but this does not alter the division of responsibility for the service pipe, with the supply pipe remaining the responsibility of the property owner, even if it also includes the water meter.

A common problem in older houses is that the water main can be remote from the property. Although the communication pipe may still be fairly short, just extending to the boundary of the street in which it is laid, the supply pipe may be very long. In these cases the private supply pipe may be laid across private land, under roads and even occasionally through other buildings on its way to the customer's property. Such complex arrangements can cause appalling problems when the supply pipe requires attention. The problems of locating the pipe and then detecting a leak or blockage is usually compounded by obtaining permission to dig up someone's lawn or driveway. It can also involve very high costs, especially if the repair involves digging up a section of local authority highway.

Before 1974 in the UK, local authorities generally held responsibility for both housing and water supply. During this period many houses were supplied with water using joint supply pipes. These serve a number of properties and are common throughout the UK (Figure 3.11). Currently between 20 and 50% of all service pipes in any supply zone are shared. Like all service pipes only the communication pipe is the responsibility of the water company; the supply pipe in this instance is owned jointly by the properties served. Because these pipes are old they are prone to problems. The most common complaint is that the houses at the end of a joint supply pipe have low pressure or insufficient water. This is often seen when terraces of houses or cottages are modernized and the water usage of the new families is much greater than when the pipe was first laid. Old pipes are much more prone to bursting or blockages, and as with long supply pipes they can be difficult to locate and repair. It was common practice to lay joint supply pipes in terraced houses through cellars, so obtaining access for maintenance and repair is not always easy.

Very often the poor supply to a property is due to defective supply pipework in which a householder may have only a part share. Water companies have specific powers in the Water Industry Act 1991 to force property owners to repair leaks, and can even insist on the installation of a new separate service connection to the main, all at the property owner's expense. In practice, the separation of joint service pipes is rarely enforced by companies. Interpretation of ownership and the degree of responsibility for service pipes can be very complex and useful guides have been published to assist in sorting out problem installations (Water Research Centre, 1991; 1992).

Service pipes can be made of mild steel, wrought iron, copper, lead, polyethylene or uPVC. Until the widespread introduction of plastic pipes, which are now universally used, the selection of the correct metal for the service pipe was very important. This is because where the water is soft, service pipes can be readily corroded. Iron and steel pipes are extremely strong and are jointed by screwing together or by flanges, so they can withstand high pressures. Mild steel is generally galvanized to protect it from corrosion. Unlike iron and steel, copper can be used with very thin walls, which makes it cheap and light to work with. Copper pipes can be easily jointed and are readily bent into shape, and can withstand high pressures. The smooth surface in copper pipes results in a very low resistance to flow so that smaller diameters can be used compared with other metal pipes. Copper and galvanized steel should not be used together, as there will be enhanced corrosion due to electrolytic action. Lead was the most widely used material for service pipes and for household plumbing before the Second World War as it is so malleable and was the best material available at that time. However, justified health concerns have resulted in lead communication pipes being replaced by the water companies whenever they are encountered. Where householders replace their lead supply pipes, the water companies will also replace the communication pipe. The problem of lead pipes and health is examined in Section 7.3.

All the service utilities now use plastic pipes or plastic ducting. In the UK these pipes are colour-coded to prevent accidents, with water pipes coloured blue, gas yellow, electricity black and telephone grey. Plastic water pipes have many advantages, especially where the soil or water conditions pose problems of freezing or corrosion. To protect service pipes from frost they should be laid at least 750 mm below the surface. Polyethylene is tough, light, cheap, flexible, easy to work and join, frost resistant and a non-conductor of electricity. uPVC, like polyethylene, is not impervious to gas and loses its strength rapidly at temperatures above 70°C. uPVC is more rigid than polyethylene and has a greater tensile strength, so that more complex piping systems can be used as a wide variety of joints are available. The greatest practical problem with using plastic pipes is that their exact location cannot be detected so readily using electromagnetic surface detectors. This can be overcome by using metal trace wire, but this is expensive. There are a number of new location techniques for plastic pipes under investigation. The most promising appear to be those based on acoustic methods. Sound waves travel readily through water, so by applying a pulse to a pipe containing water it will be transmitted along its length. In this way the exact location of the pipe can be identified by using a suitable detector to pick up the vibration (Godley and

Wilcox, 1992). Care must always be taken to accurately record the exact route of such pipes and to lay them wherever possible at right angles from the main to the boundary stoptap and then in a direct line to the house.

It was common practice in houses constructed before 1970 to earth household electrical systems to the incoming service pipe. This was phased out during the 1960s, but can still be found occasionally in houses built as late as 1968–9. Plastic pipes are non-conductors of electricity, so an alternative earthing system will be required.

One of those black magic tricks of the trade that you find in all professions, which still amazes me, is how new service pipe connections are made to the water main. This is done without interrupting the flow in the main and generally without the person making the connection getting soaking wet, although this is not always the case! A special device is used to drill the pipe and insert a ferrule onto which the communication pipe is fitted. This is carried out at the crown (top) of the pipe with the communication pipe looped over slightly in a snake shape to allow for any ground movement or thermal expansion or contraction without damaging the connection to the main.

Chapter 4
Problems with the Resource

4.1 NATURAL OR MAN-MADE?

The correct selection of water resources for supply purposes is important, as the cleaner the raw water, the cheaper the finished water is to produce, and the safer it is to drink. Any materials or chemicals that find their way into the resource may need to be removed before supply. The main parameters of concern to the water industry are those listed in the Drinking Water Directive, especially those that are potentially harmful or which affect our use of the water for purposes such as washing clothes (Section 1.3). These include nitrates, micro-organics such as pesticides, odours and taste, iron, manganese, hardness, colour and, of course, pathogens. All of these are considered in this chapter. Other problems can arise during treatment or distribution. Although the main chemical characteristics of the water remain after treatment, removing or reducing the concentration of the undesirable compounds is both expensive and technically difficult. Careful selection of resources is therefore very important.

The natural quality of water depends on the nature of the catchment area, especially its geology. Impermeable rocks such as granite result in turbid, soft waters which are slightly acidic and naturally coloured; there is no groundwater, so water supplies come directly from surface waters such as rivers or impounding reservoirs. In contrast, permeable rocks such as chalk and limestone result in clear, hard waters rich in calcium and magnesium, and slightly alkaline. These permeable rocks form aquifers so water is available from both ground and surface resources. To comply with the Drinking Water Regulations, resources are carefully selected, not only to ensure an adequate volume of water throughout the year, even during periods of drought, but also to ensure consistent quality.

Wherever humans work and live there will be an increase in toxic substances, non-toxic salts and pathogens entering the water cycle. Although industry can have serious localized effects on surface and groundwaters, the extensive nature of agriculture makes its impact on water supplies more widespread and often more difficult to control. To produce a high yield of crops, especially cereals, year after year on the same land, large amounts of chemicals are required. When the same crop is repeatedly grown in the same soil the nutrients are depleted and pests increase. Fertilizers are therefore needed to replace the nitrogen and phosphorus used by plants, and pesticides and herbicides are required to protect crops. While all these chemicals can potentially contaminate water resources, organic farming has been found to be just as polluting as conventional farming in terms of nitrate pollution, in some instances more so due to the higher levels of manure used on the land. Chemicals not only come from the large lowland arable farms, but also from other farm-based activities such as dipping sheep in upland and often quite remote farming areas.

Farmers are not the only people using chemicals on the land: a range of groups and organizations from British Rail to local authorities, the Forestry Commission to the average gardener, to name but a few, are potential polluters. This is reflected in the high concentration of essentially non-agricultural herbicides now being reported in drinking water (Section 4.3). The fish-farming industry is a major user of water, situated alongside many of our cleanest rivers. The treatment given to waste waters from such farms is extremely basic, yet they use a wide range of chemicals, many of which are toxic (Table 4.1). It is difficult to assess the overall effects of fish-farming on water quality as the dilution is so great; however, all the chemicals used are found in trace amounts in the receiving water.

The UK water industry is finding compliance with some areas of the EC Drinking Water Directive more difficult than its European partners, although nearly all European Member States have found themselves in the European Court over non-compliance with the Directive in recent years. In most European countries only a third or less of supplies come from surface waters (Table 2.2), and these are subject to extensive treatment, including storage and bankside infiltration. This compares with three-quarters of all supplies in the UK coming from surface waters, most of which only receive conventional treatment. It is therefore inevitable that supplies in the UK will be more prone to surface water problems such as colour, iron and aluminium from upland catchments, and taste, odour and THMs in lowland abstractions. High aluminium concentrations which arise naturally in some German groundwaters are legally derogatable, whereas the high aluminium waters arising during treatment of upland water in the UK using coagulation are not. Yet both are problems arising from the nature of the natural resource (Breach, 1989).

In the past we have conveniently separated the way in which we live, use our land and dispose of our waste from the quality of our water and food. But whatever we discard, whether it is in solid, gaseous or liquid form, will eventually find its way into the water cycle. Aquifers which have long retention times are like time machines, telling us 10, 20, 30 or even 40 years later what we sprayed on our land or dumped in our waste tips. Surface waters are more immediate, with chemicals flushed away eventually to be broken down in the sea. When rivers are impounded toxic materials may take years to be diluted until they are no longer detectable. Alternatively toxins may be adsorbed by sediments and plants only to reappear from time to time when they are redissolved back into solution by bacterial or chemical action. The water industry is often removing from water chemicals and pathogens that other people have added, often through neglect or poor work practices. Some of these pollutants are difficult to remove, and in practice cannot be removed satisfactorily, so in the future new management strategies that control the way in which the land is used will be required to protect drinking water as it makes its way from the sky to our taps. With dwindling water resources and an increasing demand for water it is inevitable that water will have to be used over and over again, and water from resources of poorer quality exploited.

Table 4.1. Main chemicals used in aquaculture. B = bath; A = addition to system; F = flush; D = dip; I = injection; S = spray; T = treated food; FW = freshwater; and SW = seawater. Reproduced from Nature Conservancy Council (1989) with permission of English Nature

Chemical	Use	FW/SW	Method	Remarks
Therapeutants				
Acetic acid	Ectoparasites	FW	D	Use with CuSo₄ in hard water areas 165–250 ppm up to one hour, 20 ppm four hours use in sea cages as bath is common
Formalin	Ectoparasites	FW/SW	DA	
Malachite Green	Ectoparasites and fungus	FW/SW	DFS, B	Eggs and fish, 100 ppm 30 seconds 4 ppm one hour common in FW, occasional use in cages as a dye marker
Acriflavin (or proflavine hemisulphate)	Ectoparasites, fungus and bacteria	FW	D	Mostly for surface bacteria, fish and eggs, occasional use only
Nuvan (dichlorvos)	Salmon lice	SW	B	1 ppm for one hour, canvas round sea cage
Salt	Ectoparasites	FW	DB	Occasional alternative to formalin
Buffered Iodine	Bacteriocide	FW	B	Use to disinfect eggs 10 minutes 1000 ppm
Oxytetracycline	Bacteriocide	FW/SW	T	Antibiotic widely used for systemic disease
Oxolinic acid	Bacteriocide	FW/SW	T	Antibiotic widely used for systemic disease
Romet 30 (sulphadimethoxine and orthomeprim)	Bacteriocide	FW/SW	T	Antibiotic for systemic disease
Tribrissen (trimethoprim/sulphadiazine)	Bacteriocide	FW/SW	T	Third most widely used antibiotic
Hayamine 3500	Surfactant/bacteriocide	FW	A	Quaternary ammonium compound used for treating bacterial gill diseases
Benzalkonium Chloride	Bacteriocide	FW	A	Surface antibacterial; 'Roccel' (similar to above)
Chloramine T	Bacteriocide	FW	A	As above, also effective for some protozoa
Vaccines				
Vibrio anguillarum vaccine		SW	B	Not widely used
Enteric Redmouth vaccine		FW	BSI	Widely used in trout culture
Aeromonas salmonicida Vibrio *Anguillarum* vaccine		SW	I	Not widely used

Anaesthetics			
MS222 (tricaine methane-sulphonate)	FW/SW	B	Widely used approx 1:10 000 dilution
Benzocane	FW/SW	B	Widely used, requires acetone to dissolve
Carbon dioxide	FW/SW	B	Sometimes used at harvest
Disinfectants			
Calcium hypochlorite	FW/SW	S	General disinfectant for tanks etc.
Liquid Iodophore e.g. FAM 30	FW/SW	S	For equipment and footbaths
Sodium hydroxide	FW	S	Most commonly used for earth ponds
Water treatment			
Lime	FW	A	Used in earth ponds
Potassium permanganate	FW/SW	BA	Oxidizer and detoxifier
Copper sulphate	FW/SW	A	Algicide and herbicide

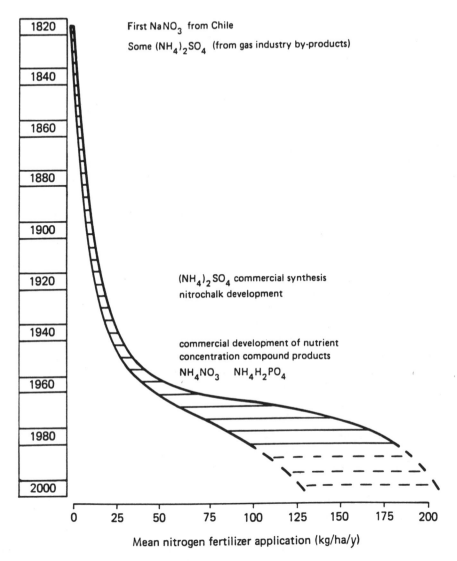

Figure 4.1. Increase in the use of fertilizer with time. Reproduced from Foster *et al.* (1986) by permission of the British Geological Survey

The most effective way to control these problems is to prevent the contamination of raw water supplies in the first place. New legislation from Europe is applying this principle with, for example, the introduction of aquifer protection zones, nitrate sensitive areas and more careful control over the disposal of all hazardous and toxic wastes. The way we dispose of our waste and use our land is the key to the quality of our raw water supplies, and more importantly to the quality of the water we drink.

4.2 NITRATES

4.2.1 Sources in Water

Nitrate fertilizer is the single most important and widely used chemical in farming. Its use on farms throughout Europe has rapidly increased over the past 20 years in particular, reaching the present phenomenal levels (Figure 4.1). In the UK the use of nitrate fertilizer increased from just 6000 tonnes/year in the late 1930s to 190 000 tonnes by the mid 1940s. This rise was due to the need to grow more food during the Second World War. However, the reasons for the increase in its use to a staggering 1 580 000 tonnes by 1985 are less clear, as is the reason for the higher application rates of such fertilizers now practised. It is estimated that application rates have doubled over the past 15 years, so that in 1992 farmers spent £600 million on nitrate fertilizer in Britain alone.

Nitrogen is an essential plant nutrient which is usually absorbed as nitrate or ammonium from the soil. It is used to form plant proteins, which in turn are used as major dietary sources of amino acids by humans and animals. The nitrogen absorbed from the soil must be replaced to maintain the fertility of the soil and therefore its long-term productivity. Farmers replace the nitrogen by either spreading manure or applying artificial fertilizers. Most of the nitrogen in the soil is in an organic form which is bound up either in plant material or organic matter and humus. As plants can only use inorganic (mineral) nitrogen this natural reserve of organic nitrogen is largely unavailable to plants for growth, unless it is broken down by microbial action from its organic form to nitrate. This slow process continuously releases mineral nitrogen in low concentrations, but not always when and in the amounts required by crops.

There are two main sources of nitrate contamination of water resources. (1) Nitrate is released when organic matter is broken down by bacteria in the soil. However, if the crops are not actively growing then the nitrate produced by microbial activity is not used by the plants and so is carried through the soil by rain-water into the aquifer to contaminate groundwater. (2) Inorganic nitrogen is added directly to fields by the farmer as artificial fertilizer. Where application exceeds plant needs or the ability of the plant to use the nitrate, then the excess will either be bound up in the soil or, more likely, washed out of the soil by rain into either surface or groundwater.

Under ideal conditions 50–70% of the nitrogen applied to land as artificial fertilizer is taken up by plants, 2–20% is lost by volatilization, 15–25% is bound up with the organic matter or clay particles in the soil, leaving between 2 and 10% to be leached directly into surface or groundwater. The percentage of nitrate leaching from the soil depends on a variety of factors, including soil structure, plant activity, temperature, rainfall, the application rate of fertilizer, the water content of the soil and many more, so it is difficult to generalize or predict accurately the amount of nitrate which will be lost from an area of agricultural land. Is it correct to conclude that the main source of nitrate pollution is artificial fertilizer? Published work on nitrate pollution of water resources clearly concludes that the leaching of agricultural fertilizers is a major source. The Department of the Environment (1988b) review on the subject, *The Nitrate Issue*, supports this view, suggesting that a cut of just 20% in the 190 kg of nitrogen now

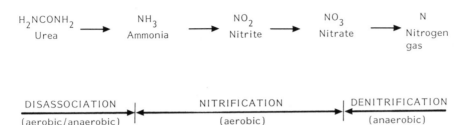

Figure 4.2. Breakdown of urea to nitrogen gas: this takes place via a number of microbial processes, both aerobic and anaerobic

typically applied to each hectare of winter wheat would reduce the nitrate leachate by up to 42%. Although there appears to have been some success in controlling nitrate by drastically reducing the application rates of nitrate fertilizers, research carried out at the Rothamsted Experimental Station indicates that the interaction between fertilizer application and nitrate leaching is far more complex (Addiscott, 1988; MacDonald *et al.*, 1989). All soils contain large amounts of organically bound nitrogen. Arable soils contain between 3000 and 8000 kg N/ha, whereas grassland can contain up to 15 000 kg N/ha, of which 90–95% is bound up with organic matter, mainly humus. Depending on factors such as the soil type, weather conditions and the previous crop grown, up to 3% of this organically bound nitrogen can be mineralized by soil bacteria into ammonia. This is a slow process, but subsequent nitrification of ammonia to nitrate occurs very rapidly (Figure 4.2). The nitrate is then available for a number of soil processes: for example, to be taken up by the crop, fixed biologically in the soil as microbial biomass, denitrified by micro-organisms to nitrogen gas which is lost to the atmosphere, or leached from the soil into water resources. It is difficult to predict whether the nitrate will be used in the soil or whether it will be lost via leaching, but tillage certainly significantly increases leaching. Where grassland is intensively fertilized by manure, due to a high stocking density of animals or the excessive disposal of animal manure or sewage sludge, then there will be an excess of nitrogen which will be readily leached as nitrate. On certain soil types, such as sandy soils, the rate of leaching may approach those levels observed for arable land. Whenever grassland is ploughed up there is a large release of nitrogen into the soil which can be leached into water resources. The actual amount depends on the age of the grass, but 280 kg of nitrogen per hectare ploughed would be an average figure.

Even when crops have no fertilizer added about 20 kg of nitrate per hectare leaches into the groundwater. Studies have shown that most of the nitrogen applied as fertilizer is used by the plant, whereas the excess nitrate, which is the most likely source of leachate, comes from the soil's own vast reserve of organically bound nitrogen, on average about 5000 kg/ha, by the action of the microbes living in the soil. The problem is that the microbes are most active when conditions are ideal for them, not when the crop needs the nitrogen. For example, it is in the autumn when the soil is warm and its moisture content is increasing that the microbes are stimulated into producing most nitrate. This is also the time that the rainfall is beginning to exceed evaporation so that water flows downwards into the groundwater, taking any soluble nitrate with it. More land in Britain is put over to growing winter cereals than any other crop and so the nitrogen applied as fertilizer is more prone to being washed out of the soil by heavy rain. When spring barley is grown, however, the soil is left bare during the winter allowing more nitrate to be leached out of the soil than when winter wheat is grown. In Europe nitrate will be mainly leached from the soil and into water resources in late autumn, winter and early spring. Water, not used by the plants nor lost by evaporation, percolates through the soil and eventually reaches the water-bearing rock, or alternatively the surface waters. Forms of inorganic nitrogen other than nitrate found in fertilizers are ammonium and urea, both of which are rapidly transformed by soil micro-organisms into nitrate—both ground and surface waters which are used for water supply can become contaminated. The problem is therefore farming practice in general rather than specifically the over-use of fertilizer.

Inorganic fertilizers also contain smaller amounts of phosphorus and potassium. These nutrients do not generally cause a problem as they are effectively bound up and held by the soil particles, and unlike nitrate are not readily leached out of the soil by rainfall. However, although phosphorus and potassium do not affect drinking water quality directly, phosphorus, along with nitrogen, can cause eutrophication in surface waters with all the associated problems of excessive plant growth (Section 4.9).

4.2.2 Effect on Consumers

Nitrate is a common component of food, with vegetables usually being the principal source in our daily diet. Intake of nitrate varies according to dietary habits, with the intake by vegetarians much higher than non-vegetarians. Some vegetables have low nitrate concentrations; for example, peas, mushrooms and potatoes all contain less than 200 mg/kg, whereas others such as beetroot, celery, lettuce and spinach are all very rich in nitrate, in excess of 2500 mg/kg. Vegetables grown out of season, or forced, generally contain higher than normal concentrations of nitrate. Therefore because our diets are so variable estimates of nitrate intake from food vary over a wide range, from 30 to 300 mg NO_3/day.

Table 4.2 shows that water can contribute significantly to the intake of nitrate; for example, when water contains the MAC limit of 50 mg/l it is contributing about half of the daily nitrate intake of an average consumer.

Breast-fed infants have a low nitrate intake whereas those fed on infant formula feeds receive nitrate from the water used in its preparation. Formula milk powder used for infant feeds contains some nitrate in its own right, equivalent to about 5 mg/l. The main source of nitrate for bottle-fed infants is the water; for example, if drinking water contains 25 mg/l nitrate, then the infant's feed will contain about 30 mg/l of nitrate.

There is little nitrite, as opposed to nitrate, in drinking water, although some foods, especially cooked meats, can contain high levels. Nitrate itself appears to be harmless at the concentrations found in water and most foodstuffs. In the body it is rapidly assimilated in the small intestine and taken up in the blood. Once absorbed, nitrate is excreted mostly unchanged in urine, but some will be reduced by bacteria to nitrite. Part of the ingested nitrate, about 25%, is recirculated by excretion in saliva. It is in

Table 4.2. Contribution of nitrate in drinking water to the daily intake of nitrate in diet. Reproduced by permission of the DHSS (1980)

Concentration of nitrate in water (mg/l)	Daily nitrate intake (mg)		Percentage derived from drinking water
	Water	Food	
10	14	57	20
50	71	57	55
75	107	57	65
100	143	57	71
150	214	57	79

saliva that most of the nitrate is reduced to nitrite by bacteria. A similar conversion to nitrite occurs in the stomach. All the health considerations relating to nitrate are related to its conversion to nitrite, which is a reactive molecule associated with a number of problems, most commonly conversion to N-nitroso compounds, the formation of methaemoglobin and cancer.

In the gastrointestinal tract, nitrite reacts with certain compounds in food under acidic conditions to produce N-nitroso compounds with amines and amides. Many of these compounds are known carcinogens. Although there is no epidemiological evidence to link nitrate directly with cancer in humans, increased concentrations of nitrite and N-nitroso compounds have been detected in people who secrete inadequate amounts of gastric acid, a group known to be particularly at risk from gastric cancer. This is further discussed by Forman et al. (1985).

The main concern associated with high nitrate concentrations in drinking water is the development of methaemoglobinaemia in infants. To cause enhanced methaemoglobin levels in blood, nitrate must first be reduced to nitrite, as nitrate itself does not cause the disorder. The nitrite combines with haemoglobin in red blood cells to form methaemoglobin, which is unable to carry oxygen and so reduces oxygen uptake in the lungs. Normal methaemoglobin levels in blood are between 0.5 and 2.0%. Research has shown that the use of water containing levels of nitrate up to double the MAC concentration, 50–100 mg/l, for infant feed preparation results in increased levels of methaemoglobin in the blood, but still within the normal physiological range. As methaemoglobin does not carry oxygen, excess levels lead to tissue anoxia (i.e. oxygen deprivation). It is only when the methaemoglobin concentration in the blood exceeds 10% that the skin takes on a blue tinge in infants, the disorder known as methaemoglobinaemia or blue-baby syndrome. The progressive symptoms resulting from oxygen deprivation are stupor, coma and eventual death. Death ensues when 45–65% of the haemoglobin has been converted. However, the disorder can be readily treated using an intravenous injection of methylene blue, which results in a rapid recovery.

Infants aged less than three months have a different respiratory pigment which combines with nitrite more readily than haemoglobin, making them especially susceptible to the syndrome. Their nitrate intake at this age is also high relative to their body weight compared with older children. Children up to 12 months of age may have an incompletely developed system for methaemoglobin reduction, being naturally deficient in two specific enzymes that convert methaemoglobin back to haemoglobin, and so are also at risk. Although methaemoglobinaemia is not usually a problem in adults, pregnant women are thought to be at risk, although the reasons are unclear (World Health Organization, 1984).

It appears unlikely that infantile methaemoglobinaemia is caused by bacteriologically pure water supplies containing nitrate concentrations up to 100 mg/l. With 98% of Europe's population using piped mains with the water treated to remove bacteriological contamination, infantile methaemoglobinaemia has become almost non-existent. The disorder has not been reported in children drinking mains piped water, except when it has been contaminated by bacteria. The remaining 2% of the population is supplied with well or spring water of variable quality and acute infantile methaemoglobinaemia has been largely associated with bottle-fed infants using high-nitrate well water. This is commonly referred to as well-water methaemoglobinaemia. Most cases have been associated with well-water nitrate concentrations in excess of 100 mg/l. Where lower concentrations of nitrate were found, then the bacterial status of the water was poor and/or the infants had gastroenteritis. Acute infantile methaemoglobinaemia can be a rare complication in gastroenteritis irrespective of nitrate intake. It is thought that this is due to the enhanced conversion of nitrate to nitrite from increased bacterial activity associated with enteritis. Like all cases of the disorder, well-water methaemoglobinaemia has become extremely rare over the past 20 years in the UK. Since 1945 there have been 35 cases, the last reported case associated with a private well in 1972. The reduction in incidence is linked with the development of public supplies to rural areas (ECETOC, 1988).

4.2.3 Nitrate Standards

Units used to express nitrate concentrations have caused much confusion. They are expressed as either milligrams of nitrate per litre (mg NO_3/l) or as milligrams of nitrogen present as nitrate per litre (mg NO_3-N/l). There is considerable difference between the two. For example, 50 mg NO_3/l is equivalent to 11.3 mg NO_3-N/l. In this book the actual nitrate concentration is used, i.e. mg NO_3/l, unless specified when the corrected value is given in parentheses. Be careful when looking at reports or data, and make sure you know which units are being used. To correct mg NO_3/l to mg NO_3-N/l multiply by 0.226; conversely, to correct mg NO_3-N/l to mg NO_3/l multiply by 4.429.

In the EC Directive on Drinking Water, nitrate is listed along with non-toxic substances (Table 1.10). The concentration of these non-toxic substances may exceed that stated in the Directive at the discretion of individual Member States, provided there is no danger to public health. This is known as a derogation. The Directive currently specifies an MAC for nitrate of 50 mg/l, which is in line with the recommended concentration specified by the WHO in 1970. The Directive also defines a target guide level of 25 mg/l. Until 1984 the WHO standard classified nitrate concentrations between 50 and 100 mg/l as conditionally acceptable, whereas concentrations in excess of 100 mg/l were not recommended. In 1984, due to the reported incidence of infant methaemoglobinaemia, the WHO reduced the recommended standard to 10 mg NO_3-N/l (45 mg NO_3/l), dispensing with the earlier conditionally acceptable range. This was revised in 1993 (Table 1.16).

4.2.4 Nitrate in Groundwater

Nitrate leaches into the water-table throughout the year, although the rate of leaching depends on factors such as geology, soil type, rainfall pattern, crop utilization rate of the nitrogen, the microbial conversion rate of nitrate, and the fertilizer application pattern. Greatest leaching occurs in the autumn and winter. Nitrate is very soluble and is dissolved by rain-water and percolates deeper into the soil where it either enters the groundwater by direct percolation or, if it meets an impermeable layer such as clay, by sideways migration through the soil until it finds a way into the groundwater. In some areas over 80% of the public supply may come from groundwater sources, whereas small community and private supplies are almost exclusively from a groundwater source via boreholes, shallow wells or springs. This means that a considerable number of people are receiving groundwater which contains increased nitrate concentrations. Where there is no intervening impermeable layer to prevent nitrate percolating into the aquifer, then in areas of intensive arable production high nitrate concentrations are inevitable. Owing to the variable migration period it is difficult to predict the rate of movement of nitrate through the aquifer or the resulting concentration in the water supply. Depending on the thickness of the overlying rock, and whether or not the rock is fissured, this migration of nitrate can take up to 40 years. As the rain-water has to percolate through unbroken rock taking the nitrate with it to reach the water-table. The extent of nitrate accumulation in groundwater depends on the climate and geology, but over the past 50 years the concentration of nitrate in rivers and waters from deep boreholes has increased steadily. The increase in some groundwaters currently being observed, especially some chalk aquifers where percolation may be less than 1.0 m/year, may reflect the intensification of agriculture during and after the Second World War when 8×10^6 acres of grassland were ploughed up and brought into arable production. The effects of peak inorganic fertilizer use may not be seen in these aquifers for another 20 years.

The 1970s and 1980s have been periods of massive nitrate use, equivalent to an eight-fold increase in fertilizer use in Britain alone. This is reflected by many aquifers showing a steady increase in nitrate concentrations. However, when and at what concentration nitrate will stabilize in groundwater supplies is unknown. Predictions for the most severely affected areas in the UK suggest that the maximum concentration in drinking water supplies will level out at 150 mg/l, assuming a continuation of current agricultural practice. In 1986 the UK Nitrate Co-ordination Group admitted that most groundwater sources were showing an overall increase in nitrate concentrations. They concluded that there will be a continuing and slow increase in groundwater nitrate concentration in most unconfined aquifers. This has indeed been so, although the rate of increase has generally been much faster than expected. The fact is that due to the time delay between the application of fertilizer and its appearance in our drinking water, in many areas the concentration of nitrate in our aquifers will continue to increase even if nitrates are banned altogether or land is taken out of production. Where the migration times are much less, however, remedial measures will have a more rapid effect on stabilizing and even reducing nitrate concentrations. As a general guide, the time lag between the reduction in nitrate leaching from the soil and a measurable reduction in groundwater nitrate concentrations will vary from a few years for limestone aquifers to 10–20 years for sandstone aquifers, and up to 40 years in chalk aquifers.

4.2.5   Nitrates in Surface Waters

Those people obtaining their drinking water from surface waters are also at risk from nitrates. Nitrates enter rivers and lakes as surface runoff or as interflow—that is, water moving sideways through the soil layer into a watercourse. Some rivers are also fed by groundwater via springs either above or below the surface of the stream. The nitrates are mainly added as runoff from agricultural land, with the highest nitrate concentrations clearly associated with areas of intensive farming. According to the Royal Society (1984) the concentrations of nitrates in British rivers have been increasing dramatically over the past 20 years; about 400% in that period alone. Rivers have a seasonal pattern of nitrate concentration, with levels at their highest in autumn and winter when drainage from the land is greatest. Many rivers in the UK have winter peaks in excess of 100 mg/l. This seasonal variation is clearly seen in the River Stour, for example, with peaks reaching way beyond the 50 mg/l MAC (Figure 4.3). Nitrogen is added from other sources apart from agriculture, such as sewage and industrial discharges, which can at times be significant. Private water sources are particularly at risk; by their very nature these tend to be in rural areas and are susceptible to surface runoff and the rapid

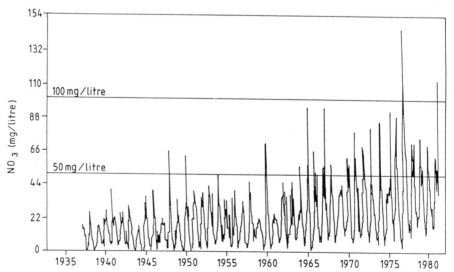

Figure 4.3. Increase in the concentration of nitrogen in the River Stour at Langham over the period 1937–82. Reproduced from the Department of the Environment (1988b) with the permission of the Controller of Her Majesty's Stationery Office

migration of pollution, not only from fertilizer and manure application, tillage and other agricultural practices, but from silage pits, slurry-holding tanks and septic tanks. Surface waters are particularly at risk from other sources of nitrate apart from ploughing up grassland and nitrate fertilizer use. In some areas land drainage has significantly increased nitrate leaching, whereas sewage treatment plant effluents, silage liquor and slurry disposal are all important sources. Dutch farmers are currently producing 94 million tonnes of animal slurry each year, although their land can only absorb about 50 million tonnes safely. This has resulted in serious contamination of drinking waters in some provinces. Water blending is now common in The Netherlands and France to reduce nitrate levels; in the UK this practice still remains largely experimental, although it is becoming increasingly widespread. In the long term, a reduction in animal stocking density appears to be the only answer in The Netherlands. Treated sewage effluents also contain nitrate and these can contribute a significant proportion of the nitrate in rivers used for supply purposes. This is especially so in rivers such as the Thames and Trent where the water is reused a number of times (Section 2.6). In these instances denitrification of effluents before they leave the treatment plant has become necessary, not only to reduce the nitrate concentration in the water being abstracted for supply, but to control eutrophication.

Increased nitrate leaching is clearly associated with the intensification of agriculture. The problem in terms of control is that nitrate from agriculture comes from diffuse sources—that is, from large undefined areas of land adjacent to rivers. In contrast, sewage and industrial nitrogen usually enters the river at point sources—that is, a single point along the river, usually a clearly defined outlet pipe which can be more easily controlled and treated.

4.2.6 Regional Nitrate Levels

In areas without intensive agriculture normal nitrate concentrations are low, between 0 and 10 mg/l. In contrast, in areas of intense agriculture such as South Yorkshire (where blending has been introduced), East Anglia and much of the Anglian, and Severn and Trent Water Company areas, nitrate concentrations are high, often breaching the EEC MAC limit. In 1987 the Department of the Environment identified 74 water supplies serving 1.6 million people which failed to meet the EEC MAC standard for nitrate. However, the Drinking Water Inspectorate, in their 1992 annual report, estimated that 5.1 million people are currently affected in England and Wales. These people live mainly in East Anglia, Lincolnshire, Nottinghamshire, Northamptonshire, Cambridgeshire, Warwickshire, Worcestershire, Bedfordshire, Buckinghamshire, Staffordshire and Shropshire. Water supply companies currently under most pressure over nitrate levels in their water supplies are the Anglia, Severn and Trent, Thames, Lee Valley, East Worcestershire and the South Staffordshire Water Companies. In 1989, 7% of the groundwater supplies in the Severn and Trent area exceeded the MAC limit, and this figure is expected to continue to increase over the next decade. In 1991 only 4% of groundwater supplies in East Anglia exceeded the limit, although the water companies themselves expect this figure to rise to 15% by 1995. The worst affected areas are listed

Table 4.3. Water supplies in England and Wales subject to nitrate concentrations in excess of 40 mg and 50 mg NO_3/l at some time during 1989 or 1990. Reproduced from Hydes *et al.* (1992) with permission of the Controller of Her Majesty's Stationery Office

Source	1989		1990	
	Number of supplies	Population affected	Number of supplies	Population affected
Supplies exceeding 50 mg NO_3/l at some time during year				
Surface	14	2 065 000	19	2 882 000
Groundwater	77	2 383 000	75	2 442 000
Supplies exceeding 40 mg NO_3/l but not 50 mg NO_3/l at some time during year				
Surface	30	8 576 000	27	7 447 000
Groundwater	129	4 913 000	124	5 291 000

in Table 4.4. In Ireland there has been a similar, although not as dramatic, increase in nitrate in surface waters, with maximum values occurring between October and February. Groundwater sources in Ireland also show similar trends of increasing nitrate levels, although other sources apart from nitrate leaching from agricultural soil have been implicated. The reason why nitrate is not such a serious problem in Ireland is due to three factors: (1) a high rainfall, in excess of 350 mm/year, which dilutes the nitrate; (2) relatively low use of inorganic fertilizers compared with other EC countries; and (3) most good lowland agricultural soils are used for grass.

In November 1992 the European Court of Justice found the British Government guilty of breaching EC limits on nitrates in 1986. The EC standards were agreed in 1980 and came into force in 1985 (Section 1.3). Twenty-eight supply zones in East Anglia were included in the action, originally instigated by Friends of the Earth. However, by the end of 1992, 13 of these zones complied with the limit, with the remaining supply areas on course for compliance by 1995.

High nitrate levels in drinking water are not a UK phenomenon. Wherever there is intensive agriculture then there are enhanced nitrate concentrations in water supplies. All European countries report problems, especially in groundwater supplies in areas of intensive agricultural activity. In Belgium, 456 000 people drink water with a nitrate concentration in excess of the MAC limit, whereas in Denmark, Germany and France 6.5, 4.0 and 2.2% of their respective populations are receiving in excess of 50 mg/l of nitrate in their drinking water. In all European countries, as well as in eastern Europe and some parts of the USA, the problem is getting worse each year (Conrad, 1990).

4.2.7 Treatment of High Nitrate Water

There are two approaches to reducing the nitrate concentrations in drinking water supplies. These are either to improve agricultural practice (prevention) or subsequently

Table 4.4. Areas of England most affected by high nitrate concentrations in drinking water. Reproduced with the permission of the Controller of Her Majesty's Stationery Office from HMSO (1989b)

Water authority	Supply area
Anglian	Barrow
	Dersingham and Snettisham
	Docking
	Heacham and Hunstanton
	Moulton and Kennett
	Wisbech
	Habrough
	Aswardby – Saltersford
	Binbrook
	Bully Hill – Barnoldby
	Bully Hill – Otby
	Clay Hill – Drove Lane
	Sleaford – Drove Lane
	Potterhanworth – Waneham Bridge
	Saltersford
	Waneham Bridge – Saltersford
	Ely and Littleport
	Habrough, Covenham
	Bowthorpe
	Heigham – Thorpe
	Otby
	Little London
	Winterton Holmes – Barrow
	Branston Booths
	Waddington – Glentham – Ulceby
	Isleham
Severn Trent Water	Parts of Leamington, Kenilworth, Warwick and Stratford
	Parts of Stourbridge, Dudley
	Parts of Workshop, Mansfield Woodhouse (Bassetlaw and Mansfield DC)
	Malvern Hills and Wyre Forest DC
	Parts of Wrekin DC (Lilleshall source)
	Parts of South Shropshire DC
	Parts of Wrekin DC (Puleston Bridge)
	Parts of Newark DC
South Staffordshire Waterworks	Parts of Lichfield DC, Sutton Coldfield, North Warwickshire CC and Walsall Metropolitan areas
South West Water Authority	St Mary's Island
Lee Valley Water	Part of Luton

to reduce their concentration in water supplies (cure). The approach selected is dictated purely by economics and curiously enough prevention of contamination may be more expensive than removing the nitrate at the water treatment plant. Ways of removing nitrates from water supplies include:

1. *Replacement*. In theory the easiest option is to replace high nitrate supplies with low nitrate supplies. However, this can be very expensive as new water mains and a suitable alternative supply are required. So, in practice, this option is usually restricted to small and isolated contaminated groundwater resources.

2. *Blending*. This is the controlled reduction of nitrate to an acceptable concentration by diluting nitrate-rich water with low nitrate water. This is currently widely practised throughout Europe. Blending becomes increasingly more expensive if the nitrate concentration continues to increase. Also, not only is a suitable alternative supply required, but also facilities to mix water at the correct proportions. Although groundwater nitrate concentrations do not fluctuate widely from month to month, surface water nitrate levels can display large seasonal variations, with particularly high concentrations occurring after heavy rainfall following severe drought periods. Most blending operations involve diluting contaminated groundwater with river water, so such seasonal variations can cause serious problems to water supply companies.

3. *Storage*. Some removal of nitrate can be achieved by storing waste for long periods of time in large reservoirs. The nitrate is reduced to nitrogen gas by bacteria under the low oxygen conditions that exist in the sediments of the reservoir, a process known as denitrification.

4. *Treatment*. The nitrate can be removed by ion exchange or microbial denitrification. Both are expensive and continuous operations. Water purification at this advanced stage is technically complex and difficult to operate continuously at a high efficiency level. Also the malfunction of nitrate reduction treatment plants would cause further water contamination. Ion-exchange systems use a resin which replaces nitrate ions with chloride ions. Although the system is efficient it does produce a concentrated brine effluent which requires safe disposal. Nitrate removal brings about other changes in drinking water composition, and the long-term implications for the health of consumers' drinking water treated by ion exchange is unknown. This is the process in which most water companies are investing. Methods of reducing nitrate concentrations by treatment within aquifers are being developed, and these, including the use of scavenging wells to intercept polluted groundwater in aquifers reaching supply wells are discussed in Section 2.4.

5. *Selective replacement*. Rather than treating the entire water supply, many water supply companies use a system of identifying those most at risk within the community from high nitrates, and supplying an alternative supply such as bottled water or home treatment units. This is by far the cheapest approach for the water industry.

Bottling plants have been used in some affected areas in Britain and Europe to supply low nitrate water for infants and pregnant women (Section 8.1). The best documented case in recent years was in villages around Ripon in the Yorkshire Dales where the local groundwater sources became severely contaminated by nitrates in 1988. The bottled water was supplied for infants under six months of age. The nitrate level was eventually reduced by blending. Similar situations are becoming increasingly common throughout Europe, but have received little media attention. Of course, the first response of a member of the public on hearing that their drinking water is contaminated is to boil it. However, as with many other inorganic chemical pollutants found in water, boiling will not remove nitrate or render it harmless, but simply increase its concentration in the water due to evaporation. In the home nitrate can only effectively be removed from drinking water by ion exchange (Section 8.2). This is further considered by German (1989).

4.2.8 Control of Agricultural Sources of Nitrate

What can farmers do to reduce nitrate contamination? The Department of the Environment has suggested a number of options: (1) cutting down on fertilizer use; (2) modifying cropping systems; and (3) light control measures. It is the last category of measures which may have the most significant effect on nitrate leaching. Among these measures are applying nitrogen fertilizer strictly according to professional advice: not applying it in autumn; leaving the soil covered over winter, perhaps by growing a 'catch crop' to mop up the nitrate; sowing winter crops as early as possible; and taking great care with manures. Specifically, good agricultural practice includes:

1. Fallow periods should be avoided. The soil should be kept covered for as long as possible, especially during winter, by the early sowing of winter cereals, intercropping or using a straw mulch.
2. Avoid increased sowing of legumes unless the subsequent crop can use the nitrogen released by mineralization of nitrogen-rich residues.
3. Grassland should not be ploughed.
4. Tillage should be minimized and avoided in the autumn. Direct drilling should be used whenever possible.
5. Slopes should, whenever possible, be cultivated transversely to minimize runoff.
6. Manure should not be spread in autumn or winter. It should be spread evenly and the amount should not exceed crop requirements.
7. The amount of nitrogen fertilizer and manure used should be applied at times and in amounts required by plants to ensure maximum uptake. The amount of available nitrogen in the soil should also be considered. Computer models are available to help the farmer calculate very accurately the amount of fertilizer required for a particular crop grown on a specific soil type; alternatively, consultants should be used to advise on fertilizer application rates. The manufacturers of the computer model and the consultants claim they can save farmers considerable amounts of money by optimizing fertilizer use. This is an example of what is good economics is also good for the environment.

Taking land out of arable cropping and putting it down to low productivity grassland would certainly reduce nitrate leaching. The set-aside policy, where arable land is taken out of production and put back to grassland, or even forestry, will reduce nitrate leaching. Of course, once this land is brought back into arable production, leaching will begin again. In those areas where aquifers are most at risk this may be the only option. Aquifer protection zones have been suggested throughout Europe. In Ireland much success has been achieved in reducing nitrate contamination of groundwater in aquifer protection zones by improving farm practice, mainly through using less nitrogen fertilizer, especially in the autumn, and by sowing winter wheat earlier. The Wessex Water Company has experimented with restricting the amount of fertilizer used by tenant farmers on its land. The 96 000 people of the City of Bath obtain 80% of their water from springs which feed several supply reservoirs. In 1984 a review of water quality indicated that the nitrate concentration would exceed the EC MAC level within a few years. Nitrate was being leached from agricultural land. It was found that in several areas there was rapid infiltration of surface water to shallow springs, and these were identified as areas of high pollution risk. The following year, when an opportunity arose to alter farming practice in parts of the catchment, Wessex Water asked tenant farmers to transfer from arable cropping to grass production, and suggested that where cereals were grown they should be autumn-sown varieties. No development of the land was allowed and the amount and timing of fertilizer application was strictly controlled. The shallow springs showed an immediate reduction in nitrate levels, with the downward trend still continuing, although the deeper springs have yet to show any reduction in nitrate concentration. The cost of this operation, in terms of land purchase and compensation to farmers for loss of income due to the restrictions, has been a fraction of the cost of advanced water treatment or replacing the resource (Knight and Tuckwell, 1988). This proved highly successful, with a significant reduction in nitrate concentrations recorded in both groundwater and surface waters.

The Water Act 1989 includes new control measures. It enables the creation not only of nitrate sensitive areas to reduce the amount of nitrate leaching from agricultural land, but also enables water protection zones to be created. These zones are different to the 800 000 ha currently designated as environmentally sensitive areas in the UK. These two new designations differ in three fundamental respects. (1) Water protection zones are restricted to prohibiting and restricting activities, whereas nitrate sensitive areas provide for positive action to be required of farmers—for example, the construction of silage facilities—in addition to prohibitions and restrictions. (2) Financial compensation may be payable in respect of obligations resulting from the designation of a nitrate sensitive area. (3) In England, it is the Minister of Agriculture, Fisheries and Food who designates nitrate sensitive areas with the consent of the Treasury, whereas water protection zones are designated by the Secretary of State for the Environment. The designation procedures are complicated and are fully explained by Macrory (1989). There are three types of designation for nitrate sensitive areas: voluntary areas, mandatory areas without compensation, and mandatory areas with compensation. So far 10 areas have been designated as nitrate sensitive areas and it has been recommended that intensive advisory campaigns should be conducted in a further nine areas by ADAS (so-called voluntary areas) (Table 4.5). Those in the designated areas will receive compensation where they voluntarily undertake to observe restrictions. Farmers who participate in the basic scheme can receive additional payments by converting some or all of their arable land to grassland. The payments vary according to the options chosen, the proportion of farm affected and the particular nitrate

Table 4.5. Nitrate sensitive areas and nitrate advisory areas in England and Wales

Nitrate sensitive areas	Nitrate advisory areas
Sleaford (Lincolnshire)	The Swells (Gloucestershire)
Branston Booths (Lincolnshire)	Bircham and Fring (Norfolk)
Ogbourne St George (Wiltshire)	Hillington, Gayton and Congham (Norfolk)
Old Chalford (Oxfordshire)	Sedgeford (Norfolk)
Egford (Somerset)	Fowlmere (Cambridgeshire)
Boughton (Nottinghamshire)	Far Bauker (Nottinghamshire)
Wildmoor (Hereford and Worcs)	Dotton and Colaton (Devon)
Wellings (Staffs and Shropshire)	Cringle Brook (Lincolnshire and Leicestershire)
Tom Hill (Staffordshire)	Bourne Brook (Warwickshire)
Kilham (Humberside)	

sensitive area. Intensive pig and poultry farmers can obtain payments towards the cost of the storage and disposal of their wastes. The advisory areas are not eligible for grant payments, but they will be encouraged to follow practices designed to reduce the risk of nitrate leaching at little or no cost to themselves. The introduction of more efficient farming practices could in fact result in an increase in profits.

Tradionally, sewage sludge has been applied to arable land in the autumn. It is clear that this coincides with the period of greatest potential risk for nitrate leaching. In nitrate sensitive areas the spreading of slurry or liquid sewage sludge is no longer permitted between 31 August and 1 November to grassland, or between 30 June and 1 November to land under arable cultivation. The maximum application of slurry or sewage sludge to agricultural land is restricted to 175 kg N/ha/year. The levels of inorganic nitrogen fertilizer must be reduced to below the economic optimum (i.e. 25 kg N/ha/year below the optimum for winter wheat, barley and forage catch crops, and 50 kg/ha/year below that for oilseed rape). These levels have to be adjusted to take into account any additional nitrogen supplied by the application of organic manures, including sewage sludge.

There are many existing and proposed EC Directives protecting groundwater resources; for example, the Directive on the Protection of Groundwater Against Pollution by Dangerous Substances (80/68/EEC). This was introduced to protect groundwater supplies that do not usually receive treatment, which could significantly reduce the contamination by the toxic compounds listed. The concept of nitrate sensitive areas and the controls seen under the Water Act have been implemented throughout Europe by the adoption of the Directive Concerning the Protection of Waters Against Pollution Caused by Nitrates from Agricultural Sources (91/676/EEC) (EC, 1991).

In many instances a reduction in nitrate fertilizer has a poor effect on reducing nitrate concentrations in water resources for a number of reasons: (1) the organic nitrogen reservoir in the soil is so large that in many instances this will take years to decrease significantly; (2) arable farming, even with reduced fertilizer use and productivity, can still produce high nitrate levels; and finally (3) in some areas nitrate from past leaching has yet to reach the water-table. In many areas this may not happen for many years. Organic farming can also cause nitrate leaching, especially the overuse of manure on land or the use of legumes in tillage rotation.

In the eastern counties of Britain the annual drainage is low, less than 250 mm, whereas farming is predominantly arable. The water draining from the land is therefore rich in nitrate. As reducing the rate of application of fertilizer may only have a marginal effect, the only effective option may be to take several thousand square kilometres of prime arable land in the UK out of production. Conversion to grassland with modest livestock numbers, or even better to forestry, will be the only way to drastically reduce nitrate leaching. Although changes in surface water nitrate concentrations will be seen relatively quickly, changes will take much longer in groundwater.

Clearly a combination of better agricultural practices and some water treatment is the best solution to control nitrates in drinking water. However, there are some very real practical problems in monitoring fertilizer use or good agricultural practice, so that the brunt of the responsibility will inevitably have to be borne by the water supply companies with the cost being passed onto the consumer. The cost will be high. Before privatization the UK Government had plans to spend £1.4 billion to comply with the Drinking Water Directive, with an estimated £90 million going to reduce nitrate concentrations, and at least £9 million for the worst hit area of East Anglia alone.

4.3 PESTICIDES AND ORGANIC MICROPOLLUTANTS

4.3.1 Organic Micropollutants

Organic compounds found in drinking water fall into three broad categories: (1) naturally occurring organics; (2) synthetic (man-made) organic compounds; and (3) organic compounds synthesized during water treatment (chlorinated by-products). The last category is dealt with in Section 5.5, and the first two categories are discussed in the following.

A large number of organic compounds are found naturally as well as synthesized for industrial uses (as solvents, cleaners, degreasers, petroleum products, plastics manufacture and, of course, their derivatives) and for agricultural use (mainly pesticides). In a survey carried out by the Water Research Centre in the UK in the late 1970s, 324 organic compounds were identified in drinking water samples (Fielding et al., 1981). Nearly all of them are toxic and most possibly carcinogenic, even at very low concentrations.

Organic micropollutants, including pesticides, are similar to nitrates in that their toxicity cannot be attributed solely to drinking water, they are ubiquitous in the environment and are found in food and even in the air. They are also similar to nitrates in that many of them percolate into the soil and accumulate in aquifers or surface waters. They can contaminate drinking water sources via many routes: agricultural runoff to surface waters or percolation into groundwaters, industrial spillages to surface and groundwaters, runoff from roads and paved areas, industrial waste water effluents leaching from chemically treated surfaces, domestic sewage effluents, atmospheric fallout, carried in rainfall, and as leachate from industrial and domestic landfill sites.

The Water Research Centre survey showed that a high proportion of the British public is exposed to minute amounts of a wide range of organic chemicals and for most of these there is insufficient toxicological data to make an accurate assessment of their safety. There is certainly growing concern about the long-term health effects of exposure to low concentrations of these organic compounds. Analytical, toxicological and epidemiological studies on organic micropollutants are underway throughout Europe. However, to date no definite health effects have been shown for the group at the concentrations usually found in drinking water. Of course, many organic compounds are present at undetectable levels and there is evidence that repeated small doses of some organic chemicals may also lead to chronic diseases. There is a lot of controversy about the presence or absence of a non-effect level, particularly for carcinogens. However, remember that these organic chemicals are not only found in drinking water, they also accumulate in food and the atmosphere. Therefore in calculating the total intake of such chemicals, the low concentrations of organic micropollutants found in drinking water may indeed be significant in terms of the overall acceptable daily intake (ADI). Other factors such as possible increased toxicity effects when two or more organic compounds are present together, and the fact that some members of the community, especially the young and old, have a much higher water intake than others, makes the calculation of safe concentrations of these chemicals in water apparently impossible.

At present there is no simple routine technique which can be used to identify and measure all the organic compounds in drinking water. There is a vast number of organic compounds, many of which may form further complex compounds in water. The analytical methods available for organic chemicals are the most sophisticated of all the chemical analytical techniques, namely chromatography and mass spectrometry. Gas chromatography, for example, can identify less than 20% of the compounds present due to volatility limitations, and although mass spectrometry is much more effective it requires a large financial investment. In practice, very little is known about what is in drinking waters in terms of organic chemicals as we are unable to determine a large portion of them and cannot afford to monitor those we can analyse often enough, even though water undertakers are required by the Water Supply (Water Quality) Regulations 1989 to monitor pesticides. To complicate matters, factors such as temperature, pH and the hardness of the water all affect toxicity. For example, organic micropollutants tend to be more toxic in soft waters. There is no doubt that organic micropollutants are the most important and potentially the most harmful contaminants found in drinking water.

The EC Drinking Water Directive specifies groups of organic compounds under two separate sections (Table 1.10). Section C covers 'substances undesirable in excessive amounts'. These are usually industrial solvents and are categorized in the EC directive as: substance extractable in chloroform (paragraph 27); dissolved or emulsified hydrocarbons (paragraph 28); phenols (paragraph 29); and other organochloride compounds not covered by paragraph 55 (paragraph 32). Section D covers what the EEC defines as 'toxic substances', which include: pesticides and related products (paragraph 55); and polycyclic aromatic hydrocarbons (paragraph 56).

Micropollutants are present in water in infinitesimal amounts. For normal inorganic compounds such as nitrate, concentrations can be adequately expressed in milligrams per litre (mg/l). However, for micropollutants, micrograms per litre (μg/l) are often used, so 1 μg/l is the same as 0.001 mg/l (1000 μg/l = 1 mg/l). For some compounds even smaller units are required such as nanograms per litre (ng/l), so 1 ng/l is equivalent to 0.000001 mg/l (1 000 000 ng/l = 1 mg/l or 1000 ng/l = 1 μg/l).

4.3.2 Pesticides

Pesticides can be classified by use as either an insecticide, herbicide or fungicide. They are used either to kill a broad spectrum of pests or specific groups or species of pest. Occasionally they may be used together, especially for fumigation and sterilization where both a broad-acting insecticide and fungicide are required. Herbicides are often mixed with fertilizers such as lawn improvers, the idea being to kill off weeds but to encourage the growth of the desired crop. Although herbicides are by far the largest groups in terms of compounds and weights used, it has been the insecticides that have, in the past, posed the major environmental and health hazards. There are four main groups of insecticides: organochlorines; organophosphorus, carbonates; and pyrethroids.

Organochlorines (chlorinated hydrocarbons) were widely manufactured and used in the 1950s and 1960s. They were considered a major breakthrough because they were very persistent—that is, they degraded extremely slowly in the environment and so went on working for very long periods of time. They have largely been banned or their use severely restricted in most countries following the major destruction of natural wildlife, especially birds. The organochlorines are lipophilic, which means they become concentrated in fatty tissues. As these animals are eaten by predators there is a net increase of these compounds in animals up the food chain. There are three types of organochlorines.

DDT, and related substances such as DDE were widely used to control malarial mosquitoes and all flying insects. DDT is an acronym for dichlorodiphyltrichloroethane, although its chemical name is 1,1′-(2,2,2-trichloroethylidene)bis-(4-chlorobenzene). It is highly insoluble in water but soluble in organic solvents. It is an extremely stable compound which is highly resistant to breakdown in the environment. Some DDT is converted by most species to its metabolite DDE (1,1′-(2,2-trichloroethylidene)bis-(4-chlorobenzene) (World Health Organization, 1984). Its use became so ubiquitous that many insects began to develop resistance to it and so it became less effective. It was also found to be accumulating in fatty tissues of all animals, including humans, and was identified in human breast milk. Today its use is severely restricted in developed countries, although unfortunately it is still widely used in some tropical countries. DDT mainly affects the central nervous system and the liver.

Lindane (γ-hexachlorocyclohexane), also written as γ-HCH or γ-BHC, and related chemicals were also widely used at one time as broad-spectrum insecticides for a wide range of applications. Like the others in this group it is only degraded very slowly in soil and has now become ubiquitous in surface waters; more recently it has begun to turn up in groundwaters as well. This group is thought to cause birth defects, including stillbirths and has been found to be carcinogenic in laboratory tests on animals.

Aldrin, dieldrin and all the other 'drins' were used mainly for seed dressing. The full chemical names for these two pesticides are warnings in themselves that they are dangerous. Aldrin (1, 2, 3, 4, 10, 10-hexachloro-1, 4, 4a, 5, 8, 8a-hexahydro-*endo*-1, 4-*exo*-5, 8-dimethanoncephthalene) and dieldrin (1, 2, 3, 4, 10, 10-hexachloro-6, 7-epoxy-1, 4, 4a, 5, 6, 7, 8a-octahydro-*endo*-1, 4-*exo*-5, 8-dimethanonaphthalene) are persistent insecticides which accumulate in the food chain and have been found to be responsible for massive deaths of birds and other wildlife. Aldrin is readily converted to dieldrin by chemical oxidation in the soil and by metabolic oxidation in animals and plants. Only dieldrin is therefore normally found in water. Today its use is very severely limited, although it is still used in the control of termites. Owing to its persistence it is still found in surface waters in some areas of the world, although in the UK it has been found in some groundwaters. Dieldrin attacks the central nervous system.

Organophosphorus compounds used as pesticides are closely related to nerve gases. Used in a diluted and modified way, they retain the same principle by attacking the central nervous system of insects and animals. Although potentially very dangerous they are not persistent and are rapidly broken down in the environment. Commonly used organophosphorus pesticides include malathion, diazinon and dichlorvos, which is widely used under the trade name of Nuvan as a control for salmon lice in aquaculture (Section 4.1).

Carbamates are naturally occurring substances originally used in medicine but also found to be effective insecticides. They also attack the central nervous system. About 20 are widely used, and all are comparatively safe as they are reasonably degradable. Although they are much more expensive than other pesticides, they are increasingly being used, e.g. dimethoroate and aldicarb.

Pyrethroids are naturally occurring compounds although they are now made synthetically. They are relatively non-toxic and widely used by gardeners.

In general, poisoning from pesticides can occur rapidly. The acute effects are nausea, giddiness, restricted breathing and eventually unconsciousness and death. More common are the short-term effects of exposure when pesticides can act as irritants, affecting the skin, lungs, eyes and gut. Chronic effects from exposure occur some time, often years later, or as a result of long-term, low-level, exposure. These include cancer, tumour formation, birth defects, allergies, psychological disturbance and immunological damage. It is difficult to be specific, especially as most studies have concentrated on the major organochlorine compounds, many of which are now banned. Little has been done on long-term exposure to very low levels or on the new (safer) pesticides. The toxicological data that are currently available suggest that the current EC MAC values are adequate for short-term protection. A major research initiative is required to determine if constant exposure at these levels is safe in the long term, but we currently just don't know. Those pesticides still registered in the UK of greatest toxicological concern are mainly insecticides. These include chlordane, coumaphos, omethoate, phorate and triazophos. The herbicide amitrole and the fungicide chlorothalonil also pose significant health risks.

The use of pesticides has developed enormously since 1950, with many new and powerful pesticides introduced during the 1960s and 1970s. Pesticides are particularly hazardous as they are chemically developed to be toxic and to some extent persistent in the environment. The most commonly used pesticides in Europe are herbicides of the carboxy acid and phenylurea groups, which are applied to cereal-growing land (Worthing and Walker, 1983). Other pesticides such as fungicides and insecticides are also widely used, although the list of pesticides in use in the British Isles, and their respective target organisms, is vast. Details of some of the most commonly used pesticides used in the UK are given in Table 4.6.

In 1986, 26 000 000 kg of pure pesticides were used in the UK. Most are so toxic that they have to be diluted hundreds of times before they can be used, so that over a billion gallons of formulated pesticides were sprayed onto crops by British farmers in that year alone. This does not include all the other pesticides used by local authorities on road verges and in parks, by British Rail along track networks, by industry or by individuals in their homes and gardens. Typical pesticide application rates in the British Isles are currently in the range of 1–10 kg of active ingredient per hectare each year. Of this, no more than 10% of the application will reach the target area, whether it is soil insects or plant roots. So where does the remainder go?

Unlike warmer climates, evaporation losses are generally low in the UK and Ireland due to the relatively low air and soil temperatures at the normal application periods (March to May and September to November), so the major proportion of the applied pesticide remains in the soil for some time. In permeable soils they are able to infiltrate to groundwater; in more impermeable soils they find their way into surface watercourses via drainage networks. This is why pesticide levels in drinking waters taken from rivers tend to be seasonal phenomena, although they can be found at relatively high concentrations at any time of the year. In groundwaters they accumulate over the years so that the overall concentration is slowly but constantly increasing, without any discernible large seasonal variation.

Although most pesticides are fairly water-soluble and able to reach concentrations between 10 and 1000 mg/l, the potential for these compounds to be leached from the soil depends on a number of factors. Most pesticides in aqueous solution show a strong affinity for soil organic matter, whereas others become concentrated in the fatty tissues of soil organisms and are therefore bound up and unable to be leached into the water. The concentration of many of these pesticides is also substantially reduced by degradation processes in the soil before they can find their way into water resources. Degradation (breakdown to harmless end-products) is by a number of routes, but mainly by chemical hydrolysis and bacterial oxidation. However, these processes are generally slow and so where the soil is very permeable some infiltration into the aquifer is probably inevitable. Once past the soil layer and into the unsaturated zone of the aquifer degradation essentially stops. Many pesticides in aquifers have resisted degradation for decades and a particular worrying trend is the sudden appearance of many organochlorine pesticides such as DDT and aldrin, which were banned decades ago, but are only now beginning to appear in some water supplies from chalk aquifers. Some intermediate degradation products may be more toxic than the original pesticide and these make complete breakdown more complicated. A summary of the major degradation products of all the widely used European pesticides is given in Appendix 1, along with some idea of their relative toxicity.

Table 4.6. Proposed permissible values for individual pesticides in UK drinking waters. Adapted from HMSO (1989c) and reproduced with the permission of the Controller of Her Majesty's Stationery Office

Pesticide	Main type/use	Advisory value (μg/l)	Pesticide	Main type/use	Advisory value (μg/l)
Aldrin/dieldrin	Insecticide	0.03	Isoproturon	Herbicide (cereals)	4
Atrazine	Herbicide (non-agricultural	2	Linuron (mainly in mixtures)	Herbicide (cereals)	10
Bromoxynil (mainly in mixtures)	Herbicide (cereals)	10	Malathion	Insecticide	7
Carbendazim	Fungicide (cereals)	10	Mancozeb (Maneb plus zinc oxide)	Fungicide (other arable)	10
Carbetamide	Selective herbicide	5000	Maneb (see Mancozeb)	Fungicide (cereals)	10
Carbophenothion	Seed treatment (cereals); veterinary use	0.1	MCPA	Herbicide (cereals)	0.5
			MCPB	Herbicide (cereals)	0.5
Chlordane (total isomers)	Insecticide	0.1	Mecoprop	Herbicide (cereals)	10
Chloridazon	Herbicide	50	Metamitron	Herbicide (other arable)	40
Chlormequat	Growth regulator (cereals)	10	Metham sodium	Soil sterilant (other arable)	
Chlortoluron	Herbicide (cereals)	80	Methoxychlor	Insecticide	30
Clopyralid	Herbicide	100	Methylbromide	Soil sterilant (protected crops)	
2,4-D	Herbicide	1000			
DDT (total isomers)	Insecticide		Paraquat	Herbicide (cereals)	10
Dicamba	Herbicide (cereals)	4	Prometryne	Herbicide	10
Dichlorprop	Herbicide (cereals)	40	Propazine	Herbicide	20
Difenzoquat	Herbicide (cereals)	80	Propyzamide	Herbicide	
Dimethoate	Insecticide (cereals)	3	Simazine	Herbicide	10
EPTC	Pre-emergence herbicide	50	Sodium chlorate	Herbicide (non-agricultural)	
γ-HCH	Insecticide	3	Triadimefon	Fungicide	10
Glyphosate	Herbicide (cereals)	1000	Triallate	Herbicide (cereals)	1
Heptachlor/ heptachlor epoxide	Insecticide	0.1			
Hexachlorobenzene		0.2			
Ioxynil (mainly in mixtures)	Herbicide (cereals)	10			

In their undiluted form pesticides are all toxic; for example, just a few drops of the herbicide paraquat could be fatal. Spillages of pure pesticide are major pollution incidents. It has been estimated that just a few litres of undiluted pesticide could contaminate tens of millions of litres of groundwater. Yet those who handle, transport and use pesticides rarely realize just how potentially dangerous they are.

There is considerable variation and uncertainty over the permissible concentrations of pesticides in drinking water between countries, although in Europe the Drinking Water Directive sets a MAC value for total pesticides and any individual pesticide compound of 0.5 and 0.1 μg/l, respectively. The World Health Organization and the US Environmental Protection Agency make recommendations for certain individual pesticides, allowing much higher concentrations than the EC Directive for most of these compounds (Section 1.3). In the revised WHO regulations (1993) guide values range from 0.03 μg/l for aldrin and dieldrin to 100 μg/l for pyridate and dichlorprop.

In the EC Directive on drinking water, total pesticides is expressed as 'pesticide and related products', which is defined as: insecticides (persistent organochlorine compounds, organophosphorus compounds and carbamates), herbicides, fungicides, polychlorinated biphenyls (PCBs) and polychlorinated terphenyls (PCTs). The term 'total pesticides' was used in earlier directives (e.g. the Surface Waters and Bathing Waters Directives) and so the current practice is often to determine the individual concentration of just three principal substances, parathion, BHC (lindane or HCH) and dieldrin, and to report the additive total as 'total pesticides'! This term is obviously misleading as only a small fraction of the real total is measured, and of course new compounds are being constantly developed. It is often unclear exactly how values for 'pesticides and related compounds' are derived in practice, and how accurately this reflects the total concentration of all those compounds covered under this broad definition.

Numerous aquifers, especially in eastern England, have been found to exceed the MAC guidelines for total pesticides in drinking water. This is due primarily to the presence of herbicides of the carboxy acid and basic triazine groups. Levels of common agricultural herbicides such as Mecoprop (MCPP) and 2,4-D, and two non-agricultural weedkillers atrazine and simazine, have also been detected in some East Anglian chalk aquifers at concentrations of between 0.2 and 2.0 μg/l, well above EC MAC levels (Croll,1985). Knowledge of the problem of runoff of pesticides from treated fields into surface water is very poor. It is therefore difficult to predict what will happen due to the large number of factors affecting the degree of leaching; also leaching is intermittent, associated in some way with the rainfall pattern. However, research has shown that pesticides can be easily washed out of the soil by rain into nearby watercourses. Williams et al. (1991), working in Herefordshire, found that although less than 1% of the total volume of simazine applied to fields actually reached the watercourse, the resultant concentration of the pesticide in the water increased to a maximum concentration of 70 μg/l, about 700 times higher than the permitted MAC value, after the first fall of rain. Also they found that simazine continued to be leached from the field into the watercourse for months, exceeding 2.0 μg/l five months later and still above the EC MAC limit of 0.1 μg/l seven months after the last application. The frequency of sampling and the selection of sampling times means that, in reality, even with the new EC Drinking Water Directive firmly in place, we still have very little idea about the true level of pesticides in our water.

In the USA the insecticide aldicarb, which is used to control potato cyst eelworm, reached concentrations in excess of 10 μg/l in groundwater sources in Long Island, New York, resulting in its ban in the early 1980s. It has also been found in other areas of the USA, most notably in the aquifers of Wisconsin. The situation in groundwater in the USA is reviewed by Ritter (1990). In Britain aldicarb and other closely related pesticides of the carbamate group are used mainly in the areas of intensive potato and sugar beet production of eastern England, resulting in localized groundwater contamination problems. In the UK the problem areas for pesticides in drinking water are London, the Home Counties, East Anglia and the East Midlands. The two most common pesticides found in drinking water are probably the weedkillers atrazine and simazine. These two herbicides are not used widely in agriculture, and it appears that the widespread contamination of groundwaters and surface waters is the result of non-agricultural uses of these chemicals such as weed control on roadsides and industrial areas. They are, for example, widely used by British Rail to control plant development along tracks, embankments and other areas under their control. British Rail originally used 2,4-D, notorious as the compound Agent Orange used by the American Air Force to defoliate Vietnam forests during the Vietnam War in the 1960s. Although 2,4-D (2,4-dichlorophenoxyacetic acid) is fairly stable chemically, it is readily broken down in water and soil, and so should only be rarely found in water resources. Where it does occur is usually due to it being directly sprayed onto, or spilt into, surface waters or very permeable soil overlying an aquifer. Under normal circumstances concentrations of 2,4-D in drinking water should therefore be very low. It is excreted almost unchanged in urine and neither stored nor accumulated in the body, and although it is not thought to be carcinogenic, those exposed to it experience fatigue, headache, loss of appetite and a general feeling of being unwell. Studies carried out in Vietnam have indicated birth defects, miscarriage and a host of other effects from exposure to this chemical. Although now banned in the UK, the pesticide 2,4-D is still commonly found in groundwaters, especially in East Anglia. Friends of the Earth, in their 1989 report published by the *Observer Magazine*, reported that both atrazine and simazine exceeded the EC MAC limits in Grafham Water, the reservoir serving East Midland towns such as Bedford, Milton Keynes and Huntingdon. An even more potentially hazardous pesticide, 2,4,5-T, has turned up in the water supplied in the past by the East Anglian Water Company. In 1989, 550 tonnes of herbicide were used in England and Wales for non-agricultural weed control. Of this active ingredient 25% was atrazine and 14% simazine, with diuron (12%), 2,4-D (9%) and mecoprop (8%) not far behind. Overall the most widely used chemicals are triazines (e.g. atrazine, simazine and terbutryn), representing 39%, and the phenoxy compounds (e.g. 2,4-D, diclorprop, MCPA and mecoprop), representing 21% of total usage.

The new water supply companies published annual water quality reports, as they are required to do under the Water Supply (Water Quality) Regulations, for the first time in 1991 covering the operational period of 1 January to 31 December 1990. These reports give the best overview yet of drinking water quality in England and Wales. There are about 450 individual substances currently used in the UK as pesticides. Thirty-three individual pesticides were detected in drinking water above the 0.1 μg/l limit in 1989 and 34 in 1990, representing a total of 43 different pesticides over the 24

Table 4.7. The 12 pesticides most often found in UK drinking waters. All are herbicides except dimethoate which is an insecticide

Frequently occurring	Commonly occurring
Atrazine	2,4-D
Chlortoluron	Dicamba
Isoproturon	Dichlorprop
MCPA (+MCPA)	Dimethoate
Mecoprop	Linuron
Simazine	2,4,5-T

months. The quality of drinking water in relation to pesticides is summarized in Appendix 2. Atrazine, simazine, isoproturon, chlortoluron and mecoprop remain the most widespread contaminants, being present in all waters supplied by some water supply companies (Table 4.7). To give you some idea of their relative toxicities, the WHO guide value for these particular pesticides are 2, 2, 9, 30 and 10 μg/l, respectively. It is difficult to compare water supply companies in this fashion, especially when it is clear that analytical and sampling problems and variations exist. Low reported levels in some areas are unexpected. For example, the total absence of pesticides exceeding the PCV levels for drinking water in the 72 water supply zones serviced by South West Water Services Ltd, and the 66 water supply zones serviced by the Newcastle and Gateshead Water Company, is surprising when we compare it to the scale of the problems experienced elsewhere in England and Wales. Those companies who are experiencing the most problems with pesticides include Anglian, Bristol, Cambridge, Colne Valley, East Anglian, East Surrey, Essex (where all their supplies are affected), Mid-Southern, Mid-Sussex, Rickmansworth, Severn Trent, South Staffordshire, Sutton District (where all their supplies contain atrazine at some time), Tendring Hundred, Thames, West Kent and York Water Works plc. Most of the other water supply companies have isolated problem areas.

Simazine is a persistent herbicide which is only slowly broken down in the soil; although the WHO (1987) reported it not to have a high mobility in soils, it is, as we have seen, often detected in both ground and surface waters. Although not fully evaluated, it does not appear to be carcinogenic, although it does display chronic toxicity in laboratory animals. Apart from its use as a non-agricultural weedkiller, atrazine is also widely used in maize-producing areas, especially Italy, where it is often found in water supplies. Like simazine, little evaluation of its health effects have been carried out. Atrazine is very persistent and can remain in water for many years, so it is particularly a problem in groundwaters. Surprisingly, little is known about its toxicity, although it is thought to be mildly carcinogenic. Atrazine is certainly more toxic than simazine, but to date no-one has been able to determine how toxic either of them really are. The WHO has recommended that the use of both these chemicals should be carefully controlled, especially in areas where they may eventually contaminate drinking water.

Apart from spraying crops, pesticides are widely used for animal applications, including fish-farming, as insecticides. Pesticides applied to animals will be washed off or excreted, and then may find their way into water resources. The major hazard, however, comes from the disposal of large volumes of chemicals in which the animals have been immersed, or as in the case of fish-farming when the chemical is added directly to the water (Table 4.1).

A major seasonal source of insecticide in drinking water is from sheep-dipping. Organochlorine, organophosphorus and phenolic compounds are used to control a number of important parasites of sheep. Contamination of water resources by these pesticides generally occurs only at the time sheep-dipping is in progress in the catchment. For example, the organochlorine pesticide γ-BHC, which is found in rivers of the Grampian region of Scotland during July and August, results in concentrations in drinking water occasionally exceeding the MAC level of 0.1 μg/l. However, the use of γ-BHC was discontinued in the mid-1980s after this particular insecticide was found in lamb carcasses at levels which prohibited their export to certain countries. Since that date this particular pesticide has not been found in the surface waters of the area. Organophosphorus insecticides are now used in its place, most notably diazinon, propetamphos, fenchlorphos and chlorfenvinphos. Although less persistent than γ-BHC, these pesticides have also been identified in surface waters. So how does sheep dip get into our water resources? Apart from trace concentrations from the animals themselves, who are soaked in the chemical which is slowly washed out of the animal's coat by rain, the main route is through the disposal of the remaining chemical after dipping is complete. The recommended disposal method is to either spread it onto land, or to dispose of it to a soakaway. Either way the chemical eventually finds its way into groundwaters or via field drains into the nearest watercourse. Littlejohn and Melvin (1991), who carried out the Scottish study, found that chemicals disposed to a soakaway designed exactly to MAFF guidelines reached a nearby stream within three hours of the sheep dip being emptied into it, with the concentration increasing over the subsequent 18 hours. Of course, there is still the problem of improper disposal, and farmers occasionally discharge sheep dip directly into rivers; in one case in Ireland into the major supply stream feeding a very small reservoir serving the local town. Sheep-farming is predominantly an upland activity, often the only one, so it is usually small streams that are affected, which often feed supply reservoirs or join large rivers that are used for supplying drinking water to towns downstream. The water from such upland catchments is thought to be very clean, and so both quality surveillance and treatment is often minimal, leaving the pesticides untouched and on their way to the consumer's tap.

Even if EC MAC limits are not exceeded, however, the effects of mixing two or more pesticides together are unknown, even at concentrations below MAC values. There is evidence to suggest that cocktails of pesticides and other organic micropollutants could intensify their potential carcinogenic properties by several orders of magnitude (tens or even hundreds of times more). As catchment areas are large, it is inevitable that water will contain chemical residuals of different types, including pesticides. It is not unusual for a single farm to use 10 or more different chemicals, so as there are many farms, possibly hundreds or even thousands, within each catchment, the complexity and magnitude of this chemical cocktail will be enormous. Of the 324 different trace organic chemicals isolated by the Water Research Centre survey, only 19 were found in all the samples tested (Fielding *et al.*, 1981). It may be misleading to assess the toxicity of pesticides and other organic chemicals individually. Of course, there are many more organic chemicals that cannot currently be measured in water at the concentrations at which they are suspected to be present, so 324 is probably a conservative estimate, especially as the Drinking Water Inspectorate has indicated that over 450 different pesticides are currently used in the UK.

The situation in the rest of Europe is equally, if not more, worrying. Several Member States are having difficulties in complying with the MAC limit for parameter 55 (pesticides and related products) of the EC Drinking Water Directive. At a meeting in November 1988 of the European Institute for Water (founded in 1983 as a forum for representatives from Community Member States to review difficulties in implementing water and environmental directives and to consider proposals for new directives made by the European Commission), it was disclosed that the problem would not be resolved in the short to medium term. Treatment of water to remove pesticides was not considered to be a practical or reliable solution in the short term; an overall reduction in the use of pesticides and/or changes in the formulation of the compounds were thought to be the only long-term solutions.

In Germany many water supplies are unable to meet the 0.1 μg/l standard for individual pesticides. In July 1989 the Federal Health Authority issued interim regulations covering pesticide levels in drinking water. Each pesticide, including its degradation products, has been categorized according to its mobility in the soil and its toxicity, and limits in drinking water have been set accordingly. Category A pesticides must not exceed 1 μg/l, category B 3 μg/l and category C 10 μg/l. A full list of pesticides, their degradation products and toxicity categories is given in Appendix 1. If limits are exceeded the consumer and local health authority must be informed, and the water supplier must prepare improvement plans. In the previous year the Pesticide Application Regulations were published by the German Government (Miller *et al.*, 1990). These regulations severely restricted the use of certain pesticides, banning some and only allowing others to be used outside groundwater catchment areas or water protection zones. In Italy there have been very high pesticide concentrations in the groundwater in the River Po region. Interim standards for pesticides have had to be set for this area which are revised every two years (Table 4.8).

The presence of any particular contaminant in drinking water, and its possible effect on our health should not be viewed in isolation from other sources of exposure, especially food and the atmosphere. Although it is clearly difficult to assess the exact contribution of drinking water to the exposure of the consumer to pesticides, it is clearly an important route. It has always been assumed that the intake of pesticides was much higher from food than from water. In fact, the World Health Organization set standards for drinking water of 1% of the ADI value. A report by the Foundation for Water Research suggested that this may be an incorrect practice for pesticides and that the intake of such compounds from water may be of the same order and in some instances it may exceed that for food (Fielding and Packham, 1990). Drinking water may therefore be the major source of pesticide exposure for some people.

Table 4.8. Maximum admissible values (MAVs) for chemicals specified in the Italian Water Supply Regulations. Reproduced from Miller *et al.* (1990) by permission of the Foundation for Water Research

Chemical	MAV ($\mu g/l$)
Atrazine	0.8
Simazine	0.4
Bentazone	4.0
Molinate	0.3

4.3.3 Industrial Solvents

There is an enormous range of organic solvents used for industrial applications, primarily degreasing operations such as metal cleansing and textile dry-cleaning. In the UK, it was ICI who pioneered the production and use of industrial solvents. Production of trichloroethylene (TCE) began as early as 1910 but it was not until 1928 that a suitable container was developed to actually hold the solvent! It was not until the 1930s that the production of TCE and its use began to increase. There is now a large number of solvents, although only six are widely used. They are described here in descending order of importance Methylene chloride or dichloromethane (DCM) has a wide range of applications including paint stripping, metal cleaning, pharmaceuticals, aerosol propellants and in acetate films. Trichloromethane or chloroform (TCM) is used in fluorocarbon synthesis and pharmaceuticals. Methyl chloroform or 1,1,1-trichloroethane (TCA) is widely used for metal and plastic cleaning and as a general solvent. It was introduced as a substitute for trichloroethylene and tetrachloroethene as it is far less toxic (both in terms of acute and chronic toxicity). It is extremely widely used now, not only in industry but in other sectors. It is used in adhesives, aerosols, inks and is the solvent for a number of correction fluids. Tetrachloromethane or carbon tetrachloride (CTC) is also used in fluorocarbon synthesis and in fire extinguishers. Trichloroethene or trichloroethylene (TCE) is the most widely used solvent for metal cleaning, although it is still used for dry-cleaning and industrial extractions. Its use has significantly decreased over the past 25 years, being replaced by TCA. Tetrachloroethene or perchloroethylene (PCE) is mainly used in dry-cleaning processes, although it is also used for metal cleaning.

The use of industrial solvents reached a peak in the mid 1970s in the UK, with a steady decrease amounting to between 30 and 40% over the past 20 years. This was due not only to the decrease in manufacturing industry, but also to improvements in process design and operation. Only four (TCE, PCE, TCA and DCM) of these chemicals are frequently found in drinking water.

Laboratory studies on mammals have shown that all the chlorinated solvents are carcinogenic and so they must be considered as potentially hazardous chemicals. The World Health Organization has set permissible concentrations in drinking water for a wide range of solvents (Table 1.16). The EC Drinking Water Directive distinguishes organochlorine compounds as either being persistent or not. Persistent organochlorine compounds are classified along with the pesticides in paragraph 55 of the Directive, whereas those not covered by this section are dealt with in paragraph 32 as 'other organochlorine compounds'. Unfortunately, there are no MAC values set for such compounds, which include these highly carcinogenic solvents. Guide values of 1 $\mu g/l$ are set for any organochlorine compound including the chlorinated solvents, but these are not mandatory. The guide value has been used in The Netherlands, whereas in Germany an annual average limit of 25 $\mu g/l$ has been set for the more toxic solvents. In the UK, the World Health Organization limits have been introduced for TCE (30 mg/l), CTC (3 $\mu g/l$) and PCE (10 ug/l). Surprisingly, no standards have been set either for TCA or DCM. In Ireland, a national limit value of 100 $\mu g/l$ has been set for all such compounds irrespective of their relative toxicity. Clearly MAC values reflecting the toxicity of these compounds will eventually have to be set by the EC, as even a small spill of just a few litres of solvent could contaminate many millions of gallons of water.

In practice the nature of these solvents reduces the potential for the contamination of surface waters. They are all highly volatile, which means that on exposure to air they rapidly evaporate into the atmosphere. In fact, up to 85% of DCM and 60% of TCE used are lost by evaporation. Once discharged to a water course, evaporation continues. Their fate in the atmosphere is less clear. Like other organic molecules these solvents are oxidized by ultraviolet radiation to harmless end-products. In general, this takes less than three months (known as the atmospheric half-life), although TCA has an atmospheric half-life in excess of five years. Due to this moderate persistence, these solvents become concentrated in the atmosphere, often reaching concentrations in excess of 1 $\mu g/l$ above industrial areas. There is evidence to show that they can subsequently be removed by precipitation, with concentrations of 0.1 $\mu g/l$ in rainfall not uncommon. Trace amounts of industrial solvents in all drinking waters are probably inevitable.

The major pollution threat is when solvents are discharged directly into or onto the ground. This may be due to illegal disposal or accidental spillage. Since groundwater is not exposed directly to the atmosphere, the solvents are not able to escape by evaporation, so it is aquifers that are most at risk from these chemicals. Industrial solvents are now stored above ground in specifically designed and lined tanks to contain any spilt chemical. However, it was, and still is, common practice to store such compounds in underground storage tanks so that if fractures occur in the tank, pipework or pipe couplings, the chemical will be lost directly into the ground and pollute the groundwater (Lawrence and Foster, 1987).

Pollution of groundwater by industrial solvents is a very widespread problem (Loch et al., 1989; Rivett et al., 1990). For example, in The Netherlands, of 232 boreholes tested for organic micropollutants, 67% contained TCE, 60% TCM, 43% CTC and 19% PCE. In 1982 over 60% of the 621 water supply wells tested in Milan contained a total concentration of chlorinated solvent in excess of 50 $\mu g/l$, with 10 wells exceeding 500 $\mu g/l$. In Italy, TCE is the most frequently detected solvent, followed by PCE, TCM and TCA. The problem is especially widespread in the USA. For example, 20% of the 315 wells tested in New Jersey contained solvents, with TCE the most widespread and occasionally reaching in excess of 100 $\mu g/l$ in some wells. Although data are sparse, with many of the surveys conducted not published, the problem in the UK is equally bad. The Department of the Environment commissioned a survey in the mid-1980s of over 200 boreholes, of which at least three-quarters were used for public supply. The results showed that TCE exceeded 1 $\mu g/l$ at 35% of the sites with 4% of these above 100 $\mu g/l$; PCE exceeded the EC guideline value of 1 $\mu g/l$ at 17% of the sites, of which 1% exceeded 100 $\mu g/l$. However, the reported incidents of solvents contaminating boreholes are fairly high. Among those reported are TCE in boreholes at Coventry, Mildenhall, London, Norwich, Dunstable and Luton; PCE near Cambridge, Dunstable and Luton; and TCA at Andover, Dunstable and Luton. Some of these, like public supply sources in the Luton–Dunstable chalk aquifer in Bedfordshire are due to specific spills of solvents, whereas others are due to years of poor management at scores of industrial sites.

In essence, aquifers underlying urbanized areas such as Milan, Birmingham, London or New Jersey will contain high concentrations of all solvents. The stability of TCE in particular and the inability of these solvents to readily evaporate mean that such contamination will last for many decades, even if the use of solvents is discontinued. Where solvents are used then spillage to the subsurface is almost inevitable. Careful handling and correct storage is vital if such contamination is to be minimized.

Industrial solvents tend not to be very soluble and when discharged into water remain as an immiscible phase, i.e. they form a separate liquid (solvent) fraction to the aqueous (water) fraction. In this form they are more dense but less viscous than water. This can result in spilt solvents penetrating deep into aquifers at a fast rate.

Solvents spilt onto the soil are less likely to contaminate the groundwater than those discharged below the soil layer. This is because the soil can be effective in retaining solvents by sorption processes, allowing longer periods for the soil bacteria to break down the organic molecules, although most polychlorinated organic compounds, such as PCE, are recalcitrant, i.e. they are highly resistant to biodegradation. In the soil layer there may still be some volatility (evaporation) losses. Once below the soil horizon, however, volatility losses will be negligible and the solvents will move much more quickly downwards towards the water-table. Bacteriological activity is greatly reduced in the unsaturated zone and this, combined with much shorter contact times between the solvent and micro-organisms, means that losses due to biodegradation will also be extremely low. Once past the soil horizon, or if spills occur below this layer, which is often the case with leaks from buried pipes or tanks, or discharges from lagoons or soakaways, then contamination of the groundwater is almost inevitable.

As mentioned earlier, chlorinated organic solvents have higher densities but lower viscosities than water. In practice this means that they tend to accumulate in the base of aquifers, with their subsequent movement or migration within groundwater dictated largely by the slope and form of the base of the aquifer, rather than the direction of groundwater flow. Solvent therefore tends to accumulate in depressions within the aquifers forming pools. The immiscible phase is the separated solvent however, with time this gradually dissolves into the water (miscible phase) forming distinct plumes. The migration of the solvent, once in the miscible phase, is governed largely by the direction and velocity of the groundwater (Figure 4.4). Where the aquifer is highly fissured, as is the case with Jurassic and chalk aquifers, then lateral migration over many kilometres is possible.

It is interesting to compare the behaviour of chlorinated organic solvents with the aromatic hydrocarbons such as benzene, toluene and fuel oils. These have lower densities than water and are more viscous, and so behave in the opposite manner to solvents. In the immiscible phase they tend to float on the surface of the aquifer, forming characteristic pancakes on top of the water-table. Migration follows the groundwater flow, but remains at the surface of the water-table. If the water level rises, then the aromatic hydrocarbon also rises, but will be retained within the porous matrix of the unsaturated zone by surface tension once the water-table falls again. This process significantly reduces the effects of lateral migration (sideways movement) of the chemical, so that the effects of such spills can be fairly localized compared with chlorinated solvents, which can affect large areas of the aquifer.

4.3.4 Polycyclic Aromatic Hydrocarbons

Polycyclic aromatic hydrocarbons (PAHs) are synthetic compounds that occur in soot, tar, vehicle exhausts and in the combustion products of hydrocarbon fuels such as household oil. Some are also formed by bacteria, algae and plants, although the major source is the incomplete combustion of organic matter. PAHs are not very soluble but are strongly adsorbed onto particulate matter, especially clay, resulting in a high

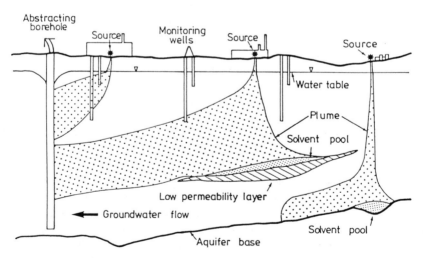

Figure 4.4. Behaviour of chlorinated solvents in groundwaters and the interactions with a pumped extraction borehole and unpumped monitoring wells. Reproduced from Rivett *et al.* (1990) by permission of *the Journal of the Institution of Water and Environmental Management*

concentration where suspended solids are present. Although PAHs are photodegradable, they do require comparatively long exposure to ultraviolet rays in sunlight to do so, so in practice most are either tied up in the subsurface soil where they are eventually degraded by soil micro-organisms, or they find their way into aquifers or surface waters (Clement International Corporation, 1990).

Nearly all PAHs are carcinogenic, although their potency varies; the most hazardous by far is 3,4-benzopyrene. PAHs are readily absorbed in the body and as they are highly lipid-soluble they become localized in fatty tissues. However, they are readily metabolized and do not generally accumulate. PAHs in drinking water are thought to cause gastrointestinal and oesophageal tumours. The annual adult intake of PAHs is between 1.0 and 10 mg, with 3,4-benzopyrene contributing between 0.1 and 1.5 mg. Drinking water contributes only a small proportion of the total PAH intake, less than 0.5% of the total. PAHs are rarely present in the environment on their own and the carcinogenic nature of individual compounds is thought to increase in the presence of other PAH compounds. There are over 100 PAH compounds, but the World Health Organization has listed just six reference compounds which should be determined. These are all combustion products of hydrocarbons and represent the most widely occurring compounds of this group found in drinking water. The EC has set a MAC volume for PAH compounds in the Drinking Water Directive of 200 ng/l, which is equivalent to 0.0002 mg/l. This is the sum of the detected concentrations of the six reference substances listed in the Directive, which are the same as originally listed by the World Health Organization, although the compound names are slightly different (Table 4.9). The MAC value for PAH compounds has been adopted by the UK Government in the Water Supply (Water Quality) Regulations 1989, although a separate maximum limit value has been set for 3, 4-benzopyrene of 10 ng/l, which is the level set by the World Health Organization and is equivalent to just 0.00001 mg/l.

A UK survey carried out in 1981 of the levels of PAH in water used for abstraction for drinking water found that groundwater sources and upland reservoirs contained low levels of PAHs, less than 50 ng/l, whereas lowland rivers used for supply contained between 40 and 300 ng/l during normal flows, increasing in excess of 1000 ng/l at high flows. This appears to be due to the increase in levels of suspended solids in fast rivers onto which the PAHs have adsorbed. Between 65 and 76% of the PAHs in surface waters are bound to particulate matter and so can effectively be removed by physical water treatment processes such as sedimentation, flocculation and filtration. The remainder can be either chemically oxidized or removed by activated carbon. Even when PAH levels in raw water are high, due to localized industrial pollution, water treatment can adequately remove PAH compounds to conform to EC limits.

Fluoranthene, which is the most soluble but least hazardous of the PAH compounds (in fact, the UK Government considers it to be non-carcinogenic), is occasionally found in high concentrations in drinking water samples collected at consumers' taps. This is due to leaching from the coal tar and bitumen linings commonly applied to iron water mains. These linings also contain other PAH compounds which can find their way into the water in the water supply mains as it flows from the treatment plant to individual homes. Conventional water treatment effectively removes the bulk of the PAH compounds adsorbed onto particulate matter, although most of the PAHs

Table 4.9. Reference compounds which are measured to give the total concentration of polycyclic aromatic hydrocarbons in drinking water

World Health Organization	European Community
Fluoranthene	Fluoranthene
3,4-Benzofluoranthene	Benzo 3,4-fluoranthene
11, 12-Benzofluoranthene	Benzo-11, 12-fluoranthene
3, 4-Benzopyrene	Benzo-3,4-pyrene
1, 12-Benzoperylene	Benzo-1,12-perylene
Indeno-(1,2,3-cd) pyrene	Indeno-(1,2,3-c-d)-pyrene

remaining are fluroanthene due to its high solubility, 240 μg/l, compared with only 4 μg/l or less for the other reference PAH compounds. Granular activated carbon can remove 99.9% of the fluroanthene and any other PAHs remaining in the water. Owing to its relatively high solubility, fluroanthene is the major PAH to accumulate in groundwater. For the consumer, most of the breaches of the MAC limit for PAHs is due to leaching in the distribution system. This problem is considered further in Section 6.5.

4.3.5 Removal of Organic Contaminants from Drinking Water

The levels of organic contaminants in water supplies vary with the catchment and are clearly linked with land use and degree of industrialization. In general, groundwaters have fewer organic micropollutants than surface waters, although urban aquifers are often locally polluted with industrial solvents. Also lowland surface waters have higher concentrations of such organic compounds compared with upland waters. Careful selection of water resources is therefore vital, and this will include an assessment of the use of organics within the catchment area. Where a problem exists, diluting supplies with uncontaminated water by blending is a widely used option to conform with standards. It may be possible to control the use of key organic micropollutants, especially pesticides, within small watersheds, although generally this will not be feasible.

Current water treatment is very limited in its ability to remove all trace amounts of organic compounds. Although membrane processes such as reverse osmosis can reduce all organic contaminants, they also remove important inorganic compounds (Baier *et al.*, 1987; Becker *et al.*, 1989). Coagulation can remove a significant proportion of colloidal material onto which organics may have complexed, although in practice this only removes a small proportion of the organic contamination (AWWA, 1979). Activated carbon is the only process able to remove pesticides, solvents and other organics (Section 3.2). PAC can be used in conjunction with coagulation and is effective in removing trace concentrations of organic compounds, whereas granular activated carbon (GAC) is used in compact filters (Becker and Wilson, 1978; Croll *et al.*, 1991). The bed-life of a GAC filter can be extended when combined with ozone before filtration to oxidize some of the organic compounds present (Ferguson *et al.*,

1991). Removal of organic micropollutants from drinking water is considered in detail elsewhere (Kruithof *et al.*, 1989; Akbar and Johari, 1990; Foster *et al.*, 1991), whereas the removal of chlorinated by-products is explored in Section 5.5.

4.4 ODOUR AND TASTE

In its purest state water is both odourless and tasteless; however, as inorganic and organic substances dissolve in the water, it begins to take on a characteristic taste and sometimes odour. Generally, the inorganic salts at the concentrations found naturally in drinking water do not adversely affect the taste—in fact, many of the bottled mineral waters are purchased because of their characteristic salty or sulphurous taste. Both tastes and odours are caused by the interaction of the many substances present. These may include soil particles, decaying vegetation, organisms (plankton, bacteria, fungi), various inorganic salts (for example, chlorides and sulphides of sodium, calcium, iron and manganese), organic compounds and gases (Cohen *et al.*, 1960; Baker, 1963). Water should be palatable rather than free from taste and odour, but with people having very different abilities to detect tastes and odours at low concentrations this is often difficult to achieve. Offensive odours and tastes account for most consumer complaints about water quality.

In 1970 it was estimated that 90% of water supplied in the UK occasionally suffered from odour and taste problems (Bays *et al.*, 1970). This is similar to the figure for the USA and Canada where 70% of water supplies are affected. However, according to the Drinking Water Inspectorate the problem is now much smaller. Odour and taste problems can be very transitory phenomena, so it is unlikely that the standardized monitoring carried out by some water supply companies involving taking just four or five samples a year will detect all but the most permanent odour and taste problems.

4.4.1 Standards and Assessment

There are two widely used measures of taste and odour. Both are based on how much of a sample must be diluted with odour- and taste-free water to give the least perceptible concentration. The threshold number (TN) is the most widely used parameter

$$TN = \frac{A+B}{A}$$

where A is the volume of the original sample and B the volume of dilutent. The value $A + B$ is the total volume of sample after dilution to achieve no perceptible odour. Therefore a sample with a TN of unity has no odour or taste.

Dilution number is used in the EC Drinking Water Directive

$$DN = \frac{B}{A}$$

The two are related as $DN = TN - 1$, so a sample with a DN of three has a TN of four.

Standards are generally the same. In the USA the maximum contaminant level (MCL) for odour is a TN of three (no temperature specified), whereas the World Health Organization only requires that the taste and odour are acceptable. The EC sets a guide DN value of zero and two MAC values. These are two at 12°C and three at 25°C. The problem is that the Directive does not specify how to interpret the standard and no methodology is specified. In the UK, the Water Supply (Water Quality) Regulations 1989 require taste and odour to be assessed at 25°C with a PCV of DN three.

It is extremely difficult to chemically analyse tastes and odours, mainly because of the very low concentrations at which many of these chemicals can be detected by humans. Also, offensive odours can be caused by a mixture of several different chemicals (ASTM, 1968). The identification of odours and tastes is normally carried out subjectively by trained technicians or by a panel of testers. Acceptable levels of odour and taste in drinking water are determined by calculating odour and taste threshold levels. This is carried out by a panel of people who taste or smell samples randomly. Odours of substances taken into the mouth can often be detected when they are not detectable by sniffing. This is due to the enclosed space of the mouth and the higher temperature, increasing the concentrations of volatile materials present. When water or food is taken into the mouth and swallowed, air is exhaled through the nose. So as the nose and sinus cavities are far more sensitive to many chemicals than the tongue, odour appears to be the primary sensation in relation to water quality. The sample is tested as near to body heat as possible. Testing is carried out by decanting the sample into a conical flask where it is shaken vigorously; the vapour is then sniffed by the panel. If an odour is detected then the panel members are asked to describe it in terms such as fishy, musty, earthy, chlorinous, oily or medicinal. A majority verdict is taken and comparisons may be made to reference samples containing known odorous compounds. If an estimate of intensity is required then the sample is diluted with odour-free water until no odour is detectable. These results can then be compared with standard samples containing odorous compounds of known concentration to give some idea of the concentration of the odour-forming compound in the water. Assessment by tasting is rarely carried out due to the potential health hazards to the tester. However, panels of testers are used to calculate at what concentrations specific compounds are no longer detectable by taste, which gives the treatment plant operator an idea of how much of a particular compound needs to be removed to maintain its wholesomeness in terms of taste and odour. There are standard methods for assessing taste and odour published in the UK (Department of the Environment, 1980) which are currently under review, and in the USA (Clesceri *et al.*, 1989a; 1989b; ASTM, 1984). This is more fully reviewed by Bartels *et al.* (1986). There is also a European standard method in preparation by the Comité Européen de Normalisation (CEN).

4.4.2 Classification

Odour problems are categorized according to the origin of the substance causing the problem. Substances can be present in the raw water, they can be added or created during water treatment (Section 5.4), or they can arise within the distribution (Section 6.1) or domestic plumbing systems (Section 7.6). Of course, the quality of the raw water will often contribute to the production of odours during treatment and distribution.

Many problems are related to transient problems with the raw water, although it is often the case that all the offending water will have passed through the treatment plant before the problem is identified by the consumer. For this reason finished water is often routinely monitored for odour at the treatment plant. Figure 4.5 outlines a scheme used by water supply companies for investigating the source of odour problems. The problems of odour and taste in raw water are due to either substances of natural origin or man-made pollutants.

It is not always straightforward to identify the source of an odour or taste in drinking water (Table 4.10). There are seven common causes:

1. *Decaying vegetation.* Algae produce fishy, grassy and musty odours as they decay, and certain species can cause serious organoleptic problems when alive.
2. *Moulds and actinomycetes.* These organisms produce musty, earthy or mouldy odours and tastes. They tend to be found where water is left standing in pipework and also when the water is warm. Usually found in the plumbing systems of large buildings such as offices and flats. They are also associated with waterlogged soil and unlined boreholes.
3. *Iron and sulphur bacteria.* Both bacteria produce deposits which release offensive odours as they decompose.
4. *Iron, manganese, copper and zinc.* The products of metallic corrosion all impart a rather bitter taste to the water.
5. *Sodium chloride.* Excessive amounts of sodium chloride will make the water taste initially flat or dull, then progressively salty or brackish.
6. *Industrial wastes.* Many wastes and by-products produced by industry can impart a strong medical or chemical taste or odour to the water. Phenolic compounds which form chlorophenols on chlorination are a particular problem.
7. *Chlorination.* Chlorine by itself does not produce a pronounced odour or taste unless the water is overdosed during disinfection. Chlorine will react with a wide variety of compounds to produce chlorinated products, many of which impart a chlorinous taste to the water.

4.4.3 Odour Causing Substances of Natural Origin

Most problems of odours in raw water treatment supplies involve organic substances of natural origin, mainly from algae and/or decaying vegetation. Large crops of algae can be present in reservoirs without causing any problems. However, there are a few uncommon algae which produce organic by-products such as essential oils, which can produce offensive odours even when the algae are present in small numbers. The odour is formed when these materials are released into the water, usually when the algae die.

There are four main groups of algae which are able to cause odour problems:

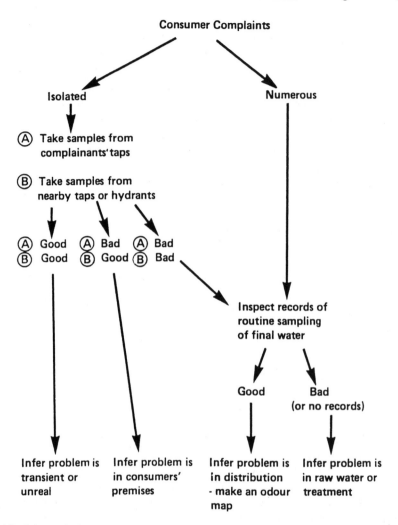

Figure 4.5. Schematic diagram of step approach of investigating an odour problem. Reproduced from Ainsworth *et al.* (1981) by permission of the Water Research Centre

1. Blue–green algae (*Cyanophyceae*). The important genera (groups) are *Anabaena, Anacystis, Aphanizomenon, Gomphosphaeria* and *Oscillatoria*. These generally give rise to grassy odours intensifying into piggy, almost septic, odours as the cells disintegrate. Some algae can also produce toxic substances (Section 4.9).
2. Diatoms (*Bacillariophyceae*). Important genera are *Asterionella, Fragillaria, Melosira, Tabellaria* and *Synedra*. They secrete oils which impart a fishy or aromatic odour. When moderate densities are present then geranium (*Asterionella* and *Tabellaria*) odours are produced.
3. Green algae (*Chlorophyceae*). Can impart a fishy or grassy odour. Important genera include *Volvox* and *Staurastrum*.

Table 4.10. Chemicals causing specific groups of odours and their possible sources

Odour	Odour-producing compounds	Possible source
Earthy/musty	Geosmin, 2-methylisoborneol, 2-isopropyl-3-methoxy pyrazine mucidine, 2-isobutyl-3-methoxy pyrazine, 2,3,6-trichloroanisole	Actinomycetes, blue-green algae
Medicinal or chlorphenolic	2-Chlorphenol, 2,4-dichlorphenol, 2,6-dichlorphenol	Chlorination products of phenols
Oily	Naphthalene, toluene	Hydrocarbons from road run off, bituminous linings in water mains
Fishy, cooked vegetables or rotten cabbage	Dimethyltrisulphide, dimethyldisulphide, methyl mercaptan	Breakdown of algae and other vegetation
Fruity and fragrant	Aldehydes	Ozonation by-products

4. Yellow–brown algae (*Chrysophyceae*). Can produce pungent odours. *Uroglenopsis* can produce a very strong fishy odour, whereas *Synura* produces a cucumber odour at low numbers, a fishy odour at moderate numbers and a piggy odour at high numbers.

It is not only algae which cause odours; rooted aquatic plants (macrophytes) can also be the source of taste and odour problems. This only occurs in shallow lakes and reservoirs when the vegetation dies back and begins to decompose. Eutrophication in rivers can cause similar problems. Odour and taste problems due to algae are generally seasonal.

The most common odour complaint in the UK is of musty or earthy odours. They are generally caused by actinomycetes and to a lesser extent by blue–green algae and fungi. They are associated with supplies taken from lowland rivers, which is probably due to the fact that such rivers are generally rich in suitable organic material on which these micro-organisms thrive. There is evidence to support the idea that some algae which do not normally produce odorous compounds do so when actinomycetes are also present, especially blue–green algae. The main organic compounds produced which cause musty/earthy odours are geosmin and 2-methylisoborneol, and consumer complaints will follow if concentrations of either of these exceed 8–10 ng/l (Table 4.10). Dimethyltrisulphide is also produced by blue–green algae and although it has a higher odour threshold than geosmin, it produces a grassy odour which intensifies into a very unpleasant septic/piggy odour with increasing concentrations of algal biomass.

Phenols can arise in water from the natural decay of organic matter, although more commonly they are associated with industrial pollution. When these are present strong medicinal (or chlorphenolic) odours are produced when the water is chlorinated.

As mentioned previously, inorganic substances rarely cause odour or taste problems in the UK, although a saline tasting water can be unpleasant if used to make tea or coffee in particular. Water will taste salty if chlorides are present at 500 mg/l or more. This can occur in groundwaters due to over-pumping boreholes near the coast, leading to saline intrusion (Section 2.4). Other commonly found salts in groundwater can also impair this taste, such as magnesium sulphate and sodium sulphate. Water that has lost its oxygen (e.g. groundwaters or the hypolimnion of lakes) (Section 2.3) allows sulphide to be formed, which has a most unpleasant odour and taste, even at concentrations below 0.5 mg/l. Iron and manganese have a characteristic bitter taste and can occur in appreciable amounts in certain underground waters and springs.

4.4.4 Man-made Odour-causing Pollutants

There is an enormous number of industrial organic pollutants that find their way into surface and groundwaters which are offensive in an organoleptic sense (e.g. alkyl benzenes, chlorobenzenes, alkanes, benzaldehydes and benzothiazole) (Lillard and Powers, 1975). As these compounds degrade their breakdown products are often even more offensive. Pesticides are highly odorous, as are many of the solvents used in their formulation. However, the odour threshold concentration is above the level at which most such pesticides are considered to be safe, so toxicity considerations are more important than organoleptic ones. Chlorphenolic compounds may also be discharged to water resources from industry, although phenolic compounds discharged mainly from petroleum and wood-processing industries (many agricultural chemicals are also phenolic) will be modified by chlorination at the treatment plant to form these odorous compounds (Section 4.3).

The removal of offensive odours and tastes is discussed in Section 5.4, whereas the problems of taste are examined further in Sections 6.1 and 7.6. Amoore (1986) has written a most interesting article on odours and our sensitivity to them. He suggests the main reason why humans have such a highly developed sensitivity to the metabolites produced by these naturally occurring micro-organisms, and which gives rise to so many complaints, is because they are biological indicators of the presence of water. Primitive tribes still display a remarkable ability to find water, even in the most arid regions, using such indicators.

4.5 IRON AND MANGANESE

4.5.1 Iron

Iron is an extremely common metal and is found in large amounts in soil and rocks, although normally in an insoluble form. However, due to a number of complex reactions which occur naturally in the ground, soluble forms of iron can be formed which can then contaminate any water passing through. Therefore excess iron is a common phenomenon of groundwaters, especially those found in soft groundwater areas.

Iron is an essential element and is very unlikely to cause a threat to health at the concentrations occasionally recorded in water supplies. It is undesirable in excessive amounts and can cause a number of problems. Iron is soluble in the ferrous state (Fe^{2+}) and is oxidized in the presence of air to the insoluble ferric form (Fe^{3+}), so when groundwaters are anaerobic, or have low dissolved oxygen concentrations, all the iron will be in a soluble form. At the treatment plant most of the iron is removed by aerating the water or by using coagulants, with the particles of insoluble iron removed by filtration. In private supplies the water will only start to become aerated as it enters the storage tank. Here the ferric iron particles will settle to the bottom of the tank and form an orange sediment which can then be resuspended when large volumes of water are being drawn from the tank. Alternatively, as the water leaves the tap it is very effectively aerated and it is here the water may become discoloured and turbid. Sediment will also form in the pipework and this will contribute to the discoloration of the water. Some stored waters can also contain iron in its soluble form, especially in deeper parts of reservoirs where the dissolved oxygen concentration will be low. In river waters, which are well aerated, iron, if present, will be in its insoluble particulate form.

Although discoloration is inconvenient, it is only an aesthetic problem. It becomes more irritating when the iron causes staining of laundry and discoloration of vegetables such as potatoes and parsnips during cooking. More importantly, iron has a fairly low taste threshold for such a common element, giving the water a strong unpleasant taste which ruins most beverages made with the tap water. High concentrations of iron in water can react with the tannin in tea causing it to turn odd colours, usually inky black.

Iron is so ubiquitous that some will find its way into nearly all supplies. The taste threshold is about 0.3 mg/l (300 μg/l), although it varies considerably between individuals. The EC Drinking Water Directive sets a MAC value of 0.2 mg/l, although the guide value (the level considered to be ideal) should not exceed 0.05 mg/l (Table 1.10). The World Health Organization guideline is set at the higher concentration of 0.3 mg/l (Table 1.18).

4.5.2 Manganese

Manganese is also widely found in ores and rocks, and like iron turns up in groundwaters due to reducing conditions in soils and rocks bringing it into a soluble form. Once it is exposed to air by aeration, manganese is oxidized into its insoluble form and deposited. It is identical to iron in all respects and indeed causes all the same problems of taste, discoloration and staining. Staining is far more severe, however, with manganese than iron, as is the unacceptable taste it imparts to the water. For that reason much stricter limits have been imposed with MAC (PCV) limits of 0.05 mg/l (50 μg/l). The EC guide level is just 0.02 mg/l (20 μg/l), the concentration at which laundry and sanitary ware become discoloured grey/black. The World Health Organization has a health-related guide value of 0.5 mg/l, but as staining occurs at much lower concentrations, a guide value of 0.1 mg/l has been set to prevent consumer complaints. Although the major source of manganese in the diet is food, all the manganese in water is readily bioavailable. However, unless the concentration in drinking water is excessively high, it is unlikely that it will significantly contribute to the daily intake of the metal. Toxicity is not a factor with manganese as it will be rejected by consumers at concentrations long before any danger threshold is reached. Excessive manganese intake does cause adverse physiological effects, particularly neurological, but not at the concentrations normally associated with water in the UK. The toxicity of manganese has been reviewed by Criddle (1992).

4.5.3 Removal by Treatment

Iron can be precipitated out of solution by a combination of aerating the water and increasing the pH before filtration. Manganese is more difficult to remove as it requires a pH of between 8.5 and 9.0, with aeration playing a minor role. The problem of using such high pH values in the removal of manganese is that any insoluble aluminium coagulant left in the water after sedimentation will be dissolved. To overcome this two filtration stages are required, the first to remove turbidity and coagulant and the second for manganese removal at the higher pH. This higher pH also means that disinfection using chlorine will require about three times the chlorine residual at pH 9 than at pH 8. Groundwaters are occasionally treated by the addition of sodium silicate to complex out the manganese and iron before aeration. Certainly for the water treatment operator, the presence of manganese in particular can produce fairly awkward operational difficulties. Iron can also be added to the water during coagulation as iron salts (Section 5.4), or can occur due to corrosion in the distribution system (Section 7.2).

4.6 HARDNESS

The hardness or softness of water varies from place to place and reflects the nature of the geology of the area with which the water has been in contact. In general, surface waters are softer than groundwaters, although this is not always the case. Hard waters are associated with chalk and limestone catchment areas, whereas soft waters are associated with impermeable rocks such as granite.

4.6.1 Chemistry of Hardness

Hardness is caused by metal cations such as calcium (Ca^{2+}), but in fact all divalent cations cause hardness (Table 4.11). They react with certain anions such as carbonate or sulphate to form a precipitate. Monovalent cations such as sodium (Na^+) do not affect hardness.

Strontium, ferrous iron (Fe^{2+}) and manganese are usually such minor components of hardness that they are generally ignored, with the total hardness taken to be the sum of the calcium and magnesium concentrations. Aluminium (Al^{2+}) and ferric iron

Table 4.11. Principal metal cations causing hardness and the major anions associated with them

Cations	Anions
Ca^{2+} (calcium)	HCO_3^- (hydrogencarbonate)
Mg^{2+} (magnesium)	SO_4^{2-} (sulphate)
Sr^{2+} (strontium)	Cl^- (chloride)
Fe^{2+} (iron)	NO_3^- (nitrate)
Mn^{2+} (manganese)	SiO_3^{2-} (silicate)

Table 4.12. Two examples of the classification used for water hardness

Classification A		Classification B	
Concentration (mg/l)	Degree of hardness	Concentration (mg/l)	Degree of hardness
0–50	Soft	0–75	Soft
50–100	Moderately soft	75–150	Moderately hard
100–150	Slightly hard	150–300	Hard
150–250	Moderately hard		
250–350	Hard	300+	Very hard
350+	Excessively hard		

(Fe^{3+}) can affect hardness but their solubility is limited at the pH of natural waters, so that ionic concentrations can be considered negligible. Barium can also cause hardness, but its concentration in water is normally extremely small. Total hardness (Ca + Mg) is expressed in mg/l of $CaCO_3$.

The classification of waters into hard and soft is arbitrary, with a number of classifications used. Two are given in Table 4.12.

There are a number of additional terms relating to hardness. *Total hardness* is the direct measurement of hardness. *Calcium hardness* is the direct measurement of calcium only. *Carbonate hardness* is the hardness derived from the solubilization of calcium or magnesium carbonate by converting the carbonate to hydrogencarbonate. This hardness can be removed by heating. *Magnesium hardness* is the hardness derived from the presence of magnesium and is calculated by subtracting the calcium hardness from the total hardness. *Non-carbonate hardness* is the hardness attributable to all cations associated with all anions except carbonate, e.g. calcium chloride and magnesium sulphate. *Permanent hardness* is equivalent to non-carbonate hardness which cannot be removed from water by heating. *Temporary hardness* is equivalent to carbonate hardness and can be removed by heating (Flanagan, 1988). All hardness values are expressed in mg $CaCO_3$/l.

In terms of water quality, hardness can have a profound effect. The hardness of water was originally measured by the ability of the water to destroy the lather of soap, as this is one of the principal problems of very hard water. Although hardness does neutralize the lathering power of soap it does not affect modern detergent formulations. Soft waters are more aggressive than hard waters, enhancing the corrosion of copper and lead pipes. Above a hardness of 150–200 mg/l scaling becomes a problem. This is only the case with calcium hardness or temporary hardness. The hydrogencarbonate is removed during heating to form calcium carbonate, which forms a thick scale over the surface of pipes, boilers and, of course, kettles. The chemical reaction that occurs is

$$2HCO_3^- \rightarrow H_2O + CO_2 + CO_3^{2-}$$

Hydrogencarbonate

$$CO_3^{2-} + Ca^{2+} \rightarrow CaCO_3$$

Calcium carbonate

The greater the degree of carbonate hardness the more severe the problem becomes, although a moderate level of hardness (150 mg/l) is useful as it forms a protective film of calcium carbonate over the inside of pipes, preventing the leaching of metals and corrosion.

4.6.2 Standards

Hardness is an important factor in the taste of water, although at above 500 mg/l the water begins to taste unpleasant. The World Health Organization sets a maximum recommended concentration of 500 mg/l in drinking water on aesthetic, not health, grounds. In the revised recommendations this value has been removed and no value set. The EC Drinking Water Directive does not set any specific standards for hardness except for water which has been softened, where a minimum concentration of 150 mg/l applies. However, G and MAC levels have been set for calcium and magnesium as well as other divalent cations which can also affect hardness such as barium, iron and manganese. Total dissolved solids dried at 180°C is closely correlated with hardness. In the USA the national secondary drinking water regulations set a non-enforceable MCL of 500 mg/l for total dissolved solids, whereas the EC sets an MAC of 1500 mg/l with no guide value (Section 1.3).

The minimum hardness in softened waters in the UK and Ireland is the MAC concentration of 150 mg/l, although in the Directive (Table 1.10) hardness is expressed solely in terms of calcium ions, so 150 mg/l as $CaCO_3$ is equivalent to 60 mg/l as Ca. To convert ions to $CaCO_3$ equivalent, multiply mg/l of Ca by 2.495, and Mg by 4.112 to give mg/l as $CaCO_3$ equivalent

$$\text{mg/l as } CaCO_3 = \text{mg/l of ion} \times \frac{\text{Equivalent weight } CaCO_3}{\text{Equivalent weight of ion}}$$

So, for example, for 30 mg/l of the ion calcium (Ca)

$$30 \text{ mg/l of Ca} \times \frac{50}{20.04} = 74.9 \text{ mg/l as } CaCO_3$$

Equivalent weights for the most important ions are Ca^{2+} 20.04, Mg^{2+} 12.16, HCO_3^- 6.1 and SO_4^{2-} 48. The equivalent weight of $CaCO_3$ is 50.

4.6.3 Health Aspects

It was observed as early as 1957 that the occurrence of cardiovascular disease in the population was related to the acidity of water (Zoetman and Brinkman, 1976). Since that time a considerable number of studies, mainly in the USA, have indicated a correlation between hardness or total dissolved solids and mortality, especially cardiovascular disease. A study carried out in the UK and reported in 1982 confirmed that mortality from cardiovascular disease was closely associated with water hardness, with mortality decreasing as hardness increased. The effect was present for both stroke and ischaemic heart disease, but not for non-cardiovascular disease. The difference in mortality rates between soft water (around 25 mg $CaCO_3$/l) areas and hard water areas ranged from 10 to 15%. This is in general agreement with an earlier figure of a 0.8% decrease in mortality with every 10 mg $CaCO_3$/l increase in hardness up to 170 mg $CaCO_3$/l (Lacey, 1981; Powell *et al.*, 1982).

It is also a general rule that the toxicity of pollutants and contaminants are significantly less in hard water than soft water. The general consensus is that moderately hard water is good for you, but causes scaling in distribution pipes and household plumbing, and reduces the efficiency of boilers, resulting in high fuel consumption. If water contains more than 300 mg/l as $CaCO_3$ then the water supply companies will tend to soften the water (Section 3.2), usually by the addition of lime or soda ash. The practice of softening has become more widespread over the past 20 years. Household water softening is considered in Section 8.2.

4.7 PATHOGENS

There are three different groups of micro-organisms that can be transmitted via drinking water: viruses, bacteria and protozoa. They are all transmitted by the faecal–oral route and so largely arise either directly or indirectly by contamination of water resources by sewage or, on occasions, animal wastes. It is theoretically possible, but unlikely, that other pathogenic organisms such as nematodes (roundworm and hookworm) and cestodes (tapeworm) may also be transmitted via drinking water.

4.7.1 Protozoa

There are two protozoa frequently found in drinking water which are known to be responsible for outbreaks of disease. These are *Cryptosporidium* and *Giardia lamblia*.

4.7.1.1 *Cryptosporidium*

This parasitic protozoan is widely distributed in nature, infecting a wide range of animal hosts including pets and farm animals. However, it was only relatively recently that it was found to be a human pathogen as well. The first recorded case of human infection occurred as recently as 1976. The problem is that the protozoa form protective stages known as oocysts which allow them to survive for long periods in water while waiting to be ingested by a host. They are also able to complete their life cycle in just a single host as well as having an auto-infection capacity (Fayer and Ungar, 1986). Once infected the host is a lifetime carrier and is subject to relapses. In normal patients the protozoa give rise to a self-limiting gastroenteritis which lasts for up to two weeks. If the patient is immunosuppressed, infection will be life-threatening. For example, it is a major cause of death among patients with AIDS (acquired immune deficiency syndrome). It is thought that 2% of all current cases of diarrhoea in the UK are due to *Cryptosporidium*, with children more at risk than adults. Two peaks in the number of infections are seen each year, one in the spring and another in the autumn.

The oocysts of *Cryptosporidium* are only 4–7 μm in diameter and so are difficult to remove from raw waters by conventional treatment. The oocysts appear to be very widespread in British drinking water resources, being found in over half the samples tested in a recent Department of the Environment study. In a separate study between 0.02 and 0.08 oocysts per litre were found in pristine streams, whereas on average 0.9–1.2 per litre were found in lakes and rivers which received waste water. The problem is therefore clearly worse in water resources receiving treated or untreated waste water, although oocysts are widespread in all types of surface waters. Low level exposure to oocysts (1–10) is capable of initiating an infection (Rose, 1990). Studies indicate that a single oocyst may be enough to cause infection (Blewett *et al.*, 1993), although outbreaks of cryptosporidiosis have been associated with gross contamination (Wilkins, 1993). The main symptoms of cryptosporidiosis are stomach cramps, nausea, dehydration and headaches. There may also be significant weight loss associated with infection. The first water-borne outbreak documented was in 1985 due to sewage contamination of a well in Texas. In 1987 an outbreak in Georgia affected 13 000 people, even though the drinking water had undergone conventional treatment (i.e. coagulation, sedimentation, filtration and disinfection).

Since 1988 two major outbreaks of water-borne cryptosporidiosis have been recorded in the UK (Table 4.13). The first in 1988 occurred in Ayrshire when 27 cases were confirmed, although many more people were thought to have been infected. Of those infected, 63% were less than eight years of age. The outbreak was thought to have been due to the finished water being contaminated with runoff from surrounding fields on which slurry had been spread. The oocysts were found in the chlorinated water in the absence of faecal indicator bacteria. A second outbreak was far more serious and affected over 400 people in the Oxford and Swindon areas early in 1989 (Richardson *et al.*, 1991a), although according to Rose (1990) as many as 5000 people may have been affected. The outbreak was quickly traced to the Farmoor Water Treatment Plant near Oxford, which takes its water from the River Thames. On investigation oocysts were found in the filters, in the filter back-wash water and in the treated water, even though it had been chlorinated and the microbiological tests had shown it to be of excellent quality. The oocysts were found in the Farmoor Reservoir and also in a tributary of the River Thames, upstream of the treatment plant. The seasonal presence of the organisms was particularly associated with the grazing of lambs and with the scouring that often arises in their early lives. Although the rapid sand filters at Farmoor were removing 79% of the oocysts after coagulation, the recycling of the back-wash water

Table 4.13. Major occurrences of *Cryptosporidium* in drinking water in 1983–9.
Reproduced from Rose (1990) by permission of *Water Engineering and Management*

Location	Year	Number of cases	Source
Cobham, Surrey, UK	1983	16	Spring and river
Cobham, Surrey, UK	1985	50	Spring and river
Braun Station, Texas, USA	1984	2000	Artesian well
Carrollton, Georgia, USA	1987	13 000	River
Ayrshire, UK	1988	27	Reservoir
Oxford and Swindon, UK	1989	400	Reservoir

had given rise to exceptionally high concentrations of oocysts (10 000 per litre), with resultant breakthrough. Disinfection with chlorine was not effective, and the first action by Thames Water was to stop recycling the back-wash water (up to then a normal practice), which brought the problem under control. There was a decrease in the number of reported cases as the number of oocysts in the water decreased. These oocysts can survive for up to 18 months depending on the temperature.

There have been further outbreaks since then, but on a much smaller scale. One was in north-west Surrey at about the same time as the Oxford outbreak. Again, water was taken from the River Thames downstream of Oxford. In April 1989 an outbreak occurred in Great Yarmouth following a change in the water supply, whereas there were 140 cases reported in Hull in January 1990. There was also an outbreak in the Loch Lomond area in 1989. In this case livestock grazing around the reservoir or in the catchment of rivers were the source of the organism. Cattle and infected humans can excrete up to 10^{10} oocysts during the course of infection, so that cattle slurry, waste water from cattle markets and sewage should all be considered potential sources of the pathogen. Cryptosporidiosis is now a reportable disease in Scotland, but unfortunately not so in the rest of the UK or in Ireland. Isolating oocysts is considered in the following.

4.7.1.2 *Giardia*

Like *Cryptosporidium*, *Giardia lamblia* is found in a wide range of animals, where it lives in the free-living (trophozoite) form in the intestines. In water resources cysts are able to survive for long periods, especially in winter. *Giardia* cysts, unlike those of *Cryptosporidium* which are round, are oval and larger (8–14 μm long and 7–10 μm wide). This intestinal protozoan causes acute diarrhoeal illness and has a world-wide distribution. It is particularly common in the USA where it is now considered to be endemic with a carrier rate of 15–20% of the population, depending on their socio-economic status, age and location. *Giardia* is the most common animal parasite of humans in the developed world, although water is probably not the most common mode of transmission. However, *Giardia* remains one of the most common causes of water-borne diseases. Between 1980 and 1985 there were 502 outbreaks of water-borne disease in the USA; 52% of these were due to *Giardia*. For example, sewage contamination of a groundwater supply at a Colorado ski resort resulted in 123 holiday-makers contracting the illness. A survey of the situation in the USA by LeChevallier (1990) showed that 81% of the raw waters and 17% of the treated waters sampled contained *Giardia*. A similar survey in Scotland showed that 48% of the raw waters and 23% of the treated waters sampled contained *Giardia* cysts. The number of reported cases in England and Wales has risen from 1000 each year in the late 1960s to over 5000 each year by the late 1980s, although these were largely associated with people travelling overseas. The number of outbreaks associated with drinking water contamination is steadily increasing in the UK. A significant outbreak of giardiasis occurred in south Bristol in the summer of 1985, when 108 cases were reported. It is thought that contamination of the supply occurred in the distribution system and was not due to any failure of the water treatment process.

The symptoms of giardiasis develop between one and four weeks after infection. These include explosive, watery, foul-smelling diarrhoea, gas in the stomach or intestines, nausea and, not surprisingly, a loss of appetite. Unlike *Cryptosporidium*, *Giardia* can be treated by a number of drugs, but there is no way of preventing infection except by adequate water treatment. Boiling water for 20 minutes will kill the cysts.

The New Safe Drinking Water Regulations in the USA require surface waters to be filtered to specifically remove cysts and be sufficiently disinfected to destroy *Giardia* (Section 1.3).

4.7.1.3 *Other*

Protozoan pathogens of humans are almost exclusively confined to tropical and subtropical areas, which is why the increased occurrence of *Cryptosporidium* and *Giardia* cysts in temperate areas is so worrying. However, with the increase in travel, carriers of all diseases are now found world-wide, and cysts of all the major protozoan pathogens occur in European sewage from time to time. All the diseases are transmitted by cysts entering the water supplies. Infection is by direct ingestion, either by drinking the water or by swimming in contaminated water.

Two other protozoan parasites which occur in the UK from time to time are *Entamoeba histolytica* which causes amoebic dysentery and *Naegleria fowleri*, which causes the fatal disease amoebic meningo-encephalitis.

Entamoeba histolytica is carried by about 10% of the population in Europe and 12% in the USA. In countries where the disease is widespread (not Europe) the highest rate of incidence occurs in those groups with unprotected water supplies, inadequate waste disposal facilities and poor personal hygiene. In Europe the number of cysts in sewage is generally low, with less than five per litre in untreated raw sewage. Because of their poor ability to settle, like all protozoan cysts they tend to pass through the sewage treatment plant and reach surface waters. Where surface waters are reused for supply purposes, any cysts present will be taken up with the water. The disease amoebic dysentery is very debilitating. The infection is centred in the large intestine and the symptoms include mid-abdominal pain, diarrhoea alternating with constipation, or chronic dysentery with a discharge of mucus and blood.

The cysts of *Naegleria fowleri* normally enter the body through the nasal cavities while swimming in infected water, although it is thought that infection is also possible in swimmers who swallow contaminated water. Once in the body the protozoa rapidly migrate to the brain, cerebrospinal fluid and bloodstream. Cases are very rare but generally fatal, and are exclusively associated with swimming in warm contaminated waters, especially hot springs and spa waters. Since it was first reported in 1965 there have only been six reported cases in the UK. The last reported case was in 1978 when an 11 year-old girl was infected while swimming in a municipal swimming pool filled with water from a natural hot spring in Bath.

4.7.1.4   *Isolating protozoa*

The incubation period for *Cryptosporidium* before the onset of symptoms is about seven days, and a little longer for *Giardia*, and so it may be several weeks before it is apparent that there is an outbreak in the community. It is often difficult to know exactly how widespread the disease is. There are technical difficulties in monitoring water for protozoa before treatment and distribution. Currently large volumes of treated water (100–500 l) must be filtered to remove oocysts and cysts. Current standard methods recommend the use of wound polypropylene cartridge filters (Department of the Environment, 1990a), although studies have shown the subsequent recovery of oocysts and cysts to be poor (Musial *et al.*, 1987). Isolation and identification techniques are discussed elsewhere (Department of the Environment, 1990a; Vesey and Slade, 1991) and are subject to intense research (Whitemore and Carrington, 1993). The removal of protozoa by water treatment processes is considered in Section 5.3.

Sample collection is normally carried out using a standard cartridge filter housing with a 1 μm pore size fibre filter cartridge plumbed into the incoming mains supply to the house. Clearly this is also an effective way of removing both *Giardia* cysts and *Cryptosporidium* oocysts from household supplies (Section 8.2).

4.7.2   Viruses

Viruses are made up of a core of nucleic acid (RNA or DNA) surrounded by a protein coat. They cannot reproduce without a host cell, but can survive in the environment for very long periods. Human enteric viruses are produced in very large amounts by infected individuals and are faecally excreted. They generally pass unaffected through the waste water treatment plant and so will be found in surface waters. Infection takes place when a virus is ingested with contaminated food or by drinking contaminated water. The viruses pass through the stomach and generally infect cells of the lower alimentary canal. Infected subjects shed large numbers of viral particles into the faeces which are then excreted. Up to 10 000 000 viral particles have been reported per gram of faeces, and can continue to be shed for long periods; for example, the mean excretion period for polio is over 50 days.

Infectious hepatitis, enteroviruses, reovirus and adenovirus are all thought to be transmitted via water. Of most concern in Britain is viral hepatitis. There are three subgroups, hepatitis A which is transmitted by water, hepatitis B which is spread by personal contact or inoculation and which is endemic in certain countries such as

Greece, and hepatitis C which is a non-A or B type hepatitis virus. Hepatitis A is spread by faecal contamination of food, drinking water and areas which are used for bathing and swimming. Epidemics have been linked to all these sources, and it appears that swimming pools and coastal areas used for bathing which receive large amounts of sewage are particular sources of infection. The virus cannot be cultivated *in vitro* so studies are confined to actual outbreaks of the disease. Hepatitis A outbreaks usually occur in a cyclic pattern within the community, as once infected the population is immune to further infection by the virus. Therefore no new cases occur for five to ten years until there is a new generation (mainly of children) which has not previously been exposed. There is no treatment for hepatitis A, with the only effective protection good personal hygiene, and the proper protection and treatment of drinking water. Immunoglobulin is often given to prevent the illness developing in possible contacts, although it is not always successful. Symptoms develop 15–45 days after exposure and include nausea, vomiting, muscle ache and jaundice. Hepatitis A virus accounts for 87% of all viral water-borne disease outbreaks in the USA (Craun, 1986).

Enteroviruses, poliomyelitis virus, coxsackievirus and echovirus, all cause respiratory infections and are present in the faeces of infected people. Poliomyelitis in particular is common in British sewage, but this is due to the vaccination programme within communities and does not indicate actual infection. Reovirus is thought to be associated with gastroenteritis whereas adenovirus 3 is associated with swimming pools and causes pharyngo-conjunctival fever.

Warm-blooded animals appear to be able to carry viruses pathogenic to humans. For example, 10% of beagle dogs have been shown to carry human enteric viruses. Therefore, there appears a danger of infection from waters not contaminated by sewage but by other sources of pollution, especially storm water from paved areas. Most viruses remain viable for several weeks in water at low temperatures, so long as there is some organic matter present. Viruses are found in surface and groundwater sources. In the USA as many as 20% of all wells and boreholes have been found to be contaminated with viruses. Two viruses which have caused recent outbreaks of illness due to drinking water contamination are Norwalk virus and rotavirus (Craun, 1991; Cubitt, 1991).

Norwalk virus results in severe diarrhoea and vomiting. It is of particular worry to the water industry in that it appears not to be affected by normal chlorination levels. Also it seems that infection by the virus only gives rise to short-term immunity, whereas life-long immunity is conferred by most other enteric viruses. In 1986 about 7000 people who stayed at a ski resort in Scotland were infected with a Norwalk-like virus. The private water supply, which was untreated, came from a stream subject to contamination from a septic tank.

Rotavirus is a major contributor to child diarrhoea syndrome. This causes the death of about six million children in developing countries each year. This is not, thankfully, a serious problem in Europe due to better hygiene, nutrition and health care. Outbreaks do occur occasionally in hospitals, and although associated with child diarrhoea, can be much more serious if contracted by an adult. Gerba and Rose (1990) have produced an excellent review on viruses in source and drinking waters.

4.7.3 Bacteria

Bacteria are the most important group in terms of frequency of isolation in drinking water and reported outbreaks of disease. The most important bacterial diseases are commonly associated with faecal contamination of water. For example, in temperate regions these include *Salmonella* (typhoid, paratyphoid), *Campylobacter*, *Shigella* (bacterial dysentery), *Vibrio cholerae* (cholera), *Escherichia coli* and *Mycobacterium* (tuberculosis).

4.7.3.1 *Salmonellosis*

The various serotypes that make up the genus *Salmonella* are now possibly the most important group of bacteria affecting the public health of humans and animals in western Europe. For humans this is undoubtedly due to the elimination of other classical bacterial diseases through better sanitation, higher living standards and the widespread availability of antibiotic treatment. *Salmonella* is commonly present in raw waters but is only occasionally isolated from finished waters, as chlorination is highly effective at controlling the bacteria. Typical symptoms of salmonellosis are acute gastroenteritis with diarrhoea, often associated with abdominal cramps, fever, nausea, vomiting, headache and, in severe cases, even collapse and possible death. Compared with farm animals the incidence of salmonellosis in humans is low and shows a distinct seasonal variation. A large number of serotypes are pathogenic to humans and their low frequency of occurrence varies annually from country to country. Low level contamination of food or water rarely results in the disease developing because between 10^5 and 10^7 organisms have to be ingested before development. Once infection has taken place then large numbers of the organisms are excreted in the faeces ($> 10^8/g$). Infection can also result in a symptomless carrier state in which the organism rapidly develops at localized sites of chronic infection, such as the gall bladder or uterus, and is excreted in the faeces or other secretions. Water becomes contaminated by raw or treated waste water.

The most serious diseases associated with specific serotypes are typhoid fever (*Salmonella typhi*) and paratyphoid (*Salmonella paratyphi* and *Salmonella schottmuelleri*). The last major outbreak of typhoid in Britain occurred in Croyden, Surrey during the autumn of 1937 when 341 cases were reported, resulting in over 40 deaths. There have been five minor outbreaks of typhoid and three of paratyphoid since then. The number of reported cases of typhoid fever in the UK has fallen to less than 200 each year, 85% of these being contracted abroad. Of the remainder, few are the result of drinking contaminated water. Although *Salmonella paratyphi* is recorded in surface waters all over the British Isles, there have been less than 100 cases of paratyphoid fever reported annually. Typhoid has also been largely eliminated from the USA, although in 1973 there was an outbreak in Dade County in which 225 people contracted the disease from contaminated well water (Craun, 1986).

4.7.3.2 *Campylobacter*

Campylobacter are spiral curved bacteria 2000–5000 nm in length consisting of two to six coils. Although discovered in the late 19th century they were not isolated from diarrhoetic stool specimens until 1972. It is only since the development in 1977 of a highly selective solid growth medium allowing culture of the bacterium that its nature has been revealed. *Campylobacter* species have been isolated from both fresh and estuarine waters with counts ranging from 10 to 230 campylobacters/100 ml in rivers in north-west England. Although the epidemiology of human campylobacter infections remains to be fully elucidated, certain sources of infection are well established. There were 27 000 reported outbreaks of campylobacter enteritis in the UK during 1987, rising to over 30 000 in 1990, causing severe acute diarrhoea. It is now thought that *Campylobacter* is the major cause of gastroenteritis in Europe, being more common than *Salmonella*. The most important reservoirs of the bacterium are meat, in particular poultry, and unpasteurized milk. Household pets, farm animals and birds are also known to be carriers of the disease. Unchlorinated water supplies have been identified as a major source of infection. For example, 3000 of a total population of 10 000 developed campylobacter enteritis from an inadequately chlorinated mains supply in Bennington, Vermont. While 2000 people who drank unchlorinated mains water contaminated with faecally polluted river water in Sweden also contracted the disease. Water is either contaminated directly by sewage which is rich in *Campylobacter* or indirectly from animal faeces. There is a definite seasonal variation in the number of campylobacters in river water, with the greatest numbers occurring in the autumn and winter. This is opposite to the seasonal variation of infection in the community, with the number of infections increasing dramatically during May and June (Jones and Telford, 1991). Serotyping of isolates has confirmed that *C. jejuni* serotypes, common in human infections, are found downstream of sewage effluent sites, confirming that sewage effluents are important sources of *C. jejuni* in the aquatic environment. Gulls are known carriers and can contaminate water supply reservoirs while they roost. Dog faeces, in particular, are rich in the bacterium. In a UK study *C. jejuni* was isolated from 4.6% of 260 specimens of dog faeces sampled, whereas *Salmonella* spp. were isolated from only 1.2%. The incidence of *C. jejuni* is low compared with other studies on dog faeces, with infection rates ranging from 7 to 49%. Dog faeces can cause contamination of surface waters during storms as surface runoff removes contaminated material from paved areas and roads. An outbreak affecting 50% of a rural community in northern Norway was traced to contaminated faecal deposits from sheep grazing the banks of a small lake that were washed into the water during a heavy storm which melted the snow on the banks. The water supply for the village came directly from the lake without chlorination. Natural aquatic systems in temperate areas are generally cool and research has shown that *Campylobacter* can remain viable for extended periods in streams and groundwaters. Survival of the bacterium decreases with increasing temperature, but at 4°C survival in excess of 12 months is possible. The incidence of *Campylobacter* in water can be estimated by an MPN technique. *Campylobacter* infections have been reviewed by Skirrow (1982).

4.7.3.3 *Shigella*

Shigella causes bacterial dysentery and is the most frequently diagnosed cause of diarrhoea in the USA, where it accounted for 19% of all the cases of water-borne diseases reported in 1973. The bacterial genus is rather similar in epidemiology to *Salmonella*, except that *Shigella* rarely infect animals and do not survive as well in the environment. When the disease is present as an epidemic it appears to be spread mainly by person-to-person contact, especially between children, shigellosis being a typical institutional disease occurring in overcrowded conditions. However, there has been a significant increase in the number of outbreaks arising from poor quality drinking water contaminated by sewage. Of the large number of serotypes (>40) *S. sonnei* and *S. flexneri* account for more than 90% of isolates. The number of people excreting *Shigella* is estimated to be 0.46% of the population in the USA, 0.33% in Britain and 2.4% in Sri Lanka. In England and Wales notifications of the disease rose to between 30 000 and 50 000 each year, falling to less than 3000 each year in the 1970s. However, in the 1980s notifications have doubled to nearly 7000 each year (Galbraith *et al.*, 1987).

4.7.3.4 *Enteropathogenic* Escherichia coli

There are 14 distinct serotypes of *Escherichia coli* which cause gastroenteritis in humans and animals, being especially serious in newborn infants and children under five years of age. It is common throughout Europe and is also thought to be the cause of 'traveller's tummy', the bout of diarrhoea that affects so many tourists who visit the warmer areas of Europe. The symptoms are profuse watery diarrhoea with little mucous, nausea and dehydration. The disease does not cause any fever and is rarely serious in adults. Up to 2.4% of children in England and Wales are thought to be carriers, although much higher percentages are found in people engaged in high risk occupations such as food handling. Enteropathogenic *E. coli* is commonly isolated from sewage but probably represents less than 1% of the total coliforms present in polluted waters. However, only 100 organisms are required to cause illness. Survival of the organism is the same as for other serotypes of *E. coli* and under warm, nutrient-rich conditions they are able to multiply in water. An outbreak of gastroenteritis in Worcester in the winter of 1965–6 affected 30 000 people and was thought to be due to contamination of the water supply as a result of flooding. In 1986 the water tanks on a cruise-liner became contaminated by sewage, resulting in an outbreak of the disease which affected 251 passengers and 51 crew.

4.7.3.5 *Cholera*

Cholera is thought to have originated in the Far East, where it has been endemic in India for many centuries. In the 19th century the disease spread throughout Europe where it was eventually eliminated by the development of uncontaminated water supplies, water treatment and better sanitation. It is still endemic in many areas of the world, especially those which do not have adequate sanitation and, in particular, in situations where the water supplies are continuously contaminated by sewage. This is the cause of the most recent epidemic raging through South America, which started in Peru at the beginning of 1992; about 400 000 people have so far contracted the disease. However, over the past 10–15 years the incidence and spread of the disease has been causing concern; this has been linked to the increasing mobility of travellers and the speed of travel. Healthy, symptomless carriers of *Vibrio cholerae* are estimated to range from 1.9 to 9.0% of the population. This estimate is now thought to be rather low, with a haemolytic strain of the disease reported as being present in up to 25% of the population. The holiday exodus of Europeans to the Far East, which has been steadily increasing since the mid-1960s, will have led to an increase in the number of carriers in their home countries and an increased risk of contamination and spread of the disease. There have been nearly 50 reported cases of cholera in the UK between 1970 and 1986, although no known cases have been water borne (Galbraith *et al.*, 1987).

Up to 10^8–10^9 organisms are required to cause the illness, so cholera is not normally spread by person-to-person contact. It is readily transmitted by drinking contaminated water or by eating food handled by a carrier, or which has been washed with contaminated water, and is regularly isolated from surface waters in the UK. It is an intestinal disease with characteristic symptoms i.e. sudden diarrhoea with copious watery faeces, vomiting, suppression of urine, rapid dehydration, lowered temperature and blood pressure, and complete collapse. Without treatment the disease has a 60% death rate, the patient dying within a few hours of first showing the symptoms, although with suitable treatment the death rate can be reduced to less than 1%.

The bacteria are rapidly inactivated under unfavourable conditions, such as high acidity or a high organic matter content in the water. In cool, unpolluted waters *V. cholerae* will survive for up to two weeks. The El-Tor type of *V. cholerae* is the predominant type today, being excreted over a longer period and surviving longer in surface waters than the classical type. Other types of *Vibrio* cause a milder form of the disease or gastroenteritis.

4.7.3.6   Tuberculosis

Mycobacterium tuberculosis, *M. balnei* (*marinum*) and *M. bovis* all cause pulmonary tuberculosis. Infection is by the inhalation or ingestion of bacilli released in the sputum, milk or other discharges, including the faeces, of infected animals. The source of infection is difficult to identify as the disease has a very long incubation period before clinical tuberculosis is diagnosed, which in some cases may be many years. However, *M. tuberculosis* is often isolated in waste water from hospitals and meat-processing plants. The bacilli are able to survive for several weeks at low temperatures in water contaminated with organic matter. Drinking contaminated water appears to be a source of infection and there is considerable circumstantial evidence to support this.

4.7.4 Unusual Sources of Contamination

There was some concern in the early 1980s over the open policy adopted by the then Regional Water Authorities in allowing more extensive use of their potable supply reservoirs for recreation. One of the potential hazards identified has been the use of bait by fishermen. It has been suggested that such bait could be a potential source of problems due to: (1) the introduction of pathogenic organisms affecting either humans or fish; (2) increasing the nutrient concentration of the water and thus causing eutrophication; and (3) that certain chrysoidine dyes which are used to colour maggots are possibly carcinogenic. In a study carried out by the Water Research Centre, there appeared to be no risk at all, although the report noted that most of the Regional Water Authorities were already adopting restrictions (Solbe, 1983). It had been shown that a botulinus toxin, which originated from maggots used as bait and contaminated with *Clostridium botulinum*, had caused the deaths of waterfowl on the River Thames in the summer of 1982. However, it was felt that chlorination was effective in destroying or inactivating the toxin, so that any risk was minimal. Restricting baiting to moderate levels, especially on small reservoirs, appears to be sensible, but the risk is negligible.

Although migratory waterfowl which roost at reservoirs have no serious deleterious effect on water quality, gulls which feed on contaminated and faecal material at refuse tips and sewage works have been shown to excrete pathogens. All five British species of *Larus* gulls have been rapidly increasing in numbers in recent years, with the herring gull (*Larus argentatus*) doubling its population size every five to six years. This has resulted in a large increase in the inland population of *Larus* gulls throughout the British Isles, both permanent and those overwintering. These birds are opportunist feeders and have taken advantage of the increase in the human population and its standard of living, feeding on contaminated waste during the day and then roosting on inland water bodies, including reservoirs, at night. Faecal bacteria, especially *Salmonella* spp., have been traced from feeding sites such as domestic wastetips to the reservoir, showing that gulls are directly responsible for the dissemination of bacteria and other human pathogens. Many reservoirs have shown a serious deterioration in bacterial quality due to contamination by roosting gulls, and those situated in upland areas where the water is of a high quality, so that treatment is minimal before being supplied to the consumer, are particularly at risk from contamination. Research has shown that numerous *Salmonella* serotypes as well as faecal coliforms, faecal streptococci and spores of *Clostridium welchii* can be isolated from gull droppings (Gould, 1977). Ova of parasitic worms have also been isolated from gull droppings and birds are thought to be a major cause of the contamination of agricultural land with the eggs of the human beef tapeworm (*Taenia saginata*). There is a clear relationship between the number of roosting gulls on reservoirs and the concentration of all the indicator bacteria. Gull droppings can also contribute significant amounts of nitrogen and phosphorus to reservoirs, which can lead to eutrophication.

Although contamination of small service reservoirs has been eliminated by covering them, this is impracticable for large upland reservoirs serving major cities and towns

(HMSO, 1989c). A considerable degree of success has been achieved by using bird-scarers and, more recently, by using techniques developed for the control of birds at airports. Roosting can be discouraged by broadcasting species-specific distress calls of *Larus* gulls, which has led to a dramatic reduction in bacterial contamination. Such a control option is far more cost-effective than installing and operating more powerful disinfection treatment systems.

Drinking water standards for microbial pathogens and the efficiency of water treatment to remove them from raw waters are considered in Section 5.3. Recent developments in drinking water microbiology have been reviewed by McFeters (1990a; 1990b) and Morris *et al.* (1993).

4.8 RADON AND RADIOACTIVITY

The main source of radiation is from naturally occurring radionuclides. In fact, the combined effect of all these natural sources accounts for 87% of the total radiation exposure to a person over their lifetime. Man-made sources only account for 13% of radiation exposure, of which 11.5% is from medical sources, 0.5% from fallout, 0.4% occupational, 0.1% from nuclear discharges, with various other miscellaneous sources accounting for the remainder. Of the natural sources, radon accounts for 32% of this total exposure.

Radon is a natural radioactive gas that has no taste, smell or colour; in fact, special equipment is required to detect its presence. Radon is formed in the ground by the decay (breakdown) of uranium, which is found in all soils and rocks to some extent. The highest levels of uranium are found in areas of granite with maximum concentrations of just under 2 ppm (parts per million) in granite from Devon. The decay of uranium results in the formation of radium-226, which subsequently decays to radon-222. The radon gas slowly migrates through the soil to the surface and is quickly dispersed into the atmosphere, where it is diluted to safe concentrations. Concern has been expressed, however, over modernized houses which are well draught-proofed. Radon percolates up through the foundations of the building, or occasionally from the granite stone used to build the house, and because there is so little air exchange in the building, the radon can accumulate to dangerous concentrations in the enclosed space of the room. Even where newly constructed buildings have adequate radon-proof membranes laid in the foundations to stop the gas from percolating through, radon gas can still accumulate in the buildings from the use of well water in the bathroom and kitchen. Radon is known to be carcinogenic, and as it is primarily inhaled as a gas it causes lung cancer. The US Public Health Service consider radon to be a major environmental health problem and evidence shows it to be the second major cause of lung cancer after cigarette smoking. The USEPA suggest that when ingested with water, radon can also increase the risk of stomach cancer (Cross *et al.*, 1985).

Radon is very soluble in water and so when the gas comes into contact with groundwater it dissolves. In a national survey of groundwaters in the USA the average radon level was 900 pCi/l (median 300 pCi/l), although in some states the mean value is considerably higher (e.g. New Hampshire 1716 pCi/l). Some groundwaters have been found to contain in excess of 100 000 pCi/l, with levels in excess of 10 000 pCi/l common in the west and north-east of the USA (Dupuy *et al.*, 1992; Helms and Rydell, 1992). Levels in European groundwaters are largely unknown.

Radon can be consumed by drinking contaminated water or through the inhalation of radon released from the water. Most of the radon is released from water when it is agitated, with up to 40 times higher concentrations recorded in the bathroom than the living room and 30 times higher than in the kitchen, due to the radon released from flushing the toilet and running large volumes of water. In the USA it is estimated that each 10 000 pCi/l of radon in drinking water results in an increase in the mean indoor radon concentration by about 1 pCi/l. It is clear that water will rarely be a major source of radon within the home overall, although increased exposure occurs in the bathroom and kitchen.

Some groundwaters in the UK may contain high levels of radon, but those that are affected will mainly be those on private supplies, especially boreholes. This water will have lost very little of its natural radiation before coming out of the tap. Having a water storage tank in a vented attic will help to reduce radon levels, and if radon is a problem, then consideration should be given to supplying all the water from the storage tank. Installing a large aquarium-type aerator in the tank will help to release the radon, whereas installing a small activated carbon filter will remove the radon from the supply to the kitchen tap which is used for drinking. Installing even a small aquarium-style aerator in the water storage tank will cause some degree of noise, especially at night; a time switch can be used to turn it off at night or the tank and motor can be insulated to reduce the vibration which causes the noise. The implications of doing this should be discussed with, and approved by, the water supply company. Of course, as the aerator will cause considerable mixing of the water, all sediment and debris must be removed from the tank before installation.

The USEPA has proposed a MCL for radon and other radionuclides of 300 pCi/l (Table 4.15). Although the Primary National Drinking Water Regulations do not apply to private water supplies, they do apply to small water schemes serving more than 25 people (or with 15 or more service connections) using groundwater. Individual supplies will require a point of entry treatment system where radon is present. In a study carried out by the USEPA, granular activated carbon (GAC) was compared with various bubble aeration systems to treat a groundwater supply containing 35 620 pCi/l radon. It was found that removal by GAC decreased over time from 99.7 to 79%. This was improved if the activated carbon was preceded by ion exchange, although the performance still decreased from 99.7 to 85% over time. Ion exchange was necessary to remove iron, which was impeding radon adsorption by fouling the surface of the GAC. In contrast, the bubble aeration systems were highly efficient (>99%) at removing radon from water (Kinner *et al.*, 1990).

There is no risk with mains supply, even if the supply is from an aquifer contaminated by high levels of radon. Water treatment of groundwaters normally includes aeration (Section 3.2), so that 99% of the radon is removed at this point. The removal efficiency varies with the aeration technique, with packed tower aeration particularly effective (>99% removal), compared with only 60–70% removal when using spray aeration. Water treatment plants also use GAC filters to remove radon and other radioactive

Table 4.14. Radionuclides found in drinking water which have a significant health risk

α-Emitting radionuclides							
^{210}Po	^{222}Rn	^{226}Ra	^{232}Th	^{234}U	^{238}U		
β-Emitting radionuclides							
^{60}Co	^{89}Sr	^{90}Sr	^{131}I	^{134}Cs	^{137}Cs	^{210}Pb*	^{210}Ra*

*Also have α-emitting daughters.

isotopes. However, there is a concern that filters could themselves become radioactive and then become a major disposal problem.

Apart from radon-222, other radionuclides are occasionally found in water; for example, the α-emitting radionuclides radium-226, which is associated with an increased risk of bone and head carcinomas, and uranium-234 and -238 associated with an increased risk of bone cancer (Longtin, 1988). With the exception of radium-228, most β-emitters are associated with human activity, whereas the α-emitters are of natural origin and so are far more likely to be detected in groundwaters (Table 4.14). There is some evidence that water supplies can become contaminated from the normal operation of nuclear power plants and reprocessing facilities. Where accidents occur, the water supplies readily become contaminated. There is understandable concern over the proposal to store radioactive waste underground at Sellafield. There are extensive aquifers in this area, widely used for water supply. Therefore any leakage would cause contamination of groundwaters.

The World Health Organization specifies guideline values for gross α and β activity of 0.1 and 1.0 Bq/l respectively (1 Ci is equivalent to 3.7×10^{10} Bq) (Table 1.17). If samples exceed these levels of activity then further radiological examination is recommended. This is a task for the national radiological authority requiring sophisticated equipment and facilities. There is currently no EC regulation covering levels of radionuclides in drinking water. A summary of current standards in the USA, and the proposed future standards, are given in Table 4.15.

4.9 ALGAE AND ALGAL TOXINS

Many impoundment and storage reservoirs are rich in nutrients from farming and the discharge from sewage treatment plants upstream. Given adequate light, algae can quickly establish themselves and become a problem. The problem is worse in the summer due to thermal stratification, with the lower zones of the reservoir (the hypolimnion) becoming anaerobic. This causes ammonia, phosphorus and silica, all algal nutrients, to be released from the bottom sediments, which encourages even more algal growth.

There are two specific problem areas for the water engineer associated with algal blooms. The first is an operational problem. The physical presence of the algae makes water treatment more difficult because the algae must be removed from the finished water. If small algae do get through the filter, then algae will enter the distribution

Table 4.15. Current and proposed maximum contaminant levels (MCLs) for radionuclides in US drinking water. Reproduced with permission from USEPA (1993)

Radioactive constituent	Primary drinking water regulation	Lowest MCL being considered by EPA	Highest MCL being considered by EPA
Radium-226, radium-228	5 pCi/l combined	2 pCi/l for each isotope	20 pCi/l for each isotope
Gross α-particle activity (including radium-226 but excluding radon and uranium)	15 pCi/l	—*	—*
Gross β-particle activity	4 millirems/year equivalent (man-made) 50 pCi/l	4 millirems/year equivalent	4 millirems/year equivalent
Uranium	—*	5 pCi/l	40 pCi/l
Radon-222	—*	200 pCi/l	4000 pCi/l

*No values set.

system and cause several problems. The water will be coloured and turbid, the algae will decay causing taste and odour problems, and finally the algae will either become the source of food for micro-organisms growing on the walls of the supply pipes, or the source of food for larger animals infesting the supply system (Section 6.6). In practice, algae reduce the rate of flow through the treatment plant by blocking microstrainers; the larger algae blocking rapid sand filters and the smaller species blocking the finer slow sand filters. In severe cases the supply may be drastically reduced or stopped altogether.

The second problem area is one of metabolism. The algae do two things. They all produce carbon dioxide during respiration, and this can cause a severe alteration of the pH of the water. This is most often caused by large crops of blue–green algae. These changes in pH severely disrupt the coagulation process, resulting in the loss of coagulant into the finished water (often aluminium sulphate), and a reduction in quality in terms of colour and turbidity (Section 5.2). Certain algae also release extracellular products, some of which are toxic, others which increase the organic matter content of the water or form halogenated by-products which are carcinogenic (Section 5.5). Taste and odour problems can cause treatment plants to close during the summer for up to two months (Section 4.4) (National Rivers Authority, 1990). There has been much interest in toxins which are released by large blooms of algae, not only in reservoirs but also in coastal waters. It is the blue–green algae which are responsible for toxins in drinking water. Despite their name, these algae are actually a group of bacteria which are capable of photosynthesis. Therefore the blue–green algae are also referred to as cyanobacteria (Hunter, 1991).

It is not a new phenomenon for algal blooms to release toxins that can kill livestock, domestic animals and even fish and birds that have drunk contaminated water (Codd et al., 1989). In fact, the earliest recorded report was over 110 years ago. In the British Isles three algae are known to produce toxins in freshwater. *Microcystis aeruginosa* produces a toxin known as microcystin-LR, a hepatotoxin which increases the level of certain enzymes in the liver causing liver damage. *Anabaena flos aquae* produces anatoxin, a neurotoxin which attacks the central nervous system. *Aphanizomenon flos aquae* also produces toxins (Falconer, 1989; 1991). When there is sunlight, a nitrogen source (e.g. nitrates from farm runoff or a sewage treatment plant effluent) and a phosphorus source (mainly from sewage treatment plants), then algal growth will develop rapidly to very high densities, i.e. eutrophication. The toxins are thought to be released on decay, rather than as extracellular products released while active. Decay can occur naturally, but massive releases of toxins only occur when the algae are killed by the addition of chemicals such as copper sulphate.

Algal toxins can affect humans and animals in three ways: through contact with contaminated water, by the consumption of fish or other species taken from such waters, or by drinking contaminated water (Hunter, 1991). A major occurrence of toxic algae took place at Rutland Water in Leicestershire, which is one of the largest reservoirs in western Europe. During late August and mid-September 1989, blooms of *Anabaena* and *Aphanizomenon* were followed by *Microcystis aeruginosa*. Twenty sheep and 15 dogs are thought to have been killed by drinking the water, which is part of the supply network for 1.5 million people. Another toxic bloom also occurred in 1989 at Rudyard Lake in Staffordshire. This time army cadets who had been canoeing and swimming through the algal bloom had influenza-type symptoms, two being admitted to hospital (Turner *et al.*, 1990).

The cyanobacterial blooms also cause strong tastes and odours. The Ardleigh Reservoir (Anglian Region) has been shut for two months every year since 1978 because of blooms of *Microcystis aeruginosa*. So at the time when water demand is at its highest and supplies are reaching their lowest, algal blooms can cause havoc with water distribution and supply (Pressdee and Hart, 1991).

There are no water quality standards set for algal (or bacterial) toxins. In The Netherlands, where this is also a serious problem, maximum acceptable algal concentrations have been suggested at 200 μg/l of chlorophyll-*a* for blue–green algae and 60 μg/l for other algal groups. These levels can be dealt with successfully by conventional treatment, but do not relate to the concentrations of the toxins. Many water supply companies now monitor reservoirs and other surface water resources daily for algae. Using an inverted microscope, they scan for the toxin-producing species and when these are identified in significant numbers action is taken. Algal blooms occur not only in lakes and reservoirs, but also in canals and coastal waters. Their occurrence has increased in recent years, although the reason for this is unknown.

At present there is no standard treatment for the removal of algal toxins from water. Conventional treatment is ineffective, except for slow sand filtration which can successfully reduce concentrations of toxins. Trials carried out by the Water Research Centre using microcystin-LR have shown that oxidation of the toxin is only effective after sand filtration using either chlorine dioxide or potassium permanganate. Chlorine and hydrogen peroxide are ineffective. Activated carbon is effective under controlled conditions, with powdered activated carbon more effective than granular (Hart *et al.*, 1992). The use of activated carbon in conjunction with ozonation is particularly promising (Himberg *et al.*, 1989). However, until there is a rapid and accurate detection method for the toxins, further progress in this area is likely to be slow.

Control options are currently directed at the blue–green algae rather than at the toxins. None of the current methods are reliable. The obvious method is to reduce the nutrient loading to the reservoir to reduce the crop of algae. Destratification of deeper lakes may eliminate some species but encourage others, whereas biological control methods such as using fish to graze the algae are difficult to operate and unreliable (Parr and Clarke, 1992). Decomposing straw has been shown to inhibit both green and blue–green algae (Gibson *et al.*, 1990), and research is currently under way to test the feasibility of using barley straw to control algae in small reservoirs.

Chapter 5

Problems Arising From Water Treatment

5.1 HOW PROBLEMS ARISE

The principle of water treatment is to make the raw water safe and palatable to drink (Section 1.3). Although water treatment normally removes or reduces unwanted chemicals and organisms from the water, during its treatment a wide variety of chemicals are used. Very occasionally excessive amounts of these chemicals are added due to poor operation or by accident and these find their way into the water supply. However, very small amounts are normally discharged with the finished water because of the nature of the processes themselves. These include coagulant residuals such as aluminium, iron and a range of organic compounds called polyelectrolytes. Operators are careful to ensure that minimum carry-over of these chemicals occurs. Other chemicals, however, are added deliberately to the water to ensure that they reach the consumer. For example, chlorine is added to disinfect water (Section 5.5) whereas fluoride is added in certain areas to protect teeth from decay (Section 5.6).

The processes involved in water treatment were outlined in Section 3.2; in this chapter the processes are examined further to see how coagulants and other chemicals get into our water supplies.

5.1.1 Coagulation and Flocculation

Coagulation-flocculation removes colloidal particles too small to settle naturally. These may include micro-organisms as well as soluble particles (Section 3.2). It is from this process that most problems of contamination of drinking water during treatment arise. The rate of addition of coagulant is governed by many factors that can alter very quickly. In practice, careful control is required over the addition of the coagulant. To understand how problems arise during coagulation it is necessary to examine the method used by operators to calculate the coagulant dosage rate.

The optimum conditions for coagulation are determined as often as possible using a simple procedure known as the jar test (Solt and Shirley, 1991). This measures the effect of different combinations of the coagulant dose and pH, which are the two important factors in the process. The jar test allows a comparison of these different combinations under standardized conditions, after which the colour, turbidity and pH of the

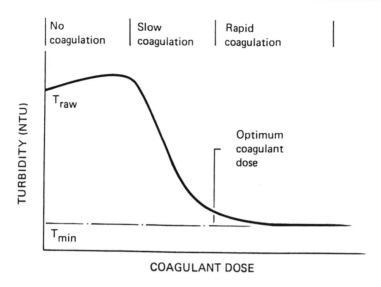

Figure 5.1. Typical coagulation curve showing the effect of the concentration of coagulant on turbidity. Reproduced from Ainsworth *et al.* (1981) by permission of the Water Research Centre

supernatant (clarified water) are measured. This involves three separate sets of jar tests, and this is how it is carried out:

1. The test is first carried out on the raw water without altering the pH. The coagulant dose is increased over a suitable range, so if five jars are used then five different doses can be tested. From this a simple curve can be drawn which shows the turbidity against the coagulant dose (Figure 5.1). This permits the best coagulant dose to be calculated.

2. The next stage is to alter the pH of the raw water by adding either an alkali or acid, and then to repeat the test using the best coagulant dose determined in the first set of tests. The range selected is normally between 5.5 and 8.5, and if possible 0.5 pH increments are used. From this a plot of final colour and turbidity is plotted against pH, allowing the optimum pH for coagulants to be selected (Figure 5.2).

3. Finally the test is repeated a third time using fresh raw water, but this time corrected to the optimum pH and tested at various coagulant doses again. This determines the exact coagulant dosage at the optimum pH. These values are then used to operate the coagulant process.

On-line, automated systems are now available, although the manual test is still widely used. The curve produced from a lowland river in Figure 5.2 is typical for hard waters of its type, and from this it is clear that few problems should occur in operational management. However, soft upland waters which are usually highly coloured as a result of humic material are far more difficult to treat. Figure 5.3 shows that there is only a very narrow band of pH at which to achieve the optimum removal of colour and turbidity. Residual colour and turbidity will always be much higher in soft upland

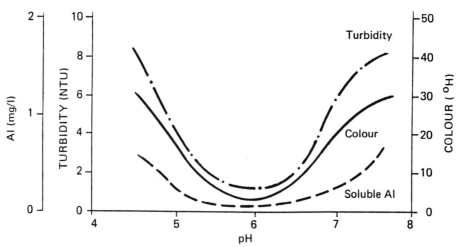

Figure 5.2. Example of a pH optimization curve for a hard lowland water showing the effect on turbidity, colour, and the soluble aluminium concentration. Reproduced from Ainsworth *et al.* (1981) by permission of the Water Research Centre

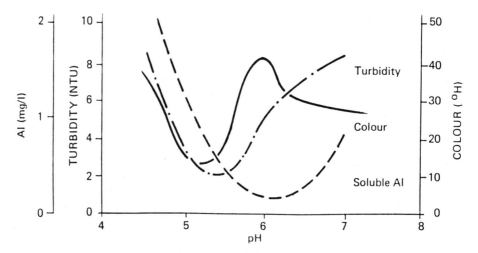

Figure 5.3. Example of a pH optimization curve for a soft, coloured water showing the effect on turbidity, colour, and the soluble aluminium concentration. Reproduced from Ainsworth *et al.* (1981) by permission of the Water Research Centre

waters than in lowland water or groundwater supplies. The hydroxide flocs formed are small and weak, causing problems at later treatment stages. Another problem is that as conditions change fairly rapidly, these tests need to be repeated, certainly daily and sometimes more often. At small plants where such waters are often treated this is just not possible, so quality in terms of turbidity and colour is more likely to fluctuate.

Changes in quality can occur rapidly, so if the operator sees a change in the colour or turbidity he or she may increase the dosage rate of coagulant, but optimum conditions may not be maintained so that excess coagulant will enter the distribution system. One of the major causes of seasonal and often daily changes is the presence of algae in reservoirs, which can significantly alter the pH of the water (Sections 4.4 and 4.8).

The jar test does not indicate how much insoluble coagulant will pass through the clarification and filtration stages, as we shall see later. It does, however, give an idea of how much soluble coagulant will be in the final water as this depends solely on pH and the dosage rate. If the curves obtained from the jar test are overlaid by the amount of aluminium that has been taken into solution, a new and more serious problem arises (Figures 5.2 and 5.3). The curve of soluble aluminium in water that is abstracted from the hard lowland river follows the same curve as both turbidity and colour, so that the amount of soluble aluminium coagulant in the treated water will be acceptably low at the optimum conditions for the removal of turbidity and colour (Figure 5.2). With soft coloured waters there is the dilemma that the best conditions for the removal of turbidity and colour do not coincide with those for minimum coagulant solubility. In the example shown the optimum removal of turbidity and colour occurred at a coagulant dose of 3 mg Al/l at pH 5.3, which results in 0.7 mg/l of soluble aluminium in the water compared with the PVC limit of 0.2 mg/l. The water is therefore clear and colourless, but contains three and a half times more than the legal limit of aluminium. Where this occurs more careful analysis of the dosage rates must be carried out to determine the optimum conditions to achieve the minimum amount of aluminium in the water. In this instance it was a coagulant dose of 5 mg Al/l, nearly twice as much as before at pH 6.6. This resulted in satisfactory aluminium (0.06 mg/l), colour (5.2 mg/l) and turbidity (1.4 FTU), although it was far more expensive and resulted in larger amounts of aluminium-rich sludge being produced. Because the operational range is so narrow, any minor change in water quality will result in a rapid deterioration in colour, turbidity and, of course, soluble aluminium. With higher residual levels of aluminium in soft waters anyway, any random adjustment by the operator, no matter how minor, could cause a significant increase in the amount of aluminium in the water. Synthetic organic polymers, in particular polyacrylamides, are used extensively to produce stronger flocs and more stable floc blankets at water treatment plants (Section 3.2). In the UK the use of polyelectrolytes is widespread, with up to 70% of all plants using such chemicals. Polyacrylamide is primarily used in the treatment of sludge, but as the polymer-rich supernatant is normally returned to the inlet of the plant it is inevitable that polyelectrolytes will find their way into drinking water.

5.1.2 Sedimentation and Filtration

It is at these stages in the process that particulate matter is removed, including the hydroxide flocs formed during coagulation. The efficiency of the sedimentation tank can be monitored continuously using turbidity meters which activate an alarm system if the turbidity exceeds the set maximum. Correct operation of sedimentation tanks is vital to minimize particulate matter passing through the plant. The most serious problem is fluctuating flow-rates, which cause the sludge blanket, through which the water flows, to expand too much. Remember that most sedimentation tanks used for water treatment operate in an upward flow direction (Section 3.2)—if the sludge blanket expands too much then particles are lost from the tank with the treated water. Infrequent desludging can also have a similar effect. The operation of filters, both rapid and slow sand filters, is complex and poor operation can lead to problems, perhaps the most serious of which is when the sand bed develops cracks, allowing the passage of unfiltered water. The polyelectrolytes used as a coagulant aid can also cause the sand in the filter to crack. The layer of micro-organisms at the surface of slow sand filters reduces the amount of organic matter in the finished water, although taste and odour problems can be caused by their activity. A similar situation can occur with micro-organisms colonizing granular activated carbon (GAC) filters (Camper *et al.*, 1985). Recycling of process water is also a major potential source of pathogen breakthrough (Section 4.7). Some of the more important water quality problems associated with water treatment are considered in the following.

5.1.3 Nitrite

Nitrogen is usually found in its fully oxidized form as nitrate (NO_3^-) or in its reduced form as ammonia (NH_3) (Figure 4.2). Nitrite is an intermediate stage in the oxidation of ammonia to nitrate, which occurs naturally in soil and water. In general, the concentration of nitrite in water is very low, but it can occasionally occur in unexpectedly high concentrations in surface waters due to industrial and sewage pollution.

It is nitrite, not nitrate, which is the actual toxicant associated with methaemoglobinaemia, the formation of *N*-nitroso compounds and carcinogenic effects, all of which have been discussed fully in Section 4.2. A very low maximum permissible value has therefore been set by the EC in their Drinking Water Directive. The EC MAC value is 0.1 mg/l nitrite (equivalent to 0.03 mg/l as nitrogen), whereas in the USA the MCL is 1 mg/l of nitrite as nitrogen. The revised World Health Organization guide value is 3 mg/l as nitrite. In their detailed report of water quality, the Drinking Water Inspectorate (DWI) (HMSO, 1991) reported increased nitrite levels in many finished waters arising from chloramination disinfection processes at water treatment plants. All the waters exceeding the PCV limit in the UK are subject to undertakings. Ammonia is not completely reacted at chlorine to ammonia ratios of 3–4:1 by weight (Section 3.2), resulting in excess ammonia reaching the distribution system. Here it acts as a source nutrient for nitrification producing nitrate or nitrite. For this reason a number of treatment works have had to replace their chloramination process due to partial nitrification leading to increased nitrite in the drinking water. For full conversion of ammonia to monochloramine a ratio of $> 5:1$ is necessary (Bryant *et al.*, 1992).

5.1.4 Discoloration

One of the most difficult problems arising from water treatment is the production of excessive colour and turbidity. Discoloration of water is a major cause of complaints, with the difficulty in maintaining optimum coagulation the prime cause. This is particularly difficult where water is abstracted from upland reservoirs and rivers where the water is usually acidic and highly coloured. With such soft water there is only a very narrow operational band in terms of pH to obtain good colour removal, so that any changes in water quality—for example, due to algal growth which affects the pH of the water quite significantly during a 24 hour period—will result in discoloration.

Particulate matter also arises from water treatment and can cause a serious deterioration in water quality, including discoloration, as can dissolved organic matter (Sections 6.3 and 6.7). Excessive amounts of suspended particulate matter and/or coagulant residual, which may be in an insoluble or soluble form, can be released from the treatment plant and enter the distribution system. These materials quickly settle out of solution by various means and accumulate in areas of low flow. If resuspended they will affect the quality of the water supplied to the consumer. These deposits are composed of aluminium coagulant, iron coagulant, manganese and iron oxides, organic matter and algae. They also harbour a variety of micro-organisms and larger animals which use them as a food source. The Department of the Environment issued guidelines to the water supply companies before privatization giving suggested values in finished water as it leaves the treatment plant to avoid discoloration problems (Table 5.1). These guidelines were in addition to the regulations arising from the EC Drinking Water Directive. All are related to the precipitation of solids, except for dissolved oxygen and pH, which are related to corrosion (Section 7.2).

5.2 ALUMINIUM

5.2.1 Sources

Aluminium is a widespread and abundant element and is found as a normal constituent of all soils, plant and animal tissue. It is especially common in food, resulting in a

Table 5.1. Guidelines issued to the UK water supply companies for avoiding discoloration in water leaving the treatment plant compared with PCV levels. Reproduced by permission of the Water Services Association

Parameter	Value to avoid discoloration	PCV
Turbidity (FTU)	<0.5	4
Colour (mg/l Pt-Co)	<10	20
Iron (μg Fe/l)	<50	200
Aluminium (μg Al/l)	<100	200
Manganese (μg Mn/l)	<50	50
Dissolved oxygen (%)	>85	75*
pH (at 25°C)	6.5–9.0	5.5–9.5

*Unless a groundwater.

typical daily intake of between 5 and 20 mg depending on individual variations in eating and drinking habits. For example, it has been suggested that as aluminium is taken up in large amounts by tea plants, drinking tea tends to enhance aluminium uptake. In fact, tea can contain anything from 20 to 200 times more aluminium than the water it is made with. Aluminium can also be leached from cooking utensils. It has been shown that cooking acidic foods such as citric fruits, rhubarb or tomatoes can lead to enhanced leaching from aluminium pots and pans. This has also been reported for spinach and other green vegetables. One interesting study looked at aluminium leaching from coffee percolators made from aluminium. In new percolators it was found that the coffee contained an aluminium concentration of 4.1 mg/l, of which 85% came from aluminium leached from the metal pot. As the percolator aged the leaching was reduced, although more than 70% of the aluminium in the coffee was still leached from the pot. Aluminium packaging materials are normally coated with lacquers to reduce leaching; however, aluminium cartons and packaging can contribute to the amount of aluminium in the diet. It is difficult to estimate the exposure to aluminium from containers and cooking utensils. In many cases they are likely to be small compared with the total dietary intake. However, aluminium pots and pans should be replaced with steel as the old ones wear out.

It has been suggested that leaching of aluminium from pots and pans is enhanced if the water has been fluoridated, or if the food contains fluoride. However, recent work has shown that this does not appear to be so, and those in fluoridated areas are no more at risk from aluminium leaching than those receiving low fluoride water, especially where the water is soft and slightly acidic.

Aluminium in water may be present due to leaching from soil and rock. Acid rain in poorly buffered regions has led to increased leaching from soils so that the surface waters have much higher aluminium concentrations. In acidic soft water areas, surface waters may contain as much as 200–300 μg/l of aluminium, increasing to 600 μg/l where these areas have been afforested. Such catchments are especially at risk if they are near the coast and subject to prevailing winds as they receive precipitation carrying higher salt burdens, which also increases acidity in the soil and leaching of aluminium.

The objective of water treatment is to provide a supply of water that is chemically and microbiologically safe for human consumption. To achieve this chemicals such as aluminium sulphate or iron(III) sulphate are added to the water during treatment to help remove fine particulate matter, including bacteria, a process known as coagulation (Section 3.2). There has been increasing concern about the amount of residual chemicals left in the water after treatment, both in a soluble and insoluble form, and the possible effects on the health of consumers. The chemical causing most concern is aluminium sulphate.

As described in Section 5.1, the dose of aluminium sulphate (alum) added to raw water depends on a number of factors, the most important being the pH, although temperature is also important. The solubility range of aluminium is shown in Figure 5.4; to achieve the EC MAC value a coagulant pH must be between 5.2 and 7.6, or 5.3 and 7.0 to achieve the stricter G level. If the pH goes outside this range the solubility increases rapidly.

The addition of alum to drinking water as a coagulant will therefore leave some

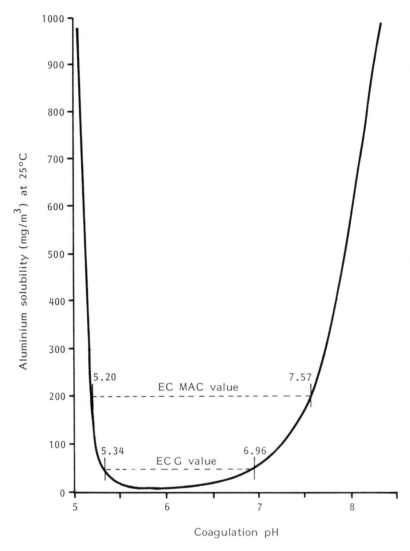

Figure 5.4. Solubility of aluminium. The pH causes severe problems during coagulation. The effective pH range to ensure PCV (MAC) values given by the EC in the Drinking Water Directive for aluminium is 5.2–7.6 . A smaller operative range of 5.3–7.0 must be used if the EC Drinking Water Directive G value is to be achieved

soluble or insoluble residual aluminium. The insoluble particulate aluminium should be removed by filtration processes, although the effectiveness of removal depends on the filtration efficiency. Where residual concentrations are high, the deposition of aluminium may occur in the distribution system. Any disturbance of these sediments will increase the concentration of aluminium in drinking water.

Reasons for increased levels of aluminium in drinking water are normally associated with operational problems at the treatment plant. However, accidents have occurred with aluminium sulphate spilt or accidentally discharged into the finished water and discharged into the distribution network. The worst such incident happened at the Sowermour Water Treatment Plant in Cornwall on 6 July 1988. This plant serves over 20 000 people around Camelford, but on that day a total delivery of alum in solution was put into the contact tank rather than a storage tank. In consequence, concentrations of aluminium several thousand times greater than the MAC limit set by the EC flowed out of the local taps, as well as killing over 30 000 fish in the rivers Camel and Allen (Clayton, 1991). Consumers who used the water before realizing that it was contaminated suffered a variety of effects such as blisters, mouth ulcers, skin irritations, sore throats, lassitude, diarrhoea and a host of other symptoms. Also, people who washed their hair in the contaminated water developed scalp rashes, whereas those with bleached hair suffered severe discoloration.

It is difficult to estimate the contribution aluminium in drinking water makes to the daily intake of the metal. If we take, for example, a daily intake of 20 mg for an adult who drinks about two litres of water each day at 200 μg Al/l, then the drinking water is only contributing 2% of the total aluminium intake for that person. Conversely, if we assume a lower daily intake of just 5 mg, then two litres of water at 200 μg Al/l would represent 8%. It is easy to see that given certain conditions, increased concentrations of aluminium in drinking water can contribute a significant amount of aluminium to our daily intake of the metal. Those who are very young or very old have a far higher intake of water. The elderly are clearly more at risk, and tend to use aluminium for packaging food and cooking utensils, because of its lightness. Any contribution from drinking water, no matter how low, is therefore significant. Evidence has shown that aluminium in drinking water is all bioavailable—that is, it is all readily absorbed by the body, although this is not so with other sources of aluminium such as food. Aluminium in drinking water may well contribute a much greater proportion of the aluminium absorbed by the body than currently thought.

5.2.2 Levels in Water

The EC Drinking Water Directive sets a MAC of 200 μg/l and a guide value of 50 μg/l. The World Health Organization no longer classifies aluminium as a chemical of significance of health in drinking water. It is now classed as a substance that may give rise to consumer complaints, and has a revised standard of 200 μg/l (Table 1.8). Both the UK and Ireland have set their national limits at 200 μg/l. These levels have been repeatedly exceeded in the past on a day to day basis in both countries. The DWI in their 1991 review of drinking water quality in England and Wales reported that 657 out of the 2577 water supply zones which make up England and Wales are covered by undertakings given by the water supply companies to the Secretary of State to take steps to comply with the current standard for aluminium (Department of the Environment, 1992).

A nationwide survey of drinking water carried out by Friends of the Earth in 1989

Table 5.2. Aluminium in drinking water in Dublin. Reproduced by permission of the Dublin Region Public Analyst. (N/A: not available)

Supply	1982–4		1985–6		1991	
	Mean(SD) (μg/l)	Range (μg/l)	Mean(SD) (μg/l)	Range (μg/l)	Mean (μg/l)	Range (μg/l)
Dodder	534(540)	30–2310	440(290)	30–1090	50	10–130
Liffey	540(722)	17–3700	210(140)	10–800	130	30–250
Vartry	149(236)	6–1220	110(110)	10–660	60	30–140
Leixlip	498(2110)	5–11 600	310(950)	10–9200	N/A	N/A

showed that high concentrations of aluminium, which exceeded the EC MAC standard, were found in most areas of the south-west, Wales (where only 62% of treatment plants were able to meet the 200 μg/l standard), the Cotswolds and north-west England.

Aluminium concentrations in Irish drinking water are also causing concern. Although mean values are generally within the MAC standard, the recorded ranges for aluminium are very wide, with supplies often exceeding the standard. For example, consumers in Dublin served by the Leixlip supply had aluminium levels which ranged from 5 to 11 600 μg/l during 1982–4 and 10–9200 μg/l during 1985–6 (Table 5.2). It is important that the number of times, or the percentage of readings, that the aluminium concentration is exceeded is published alongside mean or average values, which is not the case with the data in Table 5.2. The problems with Dublin's supplies have largely been solved by investment in new treatment facilities. However, outside Dublin most towns are served by small water treatment plants using predominately acidic waters. The potential for residual coagulant to enter the distribution system is therefore high.

This is a world-wide problem, with increased levels of aluminium found in drinking water wherever alum is used as a coagulant (Sollars et al., 1989). In the UK, Thames Water has indicated a willingness to phase out the use of alum as a coagulant, although due to technical reasons this is less likely to occur in other water supply areas where raw waters are soft. It has been suggested that iron(III) sulphate could be used instead of alum; however, it is more expensive, more difficult to handle and causes corrosion problems. Residual aluminium is tasteless, whereas residual iron is not and causes other consumer problems as well (Section 4.5). Iron salts used for coagulation are often contaminated with other metals, in particular chromium, nickel, manganese and lead, which find their way into the finished water.

5.2.3 Health Effects

Aluminium was previously regarded as non-toxic, only causing aesthetic problems if occurring in high concentrations in drinking water. However, over the past 20 years more and more evidence has come to light to link aluminium with certain neurodegenerative diseases.

Most of the information about the effects of aluminium in drinking water comes from patients undergoing kidney dialysis. It was first described in the early 1970s: a syndrome which was characterized by the onset of altered behaviour, dementia, speech disturbance, muscular twitching and convulsions. The outcome for patients with dialysis dementia, as the syndrome became known, was usually fatal. Aluminium concentrations in the water used to prepare dialysate fluid and the incidence of dialysis dementia were proved, although aluminium from other sources also contributed. People receiving long-term dialysis treatment through artificial kidneys are normally exposed to 150–200 l of dialysis fluid two or three times each week. The water used to make up the fluid must be of extremely high quality as aluminium is readily transferred from the dialysate to the patients' bloodstream during haemodialysis. Water-borne aluminium is also thought to be an important factor in the development of osteomalacia in dialysis patients.

Aluminium also appears to be an important factor in two other severe neuro-degenerative diseases: amyotrophic lateral sclerosis and parkinsonian dementia. These diseases are extremely common in the Western Pacific, especially on the island of Guam, where the drinking water and soil is low in calcium and magnesium, but high in aluminium, iron and silicon.

Several studies have reported a correlation between the concentration of aluminium in drinking water and the incidence of Alzheimer's disease, which is dementia in younger people (less than 45–50 years of age) (Vogt, 1986; Martyn et al., 1989; Neri and Hewitt, 1991; Forbes and MacAiney, 1992; McLachlan et al., 1992). Alzheimer's disease is a slowly developing disease, beginning with learning and memory defects and slowly progressing to affect all aspects of intellectual activity including judgement, calculation and language. As the terminal stages approach, urinary incontinence and the loss of useful motor activity results in the need for total nursing care. The disease may take from 18 months to 19 years to take its course, although the average is around eight years. It is a savage illness, resulting in an untreatable encephalopathy destroying the mind and body. In the USA alone, 1.2 million people have severe dementia, with another 2.5 million having mild to moderate dementia, with the cumulative lifetime risk to each individual of becoming severely demented as high as 20%. As many as 15% of all elderly people in the British Isles are suffering from dementia. Aluminium has been implicated in Alzheimer's disease because of its assumed pathogenic role in the formation of senile plaques in the brain. Certainly there are significantly increased levels of the metal in the brains of patients with Alzheimer's disease, with 1.5–2.0 times more aluminium present in the grey matter than in non-demented subjects. However, although it is clear that aluminium is associated with pathological changes in Alzheimer's disease, it is not yet clear whether its role is causal or incidental. Only time will tell.

An excellent review on aluminium in drinking water, published by the Foundation for Water Research, concludes that 'there are many questions still to be answered with regard to aluminium and Alzheimer's disease, but it is not possible to dismiss aluminium as having some kind of causal role' (Hunt and Fawell, 1990). It is obvious that aluminium should not be treated as a safe chemical and that stricter operational practice should be used to control dosage rates and handling of the chemical to prevent accidents. The MAC value is really a compromise because residual levels in excess of 100 μg/l may cause discoloration anyway. With correct operational practice aluminium levels can be kept very low indeed, so more investment is needed in water treatment in areas where the use of aluminium sulphate is made difficult by the nature of the raw water.

5.3 MONITORING AND REMOVAL OF PATHOGENS

5.3.1 Monitoring Pathogens

The isolation and identification of individual pathogenic organisms is often complex, being different for each species, and is extremely time consuming. These pathogenic micro-organisms may not be present all the time in water, and they may only be present in very small numbers. It is therefore impracticable to examine all water samples on a routine basis for the presence and absence of all pathogens. Also, the selection of individual pathogens may be misleading as each species can tolerate different environmental conditions. To routinely examine water supplies a rapid and preferably a single test is required. It is far more important to examine a water supply frequently by a simple general test, as most cases of contamination of water supplies occur infrequently, than occasionally by a series of more complicated tests. This has led to the development of the use of indicator organisms to determine the likelihood of contamination by faeces. Ideally, indicator organisms should (1) be easily detected and identified, (2) be of the same origin as the pathogens (i.e. from the intestine), (3) be present in far greater numbers than the pathogens, (4) have the same or better survival characteristics than the pathogens, and (5) be non-pathogenic themselves (Flanagan, 1988).

The non-pathogenic organisms that are always present in the intestines of humans and animals are excreted along with the pathogens, but in far greater numbers. Several of these are easily isolated and are ideal for use as indicators of faecal contamination. The most widely used are the non-pathogenic bacteria, in particular coliforms, faecal streptococci and sulphate-reducing clostridia. These three groups are able to survive for different periods of time in the aquatic environment. Faecal streptococci die fairly quickly outside the host and their presence is an indication of recent pollution. *Escherichia coli* (faecal coliforms) can survive for several weeks under ideal conditions and are far more easily detected than the other indicator bacteria. Because of this it is the most widely used test organism, although others are often used to confirm faecal contamination if *E. coli* is not detected. Sulphate-reducing clostridia (*Clostridium perfringens*) can exist indefinitely in water. When *E. coli* and faecal streptococci are absent, the presence of *C. perfringens* indicates remote or intermittent pollution. The organism is especially useful for testing lakes and reservoirs, although the spores do eventually settle out of suspension. The spores are more resistant to industrial pollutants than the other indicators and it is especially useful in waters receiving both domestic and industrial waste waters. It is assumed that these indicator organisms do not grow outside the host and, in general, this is true. However, in tropical regions *E.*

coli in particular is known to multiply in warm waters and there is increasing evidence that *E. coli* is able to reproduce in enriched waters generally, indicating an increased health risk. Great care must therefore be taken in the interpretation of results from tropical areas, so the use of bacteriological standards designed for temperate climates are inappropriate for these areas.

5.3.2 Measurement

There are two techniques widely used to enumerate indicator bacteria: the membrane filtration and multiple tube methods. The EC Directive specifies that total and faecal coliforms and faecal streptococci must be isolated using the membrane filtration method, whereas the multiple tube method is used for clostridia.

The total coliform count is a measure of all the coliforms present. *Escherichia coli* are exclusively faecal in origin and are present in fresh faeces in numbers in excess of 10^8/gram. The other coliforms are normal inhabitants of soil and water even though they can also appear in faeces. The presence of coliforms in drinking water does not imply that faecal contamination is present, although in practice it is assumed that the coliforms are of faecal origin unless proved otherwise. Therefore it is important to confirm whether *E. coli* are present. This is done alongside the total coliform count. In practice two coliform counts are reported: the total coliform count and the faecal coliform (*E. coli*) count. The membrane filtration method is now widely used for all coliform testing. Known volumes of water are passed through a sterile membrane filter with a pore size of just 0.45 μm (Figure 5.5). This retains all the bacteria present. The membrane filter is placed on a special growth medium which allows the individual

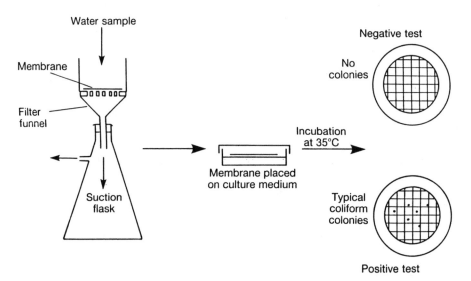

Figure 5.5. Major steps in testing for coliform bacteria using the membrane filtration technique. Reproduced from Gray (1992) by permission of Oxford University Press

bacteria to grow into colonies. Special medium which only allows specific bacteria to grow are used. For total coliforms the membrane filter is placed onto membrane-enriched Teepol medium, which contains the detergent Teepol to inhibit non-intestinal bacteria from growing. The membrane filter on the culture medium is incubated. During this incubation period, the nutrients diffuse from the culture medium through the membrane and the coliform bacteria are able to multiply and form recognizable colonies which can then be counted. Membranes are incubated at 30°C for four hours followed by 14 hours at 37°C for total coliforms or 14 hours at 44°C for *E. coli*, the colonies of which are a distinctive yellow colour. Faecal streptococci are determined using the same technique except that different culture media and incubation conditions are required (i.e. membrane enterococcus agar medium and incubated for four hours at 37°C followed by 44 hours at 45°C, with colonies appearing either red or maroon in colour).

Flanagan (1988) has summarized the interpretation of coliform results as follows: 'Where *E. coli* are present in large numbers the inference is that heavy, recent pollution by human or animal wastes has occurred. If the *E. coli* numbers are low it is inferred that pollution from the same source(s) is either recent or less severe. If coliforms not including *E. coli* are observed the indication is that either the pollution is recent and non-faecal in origin or of remote, faecal origin such that the intestinal coliforms have not survived. However, if any coliforms at all are found in a treated drinking water supply, following chlorination, it should be concluded that either inadequate treatment is being applied or else the contamination has been introduced during distribution of water, or in the sampling or handling of the sample(s). Any indication at all of contamination, however apparently mild, must be regarded as a matter of gravity and the circumstances investigated promptly.'

More information on indicator organisms and enumeration methods are found elsewhere (Mara, 1974; HMSO, 1982; Collins *et al.*, 1989). These methods are currently under review by the Standing Committee of Analysts of the UK Department of the Environment.

5.3.3 Standards

The EC Drinking Water Directive set a number of microbiological parameters using indicator organisms. Total coliforms, faecal coliforms and faecal streptococci must be absent from the finished water, and there must be less than one sulphate-reducing clostridia per 20 ml of water tested. The Directive does not allow any significant increase in the total bacteria count at either 22 or 37°C (Table 1.10).

The EC Directive does not require either viruses or protozoan cysts to be specifically measured. Studies have shown, however, that the traditional bacterial indicators will often indicate that the water is free from faecal contamination although viruses and cysts are present. The whole philosophy behind the use of coliforms as indicators is to give a very high margin of safety. The exact safety factor is dependent on the ratio of coliforms to pathogens, which in practice is never quantified. On the whole this approach is satisfactory, but problems do occur with viruses and cysts. Another problem is that many enteric bacteria are capable of surviving disinfection through a number of survival strategies. Those that are damaged (injured) cannot be cultured using standard methods, but are now known to be able to survive and eventually infect (McFeters and Singh, 1991). A lack of recovery on artificial media should not be taken to mean that the water is necessarily microbially safe (McFeters, 1990b; Rose, 1990). Coliphages and *C. perfringens* are more resistant to chlorine than other pathogens and so are useful indicators of the efficiency of treatment works in removing disinfect-ant-resistant pathogens. Research is being carried out to find better and more rapid methods of assessing the microbial quality of drinking water as the traditional indicators are not totally reliable.

There are a number of interesting alternative techniques under investigation, primarily to detect coliforms. These are focused on complex biochemical techniques such as hybridization and the polymerase chain reaction (PCR), gene probe technology and monoclonal antibody methods, which allow single bacterial cells to be detected (Richardson *et al.*, 1991b; Alvarez *et al.*, 1993). These tests will become increasingly important as there is a general swing away from standards based on microbial density to those simply based on the presence or absence of coliforms in a sample (Clark, 1990).

The USEPA set an MCL value based on the presence or absence concept for coliforms. This revised MCL came into force on 31 December 1990. For systems analysing more than 40 samples each month, <5% of the samples must be positive for total coliforms. For systems where the frequency of analysis is less than 40 samples each month, then no more than one sample each month may be positive for total coliforms. If a sample is found to be positive for total coliforms then repeat samples must be taken within 24 hours. Repeat samples must be taken at the same tap and also at adjacent taps within five service connections upstream and downstream of the original sample point. If these repeated samples are also positive for total coliforms, then the samples must be immediately tested for faecal coliforms or *E. coli*. If these are positive the public must be informed. Full details of the revised coliform rule are given by Berger (1992). The introduction of new and rapid testing systems will probably see a similar presence or absence system introduced throughout Europe.

5.3.4 Removal of Pathogens

5.3.4.1 *Bacteria*

One of the most effective ways of removing bacteria from water is by storage in reservoirs. During the spring and summer, sunlight, increased temperatures and biological factors ensure that between 90 and 99.8% reductions of *E. coli* occur. The percentage reduction is less during the autumn and winter due to the main removal mechanisms being less effective, so expected reductions fall to between 75 and 98%. The lowest reductions occur when reservoirs are mixed to prevent stratification. The greatest decrease in *E. coli* and *Salmonella* bacteria occurs over the first week, although the longer the water is stored the greater the overall reduction (Denny *et al.*, 1990).

Salmonella, faecal streptococci and *E. coli* are excreted in large numbers by gulls. The presence of gulls, especially in high numbers, on storage reservoirs may pose a serious problem either from direct faecal discharge and/or from rainfall runoff along contaminated banks (Denny, 1991) (Section 4.7).

Bacteria are removed by a number of unit processes in water treatment, especially coagulation, sand filtration and activated carbon filtration (Section 3.2). Coagulation using alum removes about 90% of faecal indicator bacteria and about 60–70% of the total plate count bacteria, although these figures vary widely from treatment plant to treatment plant. Rapid sand filtration is not dependable and is erratic unless the water is coagulated first. In contrast, slow sand filtration is able to remove up to 99.5% of coliforms, although its performance is generally worse in the winter. Activated carbon is being widely used to control taste and odour in drinking water, as well as removing a wide range of organic compounds. However, these filters can become heavily colonized by heterotrophic micro-organisms. Minute fragments are constantly breaking off the GAC and each is heavily coated with micro-organisms. These micro-organisms are not affected to any great extent by disinfection, and so can be introduced in large numbers into the distribution system (Section 4.7) (Le Chevallier and McFeters, 1990).

5.3.4.2 *Viruses*

Like bacteria, viruses are also significantly reduced when water is stored. For example, the number of viruses in River Thames water decreases from between 12 and 49 PFU/l (plaque forming units per litre of water) to 1.9 PFU/l after storage. Similar results have been reported for water from the River Lee. Temperature is also an important factor. Using water from the River Lee, researchers have found that polio virus was reduced by 99.8% in less than 15 days at 15–16°C compared with nine weeks for a similar reduction at 5–6°C. For the optimum removal of all micro-organisms of faecal origin, 10 days of retention should achieve between 75 and 99% reduction regardless of temperature.

Coagulation using alum achieves 95–>99% removal of all viruses, with other coagulants such as iron(III) chloride and iron(III) sulphate not quite as efficient. The use of polyelectrolytes as coagulant aids does not improve the removal of viruses. As with bacteria, rapid sand filtration is largely ineffective in removing viruses unless the water has been coagulated before filtration. Slow sand filtration is effective in removing the bulk of viral particles in water, with removal rates between 97 and 99.8%. However, there is always some measurable contamination left in the water and so disinfection is required (Logsdon, 1990). Activated carbon can remove viruses; these are adsorbed onto the carbon by electrostatic attraction between positively charged amino groups on the virus and negatively charged carboxyl groups on the surface of the carbon. The removal efficiency is very variable and depends on the pH (maximum removal occurs at pH 4.5), the concentration of organic compounds in the water and the time the filter has been in operation. Removal rates of 70–85% are common.

5.3.4.3 *Protozoan cysts*

Protozoan cysts are not effectively removed by storing the water in reservoirs. This is because of their small size and density. *Cryptosporidium* oocysts have a settling velocity of 0.5 μm/s. Therefore, if the reservoir is 20 m deep it will take the oocyst 463 days to settle to the bottom, assuming that there is no water current or water movement to disturb this quiescent settlement. *Giardia* cysts are much larger and have a greater settling velocity of 5.5 μm/s. In a 20 m deep reservoir it takes *Giardia* 38 days to reach the bottom. It would be feasible in a large storage reservoir to remove a percentage of the *Giardia* cysts if the retention time was greater than six weeks and if mixing and currents were minimal (Denny, 1991).

Cysts should be removed effectively by coagulation and the addition of polyelectrolyte coagulant aids should also enhance removal. Using optimum coagulant conditions (as determined by the jar test) 90–95% removal should be possible. Rapid sand filtration is not an effective barrier for cysts unless the water is coagulated before filtration. When used after coagulation effective removal of *Giardia* is achieved (99.0–99.9%). The only proved effective method of removing both *Cryptosporidium* and *Giardia* cysts is by slow sand filtration, with 99.98 and 99.99% of cysts removed, respectively (Hibler and Hancock, 1991). *Cryptosporidium* oocysts can be found in the raw water after slow sand filtration, so it must be assumed that water treatment is not able to remove all the protozoan cysts that may be present in raw water.

5.3.5 Disinfection

The efficiency of water treatment in removing pathogenic micro-organisms varies from month to month, and even when the treatment plant is achieving 99.9% removal there will always be some pathogens remaining in the water. This means that disinfection is absolutely vital to ensure that any micro-organisms arising from faecal contamination of the raw water are destroyed. Chlorination is by far the most effective disinfectant for bacteria and viruses because of the residual disinfection effect that can last throughout the water's journey through the distribution network to the consumer's tap. The most effective treatment plant design to remove pathogens is rapid and slow sand filtration, followed by chlorination or treatment by prechlorination followed by coagulation, sedimentation, rapid sand filtration and post-chlorination. Both of these systems give >99.99% removal of bacterial pathogens including *C. perfringens*. Chlorine and monochloramine are ineffective against *Cryptosporidium* oocysts, although ozone and chlorine dioxide may be suitable alternative disinfectants (Korich *et al.*, 1990). Ozone at a concentration of about 2 mg/l is able to achieve a mean reduction in the viability of oocysts of between 95 and 96% over a 10 minute exposure period. Disinfection is considered in more detail in Section 3.2. Chlorine is only effective against *Giardia* under very controlled conditions and in practice is unlikely to be an effective barrier on its own (Hibler and Hancock, 1990).

Wash-water and water treatment sludge contain all the pathogens removed from the water and so must be handled as carefully as any other microbiologically hazardous waste. It is important that the sludge is not disposed of in such a way as to recontaminate the raw water source (Section 4.7).

Conventional water treatment cannot guarantee the safety of drinking water supplies at all times. Outbreaks of water-borne diseases can and do happen, although infrequently. With correct operation the chances of pathogenic micro-organisms causing problems to consumers will be reduced even further. The greatest risks come from private supplies that are not treated. Finding dead sheep decomposing in small streams just above a water intake point for an upland cottage is not an unusual occurrence, and the risk to shallow wells from surface contamination by faeces from animals that are grazing nearby is high. It may be advisable for everyone on a private supply to install an ultraviolet sterilizer to disinfect their water as it comes into the house (Section 8.2). However, the only effective way to destroy protozoan cysts is to physically remove them. This can be done using a 1 μm pore size fibre cartridge filter which should be placed upstream of the ultraviolet sterilizer. This will also increase the efficiency of the sterilization by removing any particulate matter which may harbour and shield pathogens from the ultraviolet radiation. Those on a mains supply should be reassured that their supplies are safe.

If the sand filters at the water treatment plant are by-passed for any reason, or the disinfection system fails, then the water should not be consumed before boiling. It is the responsibility of the water supply company to ensure that notices informing the public of the need to boil supplies are issued immediately (Water Authorities Association, 1985; HMSO, 1989a).

5.4 ODOUR AND TASTE

5.4.1 Source of Odour and Taste Problems

Most odour and taste problems occur at the water treatment stage and are linked to chlorination (Levi and Jestin, 1988). Chlorine itself has a distinctive odour with a reported taste threshold of 0.16 mg/l at pH 7 and 0.45 mg/l at pH 9. Although a slight chlorine odour is generally accepted by consumers as a sign that the water is microbially safe, excessive concentrations of chlorine can make the water most objectionable. In recent years a number of outbreaks of diarrhoeal diseases have occurred when water companies have reduced the level of disinfection due to complaints of chlorinous odours. Clearly a balance must be struck between the protection of public health and whole-someness in terms of taste and odour. This is an operational problem which can be solved in a number of ways, such as using more sensitive disinfection equipment at the treatment plant, disinfecting within the supply zone or distribution network and by using alternative disinfectants.

Ammonia reacts with chlorine to produce three chloramines (monochloramine, dichloramine and trichloramine or nitrogen trichloride). These compounds are more odorous than free chlorine and become progressively more offensive with increasing numbers of chlorine atoms with trichloramine by far the worse. The proportion of each compound formed depends on the relative proportions of chlorine and ammonia present, the chlorine demand exerted by other substances and the pH (Montgomery, 1975). This is usually avoided by using breakpoint chlorination in which a high chlorine to ammonia ratio is used ($>7.6{:}1$ at pH 7–8) so that the residual chlorine present is mainly in the free form (Section 3.2). It is not only ammonia which reacts in this way to produce odorous compounds, although the reactions are considerably slower and continue within the distribution system (Section 6.1).

Phenolic compounds have already been mentioned in Section 4.4. They produce odorous compounds during chlorination. Phenol itself has very little odour, but its chlorinated forms monochlorophenol and dichlorophenol have an intense odour and are difficult to remove by subsequent treatment. Trichorophenol is the least objectionable, causing little odour, so by adjusting the chlorine to phenol ratio the formation of the more odorous compounds can be avoided. Alternative disinfectants are usually used if phenol is detected in the raw water until its source can be identified and eliminated. However, the chlorination of phenol can cause odour problems due to the formation of chlorophenols even when the phenol in the raw water is below the detection limit. For example, the phenolic compound p-cresol has an odour threshold concentration of 55 $\mu g/l$ compared with just 2 $\mu g/l$ for dichlorophenol. Many phenolic compounds occur naturally and are found in lowland rivers, making odour problems in the treated water inevitable at certain times of the year.

Algae such as *Synura* produce extracellular products which are themselves odorous. However, such odours can be enhanced by chlorination, thereby intensifying the original problem. Algae and actinomycetes accumulate and even grow on the filter media of sand filters and can impart associated odours and tastes if not kept under control. They produce a range of odorous compounds, in particular geosmin and 2-methylisoborneol (Table 4.10). Regular back-washing controls the development of these organisms, but as some odorous compounds which can be produced, such as dimethyldisulphide, have an odour threshold concentration of just 10 ng/l, the water used for back-washing should not be returned to the start of the plant for treatment if such compounds are suspected of being present as this would cause odour problems in the finished water. Recent research in this area is reported by Pearson *et al.* (1992).

5.4.2 Removing Odours and Tastes

Conventional water treatment, apart from possibly adding odours to drinking water, is not very effective at removing them, although the biological activity which occurs in slow sand filters may remove (oxidize) some odorous compounds. It may be possible to prevent odour problems arising by carefully managing the raw water source more effectively by ensuring it is kept free from industrial pollution, that phenolic and nitrogenous compounds that could react with chlorine are eliminated, and that the growth of algae and other organisms which could produce odours are controlled. Access to an alternative water resource, free of odour problems, is less likely, but would be effective.

Where odours and tastes are a problem, specific treatment methods must be used (Mallevialle and Suffet, 1987). There are three options available: (1) adsorption, either physical adsorption onto activated carbon or biological adsorption using a biofilm which grows on a natural humus material such as peat; (2) aeration, which strips any volatile odorous compounds from the water; and (3) chemical oxidation using ozone, potassium permanganate, chlorine or chlorine dioxide to chemically break down odorous compounds into non-odorous forms.

The most effective method is adsorption using activated carbon, as this removes a wide range of odours and is especially effective against the two most problematic odours. These are the musty/earthy odours produced by actinomycetes, the compounds geosmin and 2-methylisoborneol. In most instances odour problems only occur once or twice a year due to seasonal peaks in algal or actinomycete growth. Therefore activated carbon is normally used in a powdered form and added to the water stream as a slurry before sedimentation or rapid sand filtration (Najm *et al.*, 1991). The powdered activated carbon (PAC) is disposed of with the normal waterworks sludge. The use of PAC is more cost effective than using GAC in the short term as it does not require the expensive construction of a special bed or filter to house the GAC nor facilities to regenerate it (Section 3.2). However, if the odour problem is permanent then a GAC system will be required as it is more cost effective in the long term due to low operating costs, even though the initial capital costs are high. Therefore PAC systems are only suitable for short-term, intermittent use. The efficiency of activated carbon in removing odorous compounds is severely affected by the presence of other organic compounds such as natural humic substances. This can be a particular problem with GAC as it increases the operational costs significantly as the bed must be taken out of service more often.

There is currently much interest in the use of biofilms grown on natural organic polymers to remove odours. The action is the same as activated carbon, by adsorption of the odour-producing compound, but instead of then having to reactivate the carbon by heat, the biofilm, which is composed of bacteria and other micro-organisms, degrades the odorous chemicals so that the system never needs to be shut down. This makes biofilm systems much cheaper to operate than conventional activated carbon units. However, there is the possibility that odour-producing organisms may themselves colonize these biofilms.

Aeration can be successful in controlling odours caused by volatile organic compounds or dissolved gases such as hydrogen sulphide. It is also effective in controlling trichloramine. However, it is unable to remove the main odorous compounds and so is of limited use.

Chemicals can be used to oxidize organic compounds as they disinfect the water. None is effective against all odours, with each particularly effective against a specific range of odorous compounds. For example, chlorine can control sulphides in groundwater, but will generally make the odour problem worse for the reasons already discussed. It is also effective against organic sulphur compounds such as dimethyl-trisulphide, which causes a fishy odour. Chlorine dioxide can be used under certain circumstances instead of chlorine to disinfect raw water. If phenolic compounds are suspected then chlorine dioxide can be used without chlorophenolic odours being formed. It is also effective at reducing all other odours except those caused by hydrocarbons in the water. Potassium permanganate is used widely in the USA for odour control, but not in the UK. It is applied before filtration, when it is reduced to manganese oxide which is insoluble and so forms a precipitate which helps to remove odours. The precipitate is removed by filtration, but is now considered to be inefficient. Ozone can reduce or change the nature of odours fairly effectively. The problem, however, is that the ozone literally breaks up the large organic molecules into smaller ones, which are much easier for micro-organisms to use as a food source. This tends to encourage microbial growth in the distribution system. Ozonation has also been reported to produce rather intense fruity odours.

5.5 CHLORINATION BY-PRODUCTS

5.5.1 Formation

Since the beginning of the century chlorine has been used to disinfect drinking water. Although it has provided an effective barrier to the spread of water-borne diseases, chlorine is also very reactive towards the natural compounds present in water. Ammonia and the humic acids which give peaty water its clear brown colour interfere with the disinfection process, whereas other compounds such as phenol react with chlorine to affect both taste and odour (Section 5.4.1).

In the early 1970s another problem related to chlorination was identified. This coincided with the development of new analytical methods such as gas chromatography coupled with mass spectrometry, which can identify the organic compounds present in water. This new technology revealed that water could contain hundreds of natural and man-made organic compounds, most of which were present at only very low concentrations of less than 1 μg/l. It was discovered that some of these organic compounds could react with the chlorine during the disinfection process at the treatment plant to form new, complex and often dangerous chemicals. Chloroform and other trihalomethanes (THMs) were first identified in chlorinated drinking water taken from the River Rhine (Rook, 1974).

Trihalomethanes are simple, single carbon compounds which have the general formulae CHX_3, where X may be any halogen atom (i.e. chlorine, bromine, fluorine, iodine, or a combination of several of these). They are all considered to be possible carcinogens and are therefore undesirable in drinking water. There are four common THMs found in drinking water: chloroform ($CHCl_3$), bromodichloromethane ($CHBrCl_2$) dibromochloromethane ($CHBr_2Cl$) and bromoform ($CHBr_3$). The THMs and other chlorinated by-products are only found in raw or treated waters which are disinfected using chlorine. For a given chlorine dose the rate, and also the degree, of THM formation is increased at higher concentrations of humus, higher temperatures and high pH values. Where a free chlorine residual exists in the water (Section 3.2), THM formation will continue. The chlorine to substrate concentration ratio (i.e. the amount of chlorine added in relation to the amount of organic matter present) is also an important factor in determining which by-products are formed. For example, at low chlorine dose rates phenol is converted to taste-producing chlorophenols, whereas at higher dosages these are converted to tasteless chlorinated quinones.

The presence of other halogens, especially bromine, in the water is also important.

Bromide is oxidized by chlorine to hypobromous acid, which then results in brominated analogues of the chlorinated by-products. For example, bromoform is the analogue of chloroform (Fawell *et al.*, 1987). The reaction pathways are extremely complex (Peters *et al.*, 1978). Chloroform is the major THM found in chlorinated drinking waters and concentrations of 30–60 $\mu g/l$ are common, with maximum levels of up to 500 $\mu g/l$ recorded. In contrast, bromoform is typically found at much lower concentrations of < 10 $\mu g/l$. The occurrence and concentrations of by-products are reviewed by Cable and Fielding (1992).

Since the early 1970s a huge range of compounds has been identified which are formed by chlorine reacting with natural humic material and other organic compounds such as amino acids that are found in water (Table 5.3). Other commonly occurring

Table 5.3. By-products of chlorination found in drinking water

Chlorine
  Trihalomethanes
    Chloroform (trichloromethane)*
    Bromoform (tribromomethane)*
    Bromodichloromethane (BDCM)*
    Dibromochloromethane (DBCM)*
  Other trihalomethanes: chlorinated acetic acids
    Monochloroacetic acid (MCA)*
    Dichloroacetic acid (DCA)*
    Trichloroacetic acid (TCA)*
  Other trihalomethanes: haloacetonitriles
    Dichloroacetonitrile (DCAN)*
    Dibromoacetonitrile (DBAN)*
    Bromochloroacetonitrile (BCAN)*
    Trichloroacetonitrile (TCAN)*
  Other
    Chloral hydrate*
    3-Chloro-4-(dichloromethyl)-5-hydoxy-2(5H)-furanone (MX)*
    Chloropicrin*
    Chlorophenols*
    Chloropropanones*

Ozone
  Bromate*
  Formaldehyde*
  Acetaldehyde
  Non-chlorinated aldehydes
  Carboxylic acids
  Hydrogen peroxide

Chlorine dioxide
  Chlorite*
  Chlorate*

Chloramination
  Cyanogen chloride*

*Included in the 1993 World Health Organization drinking water guidelines.

chlorinated by-products include di- and trichloroacetic acids, chloral, dichloroacetonitrile, chloropicrin, chlorinated acetones, 2-chloropropenal and in most instances their brominated analogues. Although many of these by-products are mutagenic they are generally found in very low concentrations of 1 μg/l or less, with the major compounds such as the haloforms (chloroform and bromoform) and di- and trichloroacetic acids normally present in water at 1–50 μg/l. However, some by-products are very potent mutagens; for example 3-chloro-4-(dichloromethyl)-5-hydroxy-2(5H)-furanone (or MX for short). The presence of MX in drinking water at any concentration, no matter how low, is therefore undesirable. MX is currently found in UK drinking waters at concentrations ranging from 1 to 60 ng/l, depending on the humic material present. It is responsible for between 30 and 60% of the total mutagenic potential of water (Horth *et al.*, 1991).

Unlike THMs, dihaloacetonitriles (DHANs) such as dichloroacetonitrile (DCAN) are unstable and readily broken down (hydrolysed) in aqueous solutions at increased temperatures and pH values, which can occur during storage or analysis. Details of concentrations in drinking water are therefore rare. The most common DHAN is DCAN, which was first reported in drinking water in 1976 (McKinney *et al.*, 1976), with its brominated analogues bromochloroacetonitrile and dibromoacetonitrile discovered several years later. Studies in The Netherlands have shown that all chlorinated drinking waters contain DHANs as well as THMs, but in much lower concentrations within the range 0.04–1.05 μg/l. There appears to be a direct correlation between THMs and DHANs, with the average DHAN concentration about 5% of the average THM concentration (Peters *et al.*, 1990). However, in The Netherlands chlorine dosage rates are five to 10 times lower than in the USA or Canada, typically about 0.2–1.0 mg/l. This is why both THM and DHAN concentrations are so much higher in North America, with DHANs found at concentrations of 1–10 μg/l.

5.5.2 Standards

It is remarkable but there is no explicit EC standard for THMs. The problem appears to be that THMs fall outside the provision in the Drinking Water Directive for the category pesticides and related products, which only includes persistent organochlorine compounds; THMs are not persistent in the intended sense and so are excluded. There is a G value in the directive of 1 μg/l for other organochlorine compounds not included in the list of pesticides and related products, and there is also the provision that 'haloform concentrations must be as low as possible'. There is no mandatory EC limit. However, some countries have set national limits for THMs. In Britain it is 100 μg/l, whereas in Germany the national limit is 25 μg/l. The USA limit was 100 μg/l, but in 1987 the US National Research Council advised that this may be too high and recommended a reduction (Table 1.12). The World Health Organization has set guide values for the disinfection by-products indicated in Table 5.3. These are reviewed in Table 1.16.

In the UK the 100 μg/l limit is being met by most supplies; however, there are areas where THM concentrations of almost double the national limit have been recorded. In a survey carried out by Friends of the Earth and published in the *Observer Magazine* in August 1989, data on THM concentrations from water supplies throughout England and Wales showed that the limit was exceeded at some time during the previous two years in no less than 82 local supply areas. They identified a number of particular black spots including Cheltenham, Gloucester and Stroud, which had concentrations of THMs up to 165 $\mu g/l$.

5.5.3 Health Risks

Most of our information relating to the effects of THMs on human health is based largely on chloroform. However, the toxic effects of other THMs will probably be similar to chloroform. The lethal dose of chloroform is about 630 mg/kg body weight, so about 44 g is required to kill an adult man weighing approximately 70 kg. Laboratory studies on animals have shown that smaller doses are mutagenic and can induce cancer, although epidemiological studies have not positively identified a strong causal relationship between THMs and cancer. There is some evidence to suggest an association between long-term, low-level exposure to THMs in drinking water with rectal, intestinal and bladder cancer (World Health Organization, 1984). It is clear that many THMs are carcinogenic at high doses in laboratory animals and also that certain chlorinated by-products are more dangerous than others, such as MX. It therefore appears prudent to try to reduce the levels of these compounds in our drinking water to as low a concentration as possible. The health effects of disinfection by-products are reviewed by Bull and Kopfler (1991).

5.5.4 Prevention of By-product Formation

A high organic matter content in water supplies is not only due to waters rich in humic acids which drain from peaty moorlands and feed upland reservoirs and supply streams. It is also a problem in nutrient-rich lowland rivers. These supplies are most at risk from THM formation due to chlorination. The greater the chlorine dose, the greater the risk of producing THMs. Although some advances have been made in removing chlorination by-products using GAC filters (Lykins *et al.*, 1988) and other advanced techniques, it is expensive. Also, like all other water treatment processes it can result in undesirable changes in water quality.

The risk from THMs appears to be very low. The World Health Organization (1984) estimates that the risk of an additional cancer due to a lifetime exposure to a carcinogen such as chloroform in drinking water, assuming a concentration of 30 $\mu g/l$ and a daily consumption rate of two litres, is less than 1 in 100 000. Clearly the best option is to control chlorination rates as carefully as possible. In those areas where there is a high concentration of organic matter which can react with chlorine to form THMs, then the use of chlorine dioxide or chloramination should be considered, or alternative methods of disinfection such as ozonation (Jacangelo *et al.*, 1989) or ultraviolet radiation (Section 3.2) examined. The risk associated with inadequate disinfection would be much greater than that resulting from concentrations of chloroform and other THMs well in excess of the World Health Organization's guideline value of 30 $\mu g/l$.

Alternative disinfectants to chlorine include ozone, chlorine dioxide and chloramine; however, they also produce various by-products (Table 5.3). Ozone does not form THMs, but does not have a residual disinfection potential. It is so reactive that oxidation products are also formed. Those causing most concern are formaldehyde, acetaldehyde, other non-chlorinated aldehydes and carboxylic acids. Bromide, if present, will be oxidized during ozonation to bromate, which is thought to be carcinogenic, and other brominated compounds such as bromoform (Jancangelo et al., 1989). So if chlorination is used after ozonation then other THMs will be formed such as BDCM and DBCM. Chlorine dioxide is an excellent disinfectant and does not generally react with organic matter to form chlorinated by-products (Lykins and Griese, 1986). However, it is often contaminated with chlorine, which forms THMs. In practice, as the chlorine dioxide oxidizes any organic matter in the water it is reduced to chlorite and chlorate, both of which are toxic. The same by-products are formed as with chlorine, except as chlorine dioxide is far less reactive significantly lower concentrations of by-product are formed. Although less efficient than free chlorine, chloramination is becoming more widely adopted as a disinfection method. It uses chlorine and ammonia to produce primarily monochloramine. A significant problem is the excessive use of ammonia to form monochloramine, resulting in the excess ammonia being converted to nitrite in the distribution system (Section 5.1). Cyanogen chlorine is possibly the major by-product of chloramination. It is metabolized rapidly in the body to cyanide and thiocyanate (Krasner et al., 1989).

As by-products are not readily removed by existing treatment processes, attention is being paid to their formation by the more effective removal of dissolved organic matter before disinfection. This is primarily achieved by coagulation, which can remove substantial amounts of organic matter.

5.6 FLUORIDATION

The fluoridation of water supplies has been as controversial in the past as aluminium in water is today. It was introduced in the 1940s to reduce the incidence of tooth decay in the population after a number of surveys in the USA had shown that it had a beneficial effect. Three relationships were identified: (1) fluoride levels in excess of 1.5 mg/l lead to an increase in the occurrence and severity of dental fluorosis (teeth become mottled and brittle) without decreasing the incidence of decay, missing or filled teeth; (2) at 1.0 mg/l there was the maximum reduction of decay with no fluorosis; and (3) at concentrations less than 1.0 mg/l some benefit was observed. With decreasing concentrations of fluoride in water there was an increase in the incidence of tooth decay.

These studies therefore showed that the addition of fluoride to water supplies to bring the level above 0.6 mg/l led to a reduction in tooth decay in growing children, and that the optimum beneficial effect occurred around 1.0 mg/l. Other studies have indicated that fluoride is also beneficial to older people in reducing the hardening of the arteries and, as fluoride stimulates bone formation, in the treatment of osteoporosis,

Table 5.4. Areas in England which had fluoridated water supplies in 1988. Reproduced with permission from the Water Services Association

Some of the population in district		All population in district
East Cumbria	Doncaster	North West Durham
West Cumbria	Dudley	Gateshead
Durham	Wolverhampton	Newcastle
South West Durham	Worcester	Bassetlaw
Northumberland	Bromsgrove and Redditch	Sandwell
North Tyneside	North Warwickshire	Walsall
Grimsby	South Warwickshire	Solihull
Scunthorpe	Rugby	North Birmingham
Huddersfield	Coventry	East Birmingham
Chester	Mid Staffordshire	Central Birmingham
Crewe	East Staffordshire	West Birmingham
North Derbyshire	Oxford	South Birmingham
South Derbyshire	Wycombe	
North Lincolnshire	North Bedfordshire	
South Lincolnshire	South Bedfordshire	
Central Nottingham	South West Hertfordshire	

although the most recent evidence contradicts this (Cooper *et al.*, 1990). These early studies were so convincing that fluoridation was adopted around the world, so that today over 250 million people drink artificially fluoridated water. In the USA, for example, over half of the water supplies are fluoridated. It is widely used in New Zealand, Canada and Australia; in fact, it is widespread in most English-speaking countries. Brazil and the former Soviet Union are two other countries with a strict fluoridation policy. In the Republic of Ireland all water supplies are fluoridated where the natural levels are less than 1.0 mg/l, which is also the case in Greece. None are fluoridated in Northern Ireland and only about 10% of supplies are fluoridated in the UK (Table 5.4). In western continental Europe its use has been terminated or never implemented, primarily over health concerns and less than 1% of Europeans drink artificially fluoridated water. For example, in The Netherlands fluoridation was recently phased out and it is also banned in Sweden. In contrast, Spain has been gradually introducing fluoridation since 1985 starting in Andalucia, Pais Vasco and Gerona City (1988), and most recently in Galicia in 1990.

Many areas have naturally occurring fluoride in water (Figure 5.6). In some areas high natural levels of fluoride have to be reduced by mixing with low fluoride water to bring the concentration to less than 0.9 mg/l.

The EC Drinking Water Directive sets MAC values for fluoride at 1.5 mg/l at 8–12°C, or 0.7 mg/l at 25–30°C. This MAC value of 1.5 mg/l has been adopted in the UK, although in the Republic of Ireland a national limit of 1.0 mg/l has been set. The effects of fluoride vary with temperature, so as the temperature increases the concentration permitted in drinking water decreases. Some bottled mineral waters may also contain high levels of fluoride (Section 8.1). The USEPA has set two standards for fluoride: a Primary Drinking Water Standard of 4 mg/l to protect against skeletal fluorosis, and a non-mandatory Secondary Drinking Water Standard of 2.0 mg/l to

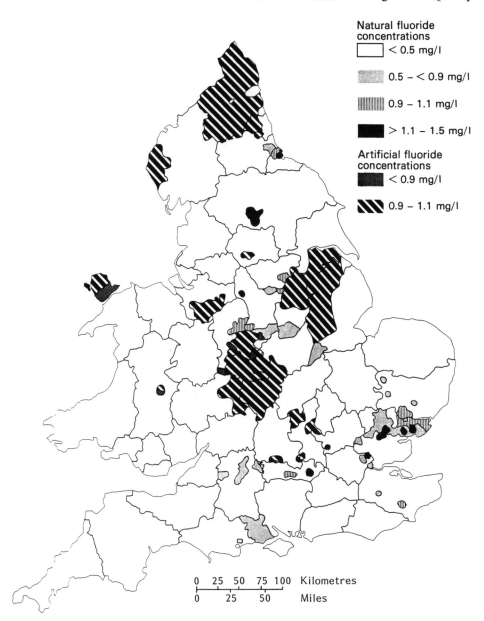

Figure 5.6. Natural and artificial fluoride concentrations in water supplies in England and Wales for the year 1988. Reproduced from the Department of the Environment (1990b) with the permission of the Controller of Her Majesty's Stationery Office. More detailed information is given in Department of the Environment (1993)

prevent dental fluorosis (Table 1.12), both of which are under review. The World Health Organization has set a revised guide value of 1.5 mg/l (Table 1.16).

5.6.1 Fluoride Addition

Fluoride must be added after coagulation, lime softening and activated carbon treatment as it can be lost during any of these unit processes. Filtration will not reduce the fluoride concentration provided it is fully dissolved before reaching the filter. As chlorination has no effect on fluoride, fluoridation is normally carried out on the finished water either before or after chlorination.

Fluorine in its pure elemental state is very reactive and so is always used as a compound. Those most widely used for the fluoridation of supplies are: ammonium fluorosilicate [$(NH_4)_2SiF_6$], which is supplied as white crystals; fluorspar or calcium fluoride (CaF_2) as a powder; fluorosilicic acid (H_2SF_6) as a highly corrosive liquid which must be kept in rubber-lined drums or tanks; sodium fluoride (NaF), which is either a blue or white powder; and sodium silicofluoride (Na_2SiF_6), which is also either a white or blue powder. All these compounds dissociate (break up) in water to yield fluoride ions, with the release of the fluoride being almost complete. Silicofluorides are generally used in water treatment in the British Isles, with sodium silicofluoride the cheapest and most convenient to use. The reaction which takes place when it is added to water is:

$$SiF_6 + 3H_2O \rightleftharpoons 6F^- + 6H^+ + SiO_3$$
$$\text{Silicofluoride} \qquad \text{Fluoride} \qquad \text{Silicate}$$

The compound is emptied into hoppers and then slowly discharged into a tank by a screw feed where it is mixed with water to dissolve it. After correction of the pH the fluoride solution is added to the finished water. Fluoride is removed by precipitation with calcium and excess aluminium coagulant in the finished water, with the precipitate forming in the distribution system. Care is taken to ensure that the dosage is kept at the correct level. This is done by monitoring the weight of chemical used each day and the volume of water supplied. Also fluoride concentration is tested daily by the operator, normally using a simple test kit, or continuously using an ion-specific electrode and meter to ensure the final concentration is as near to 1.0 mg/l as possible.

5.6.2 The Case Against Fluoridation

Excess fluoride causes teeth to become discoloured (fluorosis) and long-term exposure results in permanent grey to black discoloration of the enamel. Children who drink water containing fluoride in excess of 5 mg/l also develop severe pitting of the enamel. Where excess fluoride is a problem, defluoridation is required. Leaks and spillages of fluoride can be very serious and even fatal.

Fluoridation has been universally heralded as being a useful and positive process. However, new studies have shown that when similar communities in developed countries are compared there is little difference in the levels of tooth decay between communities using fluoridated and unfluoridated water. This has been supported by a study carried out in the UK and published in 1990 which looked at 105 292 children in 170 health districts. The primary explanation given is the now widespread use of fluoridated toothpaste and better dental hygiene.

The major force gainst fluoridation in the UK was the National Anti-fluoridation Campaign, which was formed in 1960, although they changed their name in 1990 to the National Pure Water Association and are now concerned with all aspects of water quality, although fluoride is still one of their major issues. Their opposition to fluoridation focuses on the following points: (1) fluoride is one of the most toxic inorganic poisons known; (2) it is a precipitative and cumulative drug as well as being an industrial waste product; (3) as fluoride is in the drinking water you have no control over your daily intake; (4) it does not boil off and so it is both difficult and expensive to avoid; (5) fluoride has been rejected because of its damage to health by 13 western European countries; (6) fluoride does not reduce the rate of tooth decay, which can be attributed to other factors including a better diet; (7) the function of public water is to provide safe drinking water, not to serve as a vehicle for dispensing drugs to treat people.

You must make up your own mind about these claims. What is certain is that there is evidence to show that fluoridation does not make a significant difference to the incidence of tooth decay (USPHS, 1991); the incidence of dental fluorosis has increased since fluoridation in some areas; sensitivity to fluoride, which is well known from patients receiving fluoride treatment for osteoporosis, is also seen occasionally in people who drink large amounts of fluoridated water (e.g. avid tea and coffee drinkers); and preliminary studies carried out by the USEPA have indicated that fluoride may be a carcinogen, although the available epidemiological evidence has not found a causal link between the fluoridation of drinking water and cancer (Chilvers and Conway, 1985). Fluoride is cumulative and so any long-term effects of low-dose exposure may not yet be realized. All these findings need more research, as do the claims of the anti-fluoridation campaigners who are found all around the world.

Those most at risk are formulae-fed infants and children who drink mainly tap water. Other people such as outdoor workers, long-distance runners, people with diabetes insipidus, and in fact anyone who drinks large amounts of tap water are at risk. Other groups who are at even greater risk are those with malfunctioning kidneys who store excess fluoride in their bones, a condition known as skeletal fluorosis. The reason for the fluoridation of drinking water was to reduce the level of dental decay in children. However, after the age of 12 years calcification of teeth stops so that fluoride is of no further benefit. Therefore it does seem more appropriate to prescribe fluoride drugs solely for children rather than this mass-medication approach. Fluoride is widely available in toothpaste, as well as in other forms, so there is now a ready source of fluoride for children.

An incident in Glasgow highlighted the problem of giving children fluoride tablets and then using fluoride toothpaste. This resulted in dental fluorosis. Some medical officers have indicated that fluoride toothpaste is not necessary when water has been fluoridated, or if it is used then only small amounts of toothpaste should be placed on brushes to avoid the possibility of dental fluorosis, albeit unlikely. However, manufacturers of toothpaste are advising parents to use only a pea sized amount of fluoride toothpaste when fluoride supplements are taken.

Accidents have occurred resulting in overdosing of fluoride, with reported levels at the consumers' tap of between 30 and 1000 mg/l. Excessive fluoride will lower the pH of the water increasing its corrosiveness and thereby increasing the concentration of lead and copper in the water. The general effect of moderately increased fluoride (30–50 mg/l) will be mild gastroenteritis and possible skin irritation (Peterson *et al.*, 1988).

In the UK the Water (Fluoridation) Act 1985, now incorporated into the Water Industry Act 1991, does not impose any obligation on water companies to fluoridate water supplies when requested to do so by health authorities. However, many of the agreements to fluoridate water supplies were made before the present Act was passed, and it appears that the water supply companies may be forced to comply.

Chapter 6
Problems Arising in the Distribution System

6.1 ODOUR AND TASTE

Considerable odour and taste problems develop while the water is in the distribution system due to the material from which the mains are constructed, or the effect of biological growths on the walls of the pipes. However, most problems originate either in the raw water (Section 4.4) or from the treatment plant (Section 5.4). The water supply company is careful to eliminate these as possible sources of odour before testing the water within the distribution system to isolate the problem area (Figure 4.5).

The major complaint arising from the distribution system is that of musty/earthy odours. These are due to the development of micro-organisms on the walls of the distribution network. Although actinomycetes and fungi are generally responsible, heterotrophic bacteria can also cause similar problems if present in large numbers. This suggests that a high activity of any micro-organism, regardless of its identity, may result in odours. A number of steps can be taken to reduce microbial growth within the distribution system, the main one being to ensure adequate disinfection with sufficient residual chlorine. Other methods include reducing the amount of organic matter and nutrients in the water by more effective treatment (Section 6.7) and using pipe materials which do not encourage microbial development. The growth of actinomycetes and fungi is controlled primarily by temperature, with optimum growth occurring at 25°C. At temperatures below 16°C growth is so reduced that complaints due to odour are generally eliminated. It is therefore essential to prevent water in distribution systems from standing for long periods and warming up. During warm weather conditions it is possible for many supplies to reach ambient temperatures as a consequence of this long residence time, thus encouraging the unwanted growth of micro-organisms. Long residence times also encourage organic material to flocculate and settle, which then acts as a source of food for micro-organisms. The water supply companies are careful not to over-design distribution systems to ensure constant and rapid movement of water, and avoid spur mains wherever possible (Section 3.3).

Free chlorine will slowly convert ammonia or monochloramine into odorous dichloramine, a process known as disproportionation (Section 3.2). As a consequence dichloramine concentrations tend to increase towards the end of, or the extremities of, the distribution network. The reaction is considerably enhanced by lower pH values so that chloramine odours are often produced where waters from different origins are mixed, which often occurs in larger cities where the water may be coming from several different sources.

Iron is a major problem for a number of the larger water supply companies. It can originate from the raw water, from the use of iron salts as coagulants during treatment, or more commonly due to the corrosion of old iron mains. Iron imparts an unpleasant taste to the water and has a mean taste threshold concentration of 3 mg/l, although for the most sensitive 5% of the population this threshold concentration decreases to a staggering 40 μg/l. Clearly what is potable to the majority will not be to those who are more sensitive to tastes and odours. In response to this particular problem there is a major programme of mains replacement and renovation (lining with a plastic coating). For example, Severn Trent Water are currently upgrading over 1000 km of iron mains each year, and similar improvement programmes are being undertaken by all the water supply companies. In 1991, 31% of the 2577 water supply zones in England and Wales failed to comply with the PCV for iron. The number of zones covered by undertakings to the Drinking Water Inspectorate is currently 2388.

Iron mains are not the only material to cause problems. Odours are occasionally released from older lining materials, especially bituminous-based compounds which release naphthalene. This gives the water a strong oily odour (Section 6.5).

6.2 DISCOLORATION AND IRON

Poor treatment plant operation can result in particulate matter getting into the distribution system and causing discoloration and other problems. The main cause is the use of excessive amounts of coagulant, which results in either iron or aluminium entering the system, although manganese, algae and organic matter can all arise from poor treatment. As the contaminated water passes through the distribution system, deposition and sedimentation take place, thereby reducing the concentration of these substances downstream. Examples are given in the following to illustrate this from different distribution networks for three metals: aluminium, manganese and silica (Figure 6.1).

Even the most efficient treatment plant can only remove colloidal and suspended organic matter, with dissolved organic material only slightly reduced (20–25%) by conventional treatment. It would not be uncommon to have 10 mg/l of organic matter, all in soluble form, entering the distribution system (Section 6.7). When organic matter enters the distribution system it is fairly rapidly reduced. This is due to a number of mechanisms such as co-precipitation with iron, manganese or aluminium, adsorption onto corrosion products or utilization by micro-organisms. Whichever mechanism is responsible, the organic matter is taken from solution and concentrated in the deposits forming a thin layer of loose debris; this is easily flushed out at higher flow-rates. This debris is the food source for a variety of animals and micro-organisms which find a home in the water mains (Section 6.6). The actual amount of organic matter will depend on the water source. For example, groundwaters and very pure surface waters will contain very little organic matter in solution so there will be little or no related activity in the mains. Upland sources from peaty moorland areas are rich in humic

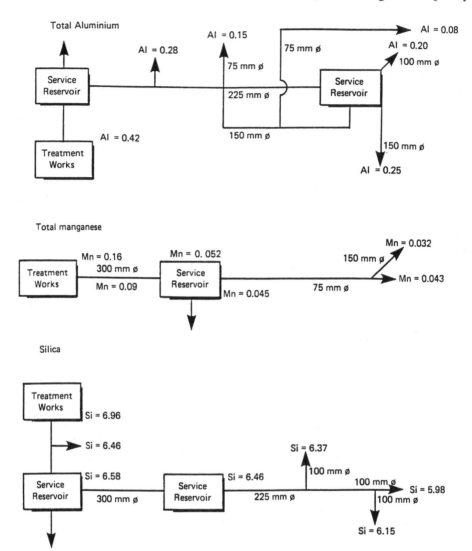

Figure 6.1. Schematic diagram showing variation in quality in three cast iron distribution water mains networks. Concentrations of aluminium, silica and manganese in mg/l and pipe diameters in mm. Reproduced from Ainsworth *et al.* (1981) by permission of the Water Research Centre

compounds, which gives the water a moderate organic matter concentration. This water shows a slight decrease in organic matter content as it passes through the distribution system, suggesting some activity. Lowland surface waters contain high concentrations of dissolved organic matter which is significantly reduced as it passes through the mains. If this organic matter supports excessive microbial growth, normally as a slime which covers the walls of the pipework, discoloration as well as taste and odour problems can result (Section 6.1). Micro-organisms are likely to cause discoloration if counts exceed 10^3/ml, although this is unlikely in either groundwaters or high quality surface waters. The main source of bacteria will be lowland surface water supplies, where counts can exceed 10^6/ml. This high level of bacterial activity can cause a number of other related problems (Section 6.7).

Although iron may originate from the use of coagulants, increases in the total iron content of a water supply as it passes through the mains is generally caused by corrosion (Section 4.5). Temporary increases in iron are caused by mechanical activity and disturbing existing deposits. Corrosion is a major source of consumer complaints. Galvanic corrosion is explained in Section 7.2 and occurs where different metals or alloys are coupled together in water. Most water mains are made of cast iron, and these are also susceptible to corrosion, although this proceeds more slowly. As it corrodes, iron (Fe^{2+}) is released into solution (De Rosa and Parkinson, 1986).

The major factors controlling corrosion rates are those related to the nature of the water itself. These include: (1) increases in alkalinity or hardness (concentration of calcium and magnesium ions) of the water, which results in a decrease in the corrosion of iron; (2) increases in the concentration of chloride or sulphate, which increases the corrosion rate, although in the case of chloride the presence of hydrogencarbonate (HCO_3^-) significantly reduces its corrosivity; (3) the deposition of sediment and the growth of microbial organisms on the pipe, leading to oxygen depletion which results in localized corrosion under such deposits and growths; (4) corrosion caused by bacterial action (e.g. sulphate-reducing bacteria)—this is more common in lowland surface supplies which contain the significant amounts of organic matter needed to maintain such populations of micro-organisms; and (5) water with a low dissolved oxygen concentration (<4 mg/l) can also produce enhanced corrosion.

Soft moorland water is therefore more corrosive than hard lowland water. This is supported by measurements of the depth that corrosion penetrates into iron mains. The mean depth that soft, acidic water penetrated was 60 μm/year, which is four times greater than the 15 μg/year recorded for hard groundwaters. Most iron pipes are protected by a lining material. Bituminous materials such as coal tar were commonly used in the past at a thickness of about 100 μm. However, if the water is very aggressive, then this material only offers temporary protection. Plastic liners are now used. Any holes in the lining, or any cracks in the inner skin of iron oxides, silicates or aluminosilicates which is formed during casting and acts as a protective barrier against corrosion of the metal underneath, or any abraded areas, will become areas of corrosion.

Three types of water resources have been identified as being more likely to corrode metallic pipes and therefore cause discoloration problems. These are: (1) soft waters, which are normally derived from upland catchment areas and have a hardness and alkalinity of less than 50 mg/l—these currently represent a third of the total volume of UK supplies; (2) chloride- and sulphate-rich waters—owing to interactions with other ions, especially hydrogencarbonate, no fixed concentration can be set for corrosivity, although 50 mg/l of either chloride or sulphate should be considered as possibly corrosive (this is a particular problem where saline intrusion occurs in boreholes in coastal areas); and (3) anaerobic waters—water with less than 4 mg/l dissolved oxygen can become corrosive and cause discoloration complaints. Some groundwaters may fall into this category.

If discoloration is not being caused by problems at the water treatment plant or by microbial action within the distribution mains, discoloration due to rusty deposits is most likely to be due to corrosion where the above types of water supplies are involved. However, most supplies suffer from corrosion and even hard waters will corrode iron mains if they contain high levels of organic matter; this is especially true if they contain high concentrations of chlorides and/or sulphates, which is possible in some coastal locations or if the water is being reused. Chloride is not removed during sewage or water treatment. Every packet of salt we buy ends up down the drain because chloride is not retained to any significant extent by the body. Urine contains 1% chloride, so as the water is used and reused for drinking, the level of chloride builds up significantly, which can increase the corrosion potential of the water.

Population centres in mainland Europe are generally more dispersed, with water distribution systems not as extensive or integrated as in England. In northern Europe in particular, the distribution systems are much newer than those in the UK, so that mains causing the discoloration of water due to iron are more common in Britain. This problem is not naturally occurring, so although high natural levels of iron may be a cause of derogation, problems caused by distribution are not.

6.3 SEDIMENT AND TURBIDITY

Consumers often find fine sediment in the water from their taps. It is normally only noticed when a bath has been run or a glass of water left to stand. Sediment can consist of either organic matter, including micro-organisms, or insoluble material, mainly iron and manganese. Sediment is often the extreme problem of particulate matter being present in water, as usually only a slight cloudiness of the water (turbidity) will be noticed. The origins of these sediments have already been discussed elsewhere in this book, so only a summary is given here.

Particulate inorganic and organic matter, micro-organisms and algae in the raw water are generally removed at the treatment stage. However, due to operational problems it is possible for unfiltered water to occasionally pass into the distribution system. The presence of fine clay or silt particles and algae will increase the turbidity. If the turbidity exceeds 5 FTU then it is clearly visible in a glass of water and usually rejected by the consumer on aesthetic grounds. The UK water quality PCV standard for turbidity is 4 FTU, which includes suspended solids. This is the EC MAC value; the EC also has a requirement that there is no persistently visible suspended solids in the water (Table 1.10). Although inorganic particles such as clay do not usually affect health, taste may be effected. High turbidity may also warn the consumer of a treatment plant malfunction, which may mean that the water has not been fully treated or disinfected. Turbid waters should always be treated with suspicion. In some instances, such as with groundwaters, soluble iron and manganese may be allowed into the mains; they will then form insoluble particulate iron and manganese as the water is aerated and the pH increases. This especially occurs within the distribution system where water from another source is mixed, thus altering the chemical nature of the water and causing precipitation of the metals. An increase in the turbidity of groundwater after rain may indicate that surface runoff is getting into the borehole.

The solids, which are normally either brown or orange in colour, either settle in the distribution mains forming a deposit or are carried on to the consumer. The settled deposit in the mains will eventually be resuspended and also reach the consumers. Corrosion of water mains results in soluble ferrous iron (Fe^{2+}) entering the water. This may rapidly precipitate out in the mains adding to the pipe deposits, or more likely it will only form the insoluble ferric (Fe^{3+}) form as it enters the consumers' water storage tank or comes out of the tap. Corrosion may result in thick encrustations on the surface of the iron pipes from which small flakes and particles are constantly being broken off. Soluble organic matter is not fully removed at the treatment plant and is adsorbed by the existing sediment in the distribution mains and also by micro-organisms living on the walls of the mains and in the debris itself. This source of material is a particular problem in organically enriched lowland rivers, which are often used for supplies (Section 6.2). Sediment problems tend to occur at the end of distribution systems, especially at dead-ends—in fact, anywhere that the solids can settle and accumulate. This results in a brownish sediment in the water not unlike fine soil, and occasionally animals may also be associated with the sediment. This problem is exacerbated when repair and maintenance work is carried out on the distribution system, or if sections are cleaned due to low pressure problems. Cleaning is often carried out to remove sediment and slime accumulations on the walls of the pipes. It can be carried out using a variety of methods, including flushing out debris by opening hydrants and allowing large flows to scour out any material inside, by pushing a tight-fitting swab through the pipe by water pressure to wipe the pipe surfaces clear of slime and accumulated solids, or by air-scouring. The latter is achieved by pumping air into the water stream, which causes so much turbulence that all the debris is scoured from the pipe. The dirty water is allowed to run to waste at the next hydrant. Corrosion encrustations require a different and more erosive technique to scrape the surface of the pipe clean. This is usually only carried out before relining the pipe with epoxy resin, cement mortar or other permissible materials.

Within household plumbing systems sediment only really occurs from corrosion or by soluble iron and manganese being oxidized into their insoluble forms. These problems are more common in households receiving private groundwater supplies. Occasionally water can have a sandy sediment present. This is usually associated with the corrosion of galvanized storage tanks or galvanized pipework (Section 7.5).

6.4 ASBESTOS

6.4.1 Nature and Standards

Asbestos is the name given to a class of naturally occurring inorganic silicates. Its name is derived from the Greek meaning unquenchable or inextinguishable, which refers to

Table 6.1. Most important types of asbestos mined and used and the main cations linking the silicate structure

Mineral	Main cation(s) linking silicate structures
Serpentine group	
Chrysolite (white asbestos)	Mg^{2+}
Amphibole group	
Amosite	Fe^{2+}, Mg^{2+}
Crocidolite (blue asbestos)	Na^{2+}, Fe^{2+}, Mg^{2+}
Anthophyllite	Mg^{2+}, Fe^{2+}
Tremolite	Ca^{2+}, Mg^{2+}
Actinolite	Ca^{2+}, Mg^{2+}, Fe^{2+}

the fact that it cannot be destroyed by fire. It is a widely used material, with over 5×10^6 tonnes mined annually. Among its important properties are its resistance to fire, its thermal and electrical insulation properties and its binding capacity, which provides strength and stability when it is incorporated with cement or other materials or used in the formulation of brake linings and even floor tiles (Commins, 1979).

There are a number of different types of asbestos; the most important are listed in Table 6.1. Those most widely used are chrysolite, amosite and, in the past, crocidolite. These are the forms most commonly used in asbestos cement products including water pipes. Asbestos is found widely throughout the world, although the largest deposits mined are found in Canada, Russia, Africa, Australia and the USA.

Asbestos fibres in water supplies originate from a number of sources; for example, from the dissolution of natural asbestos mineral, industrial effluents, atmospheric fallout and the effects of aggressive and corrosive water on asbestos cement water distribution pipes and storage tanks. Once asbestos is released into the atmosphere it is inevitable that it will find its way into water, which acts as an irreversible sink for the fibres.

In Britain outcrops of chrysolite and tremolite are found in Snowdonia, Caernarvonshire, Cornwall, Hereford, Aberdeenshire, Banffshire, Inverness-shire, Skye, Sutherland, Jersey and Guernsey. Although none of these are mined commercially, where asbestos is mined on a large scale up to 1200×10^6 fibres/l can find their way into drinking waters. In the UK the major outcrops occur at Lochnager in Scotland and on the eastern slope of Moel Yr Ogof in Snowdonia, North Wales. Neither of these areas drains into water supply catchments, and so no studies have been carried out to determine if there is any significant natural runoff of asbestos.

Interest in asbestos was triggered in the USA and Canada by the discovery of relatively high levels of fibres in Lake Superior and in a number of water supplies in mining areas. This coincided with studies on the inhalation and ingestion of asbestos dust in occupational situations. Asbestos fibres found in water are generally very small. For example, chrysolite fibres are typically 0.5–2.0 μm long and only 0.03–0.1 um in diameter. Optical (light) microscopy is therefore inadequate for identifying or counting fibres in water. Asbestos analysis is complex and time consuming, involving a combination of transmission electron microscopy with either selected area electron diffraction or energy dispersive X-ray analysis. Routine monitoring of water supplies for asbestos is not carried out. No-one is looking at the specific problems of monitoring asbestos in drinking water nor automated methods of analysing fibre concentrations in water. The problems of sampling, sample preparation and analysis are reviewed by Commins (1988).

Very little work has been carried out on asbestos in drinking water, with most studies performed on inhalation effects. The effects of inhaling asbestos fibres are associated with lung disorders such as asbestosis, which is incapacitating, causing cancer of the lung and pleura, and cancer of the gastrointestinal tract. The lung cancer takes the form of bronchial lung carcinoma, whereas cancer of the pleura, which is the membrane surrounding the lung, produces a tumour known as mesothelioma. Gastrointestinal cancers are usually in the form of peritoneal tumours, the peritoneum being a membrane associated with the gastrointestinal region. All types of asbestos are potentially carcinogenic, although it is considered by most experts that crocidolite (blue asbestos) is the most dangerous. In fact, the importation of crocidolite into the UK has been banned since 1970, although it is still commonly encountered. Owing to the risk to health, atmospheric quality standards have been set for chrysolite at 2 fibres/ml of air and for crocidolite at 0.2 fibres/ml, although lower standards are expected to be set in the near future. Asbestos is ubiquitous in the environment and almost everyone is exposed to air-borne asbestos, albeit in very small amounts compared with those employed in mining or associated industries where asbestos is used. However, very few people develop mesothelioma—about 200 each year in the UK. Inhaled fibres can pass through the lung tissue into the bloodstream to other parts of the body such as the gastrointestinal tract. Inhaled fibres can also be subsequently ingested as they are brought up by ciliary activity from the lungs; also asbestos dust is rich in fibres. There is no consensus in the medical profession as to whether it is inhaled or ingested fibres that are responsible for the increased incidence of gastrointestinal cancer. This makes the present position on the health effects of asbestos in drinking water inconclusive, which is why no guidelines have been set by the EC specifically in relation to drinking water (Commins, 1988). However, the USEPA has included asbestos in the National Primary Drinking Water Regulations. The current maximum contaminant level set is 7×10^6 fibres/l (Table 1.13). However, the scientific basis for the selection of this limit is unclear. In the revised World Health Organization guidelines, asbestos has been included for the first time. It is categorized under 'chemicals not of health significance at concentrations normally found in drinking water'. It is grouped with silver and tin, for which no guide values have been set.

Because asbestos is so ubiquitous it can enter the water cycle by a number of different routes; most water supplies are found to contain asbestos fibres. Studies carried out in the UK, The Netherlands and Germany have indicated that levels in drinking water range from 0.2 to 2.0×10^6 fibres/l, with an average concentration of 1.0×10^6 fibres/l (Conway and Lacey, 1984). This is similar to findings in the USA and Canada, but in these countries asbestos is widely mined and so more widely used in allied industries. This results in concentrations in excess of 10×10^6 fibres/l not being uncommon (Chatfield and Dillon, 1979; Millette et al., 1979).

6.4.2 Asbestos Cement Distribution Pipes

Asbestos cement has been widely used to make low cost water pipes and water-holding tanks. The fibres act as a binder or filler in the cement and have been widely used in the UK for water distribution mains and in the construction of service reservoirs. It is not understood exactly what factors cause the fibres to be released from the pipes, whether it is mechanical or chemical, but increased levels of fibres were recorded by a study carried out in the UK (Conway and Lacey, 1984). Eighty-two potable supplies were examined after they had passed through asbestos cement pipes. All the waters were considered to be either aggressive or moderately aggressive (i.e. soft and acidic). The pipe lengths examined ranged from 375 to 10 500 m, with the age of the pipes ranging from seven to 35 years. After passing through the pipes, 17 samples contained fibre concentrations greater than 1×10^6 fibres/l, and four in excess of 3×10^6/l. As the fibre levels could not be attributed to natural sources it showed conclusively that they were from the pipework. The study showed that pipe length, age and the aggressiveness of the water were all important factors affecting the concentration of asbestos fibres in drinking water. After flushing the mains by opening hydrants, or if the fire brigade used large amounts of water, then up to 14×10^6 fibres/l were found. The problem was worst in spur mains where fibres tended to accumulate along with other debris. In the UK about 10% of the water mains are asbestos cement, containing on average between 10 and 15% asbestos by weight. Asbestos is also widely used in gaskets, packing for pumps, glands and joints.

Asbestos fibres are extremely small, often less than 1 μm long and less than 0.1 μm in diameter. They are much smaller than many bacteria and can readily pass through internal membranes within the body. They are transparent in water, so even when the water is grossly contaminated, as has happened in some regions of Canada, the USA and Africa, where asbestos is mined, the water supplies remain clear to the naked eye or only slightly turbid (cloudy). Research on the use of contaminated water in showers and portable home humidifiers have shown that the fibres can be readily converted to an aerosol and inhaled. Of particular concern is the use of ultrasonic humidifiers, which produce airborne fibre concentrations directly proportional to the concentration of asbestos in water. In some homes these units were found to produce airborne asbestos concentrations well in excess of the exposure level of 0.2 fibres/cm^3 air (Hardy et al., 1992).

In a detailed review, Commins (1988) concludes that asbestos in drinking water is 'a non-problem.' However, the evidence is far from clear, leaving several unanswered questions. Is there a causal link between asbestos in drinking water and cancer in humans? At what levels can exposure to asbestos in drinking water be considered safe? How much is there in drinking water supplies, especially at the end of distribution pipes, or where supplies come from areas served by asbestos cement pipes? Which supplies are receiving high levels of asbestos from fallout? Asbestos is primarily a distribution problem and so the removal of fibres can only be achieved by home treatment systems (Section 8.2). The use of asbestos cement for water mains and tanks has largely been phased out in the developed world. In the UK there are no plans to replace existing asbestos cement pipes unless they show significant signs of deterioration, when they are replaced by plastic mains.

6.5   COAL TAR LININGS AND POLYCYCLIC AROMATIC HYDROCARBONS

In the UK, 60% by length of the distribution pipes are ductile iron. These were coated internally by immersing the pipe sections in a batch of coal tar preparation held at a high temperature. The coal tar is intended to protect the pipe. Asbestos cement pipes are also occasionally coated internally with bitumen materials, as are water-storage tanks used by the water companies. Finished water is often distributed over considerable distances through ductile iron mains lined with bitumen and related coal tar products, which are derived from petroleum sources. This means that the water may have a long contact time with such tar-based materials between treatment at the water treatment plant and being used by the consumer. The length of contact may be significantly increased where the water remains static or flows very slowly towards the end of the water mains.

The coal tar lining usually applied to iron pipes contains massive levels of polycyclic aromatic hydrocarbons (PAHs) and these readily leach into the water supply by diffusing into the water itself. Fluoranthene is the most soluble of the PAH compounds and so will comprise the bulk of the total PAHs present, although PAH compounds can also be found in drinking water due to small fragments of the lining becoming detached. As coal tar pitch ages, small particles can be shredded from the lining by the flow. Exfoliated particles are dense black in colour and are commonly seen in mains deposits and when the mains are flushed. The particles contain high concentrations of all the PAH compounds listed in Table 4.8, up to 50% PAH by weight. Slime growth on the inside of the pipe will also contain PAHs leached from the liner, and together with the exfoliated fragments this will also carry PAHs to consumers' taps in the form of fine particulate matter. There are sufficient PAHs in coal tar linings to produce water containing PAHs at 0.2 μg/l for well over 100 years.

Under normal circumstances the increase in PAHs will be within the EC limit of 0.2 μg/l, although studies have shown that increased levels are common in some distribution systems, often exceeding the MAC limit. Groundwater appears to leach PAHs from linings more readily than surface waters, whereas temporarily high levels are associated with repair work or where new pipes have been laid. Leaching experiments where water was kept in a section of new main pipe for 24 hours resulted in PAH levels of 49 μg/l (49 000 ng/l), which only fell to around the MAC level of 0.2 μg/l (200 ng/l) after 20 weeks, when they stabilized at that concentration.

Coal tar contains higher concentrations of PAHs than bitumen linings. The National Water Council in 1976 agreed that coal tar was no longer suitable for lining drinking water pipes and recommended alternative linings. Since 1977 the use of coal tar has been drastically reduced, although there are still many thousands of miles of coal tar lined water mains in the UK supplying water to consumers. The alternative bitumen lining materials now used also contain PAHs, although these leach out at far slower rates.

It was common many years ago to line household plumbing pipes, fittings and storage cisterns with coal tar. Repairs were often carried out by plugging holes and painting over them with coal tar pitch. It is now illegal to use coal tar or any bitumen based material in household plumbing systems (Water Bylaw No. 8). The presence of PAHs in drinking water is considered in greater detail in Section 4.3 and their toxicity has been reviewed in depth elsewhere (Clement International Corporation, 1990).

6.6 ANIMALS ON TAP

6.6.1 Microbial Slimes in Distribution Pipes

Microscopic organisms such as bacteria and fungi are common in water mains. They grow freely in the water, and more importantly form films or slime growths on the side of the pipe wall, which makes them far more resistant to attack from residual chlorination (Section 6.7). In an operational sense water supply companies find slime formation undesirable as it increases the frictional resistance in the pipes, thereby increasing the cost of pumping water through the system. Certain bacteria attack iron pipes increasing the rate of corrosion, and can also affect other pipe materials. In terms of water quality, microbial slimes can alter the chemical nature through microbial metabolism, reducing dissolved oxygen levels and producing end-products such as nitrates and sulphides. Odour and taste problems have been associated with high microbial activity in distribution systems (Section 6.1), as have increased concentrations of particulate matter in drinking water (Section 6.3). The microbial slimes are also the major food source for larger organisms that are normally found in the bottom or at the margins of reservoirs, lakes and rivers.

Many upland rivers contain water of an exceptionally high quality, with a low density of suspended solids and good microbial quality. Such supplies may only receive rudimentary treatment, so very small animal species can enter the mains in large numbers. Free swimming (planktonic) species such as *Daphnia* (water flea) are unable to colonize the mains, whereas many of the naturally occurring bottom sediment dwelling species (benthic) such as *Cyclops*, *Asloma* and *Nais* easily adapt to live in the mains and even form breeding populations. The species which give rise to most complaints are not necessarily those species which enter the mains in the greatest numbers from the resource, rather it is those which are able to successfully colonize and reproduce within the distribution system.

So how do these species survive in what is, after all, pure water? Most treated water contains particulate organic or plant material in the form of algae, although most of the organic matter present will be in a dissolved form (>85%). Most of the animals present are either feeding off the particulate organic matter or the microbial slime, which is itself feeding off the dissolved organic matter. These animals are eaten by carnivores, forming very basic food chains. The density or number of animals in the mains will increase if there is an increase in organic matter or algae entering the system from the source, whereas the accumulation of organic material, including dead animals, provides a rich source of food and an ideal habitat for many animals. For this reason animals are often associated with discoloured water and often cause discoloration. Also, as organic particles tend to accumulate towards the end of spur mains, in cul de sacs, for example, the problem of discoloration, sediment and animals tends to increase towards the end of the pipework.

The source of water is important in promoting animal growth. Groundwaters contain less dissolved and particulate organic matter than surface waters, as it is removed by filtration as the water moves through the aquifer. Also, far fewer animals naturally occur in such waters to replenish or inoculate the mains. The lack of light in aquifers prevents algal development, which is an important source of organic matter in many other sources. It is the eutrophic lowland reservoirs or rivers which have the greatest potential for supporting large animal populations, but these particular sources usually receive considerable and often complex treatment before entering the mains (Section 3.2). In practice it is upland sources, which receive basic treatment only, which give rise to major consumer complaints of animals in drinking water. The greater the level of organic matter in supplies, the greater the problems from excessive slime growths in distribution systems, often made worse by not having animals present to graze and control the development of slime.

One of the most common complaints about water quality is the presence of animals. Although surprisingly large animals miraculously plop out of the tap on rare occasions, such as fish (the slim ten-spined stickleback), tadpoles and even baby frogs, the vast majority of complaints are about smaller species. Animals in tap water are due to two possible causes. If it is a mains supply then the animals are living in the distribution system; if it is a private water supply then the animals may be making their way straight down the pipe.

6.6.2 Private Supplies

Problems occur mainly because the water is being taken either directly from a stream or spring, or from a wide borehole or old type of well. In my experience the larger beasts come from these types of sources. Where wells are used then covering them to exclude light will ensure that no algae will develop and, of course, will ensure that no frogs can enter to lay eggs. In streams where the water is taken from a small chamber built into the stream bed it is inevitable that some animals will be sucked into the pipe when the water is being used in the house. In both instances it is advisable to fit a series of mesh screens over the inlet, starting with a broad grade and ending with a narrower grade to ensure that animals and silt are both excluded. Mesh screens on the inlet pipes of wide shallow wells are also advisable. These screens will require periodic cleaning, but they will ensure that the larger species are kept out of the pipework. Periodic examination of the water-storage tank should also be made for animals and, if present, it should be drained and carefully cleaned out, ensuring that all the silt and any surface slime growths in the tank are also removed using a weak solution of bleach. Elvers, baby eels, migrate up all major rivers each year, especially those on the west coast. They swarm in their millions against the flow of water until they reach the smaller upland streams from which isolated dwellings often take their water supply. Unless there is a fine screen over the intake, then the occasional elver will get through. Elvers often get into domestic water-storage tanks and once there may grow into fully grown eels, living off whatever they can find. At the worst they will starve to death and decay, which is the fate of most of the other creatures.

6.6.3 Mains Supplies

All water mains contain animals and to date about 150 different species have been identified in British water distribution networks. Most of these species are aquatic and enter the mains from the treatment plant, either during construction or maintenance work, or during a temporary breakdown. Other species which are either terrestrial such as earthworms and slugs, or are aquatic for only part of their life cycle, enter the mains during construction, maintenance or via poorly fitting manholes and vents. Obviously terrestrial species, or those that are not aquatic for the whole of their life cycle will eventually die out, so problems arising from these animals are transient. The major sources of complaints are due to truly aquatic species which are able to colonize the mains and breed successfully, although flying insects may enter service or storage reservoirs via unprotected vents and lay eggs at the water surface. This results in a seasonal problem of aquatic fly larvae. This is normally overcome by the use of 0.5 mm mesh screens placed over all access points.

So what kind of animals can we expect to find? The largest beasts in British water mains will be the water louse (*Asellus*) and the water shrimp (*Gammarus*). Once the animal has been identified, then the water supply company may take the following action depending on the species. The isolated appearance of a single organism, although potentially distressing, should not be taken as a serious indication of a problem. Often animals such as *Asellus* or mussels are present in very large numbers in water mains without ever appearing at the consumer's tap and without causing any reduction in water quality. Conversely, the creature which falls into the sink from the tap may have been the only one. Animals usually appear at the kitchen tap, or less often in the storage tank. If animals appear in taps not directly connected to the rising main, then the water storage tank should be examined immediately. Whenever possible invertebrates should be eliminated before they pass into the distribution system. Once in the distribution network several control options are available. Chemical control includes disinfectants (chlorine and chloramines) and pesticides such as pyrethrins and copper sulphate. Physical control includes hydrant flushing, cleaning and relining pipes, and the elimination of dead ends and areas of low flow.

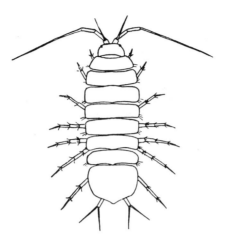

Figure 6.2. *Asellus* sp., an isopod. This invertebrate is on average 8–10 mm long. Reproduced from Macan (1970) by permission of Longman Education

6.6.4   Common Species

The invertebrates commonly found in USA distribution systems are reviewed by Levy (1990), whereas the main groups of animals found in tap water in the British Isles are described in the following.

6.6.4.1   *Isopoda, e.g.* Asellus aquaticus

These are the largest animals that most of us will come across emerging from our taps. They are up to 15 mm long and quite obvious (Figure 6.2). They usually live among the bottom debris in streams and lakes feeding on organic debris and therefore readily adapt to the conditions within the distribution system. They cling tightly to the pipe wall and can withstand adverse conditions such as low dissolved oxygen concentrations, and therefore usually survive once they get into the mains and can reproduce fairly successfully. Most complaints occur when the adults die after reproduction and are swept away in the flow of water. For this reason few will be seen during the winter, although they are still in the mains. If they have to be removed then the water supply company will use an aggressive flushing technique (water or air-scouring) to dislodge them, or even swabs. Swabs are made from polyurethane foam and are forced by water pressure through the mains, gently scouring the sides of the pipe as they move forward. In extreme cases the mains may have to be treated with a pesticide such as pyrethrin, which would involve shutting the system down (Sands, 1969). This is particularly so where the mains are old and heavily encrusted, allowing *Asellus* a better surface on which to hold. Pyrethrins have a low toxicity to mammals and are extracted from the flowers of *Chrysanthemum cinerariaefolium*. There are now a number of synthetic

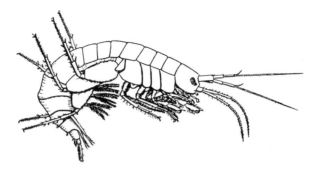

Figure 6.3. *Gammarus pulex*, an amphipod. This species varies considerably in length but is usually less than 12 mm long in in tap water. Reproduced from Macan (1970) by permission of Longman Education

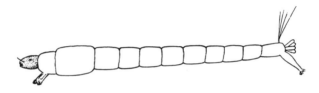

Figure 6.4. *Chironomid* larvae. These dipteran larvae vary widely in length. Reproduced from Macan (1970) by permission of Longman Education

analogues of natural pyrethrins, pyrethroids, that appear to have an equally low toxicity to humans and yet have a high insecticidal activity.

6.6.4.2   *Amphipoda, e.g.* Gammarus pulex

These are similar in habit to *Asellus*, although slightly smaller, reaching 13 mm (Figure 6.3). They are unable to cling onto the sides of the water main and so actively swim, resting in crevices and the debris at the base of the pipe. *Gammarus* is more easy to remove because it is unable to cling, so flushing should be adequate. Swabbing is particularly successful in new mains, but in older mains dosing with pyrethrins may have to be considered.

6.6.4.3   *Insecta, e.g. chironomid larvae*

Many flying insects have freshwater larval stages. Among the most often encountered in water mains are the chironomid larvae (Figure 6.4). These can be up to 25 mm in length and may be highly coloured, although most are considerably smaller, 5–10 mm, and white in colour and so are inconspicuous. One or two species can complete their life cycles within the distribution system and so need to be dealt with using an insecticide. However, most occurrences are due to small larvae getting through the treatment plant, or adults entering vents and laying eggs in service reservoirs.

6.6.4.4 *Oligochaeta (segmented worms), e.g.* Nais communis

Earthworms sometimes appear, usually after maintenance or repair work. The most common group of worms in the water mains are very small and inconspicuous in comparison, 5–7 mm long and only 0.3 mm across. They usually go unnoticed by consumers, but when they wriggle or swim then they may catch the eye, especially when the water is in a clear glass. These species are usually controlled by the residual chlorine in the water and so can be eliminated by raising the free chlorine concentration to between 0.5 and 1.0 mg/l throughout the system. This has to be kept at that level for several weeks and will itself be the cause of consumer complaints. Usually the free chlorine concentration of 0.2 mg/l throughout the distribution system will prevent a recurrence of this species. Flushing and swabbing can also be used, but infestations quickly reappear unless chlorination is adequate.

6.6.4.5 *Hirudinea (leeches), e.g.* Erpobdella octoculata

Leeches are very unusual in water mains, but if they do appear then complaints are certain and justified. The maximum size you can expect to encounter from your tap will be 35–40 mm in length. They are difficult to dislodge from the mains and require the consumers to be disconnected while the system is heavily dosed with chlorine for at least six hours before being flushed out.

6.6.4.6 *Mollusca (snails), e.g.* Potamopyrgus jenkinsi

This is a very common animal in water mains, reaching 5 or 6 mm in length, although within the mains much larger specimens will be found. They cling to the wall of the pipe and are not readily removed by flushing. Swabbing is effective in newer mains. In practice, consumers rarely come across them in the water; however, if they have an aquarium the snails rapidly colonize it and grow, eventually taking over the tank, encrusting the sides and all the water circulation pipework. Like most of the animals in your water, snails are harmless to consumers in temperate areas such as the British Isles.

6.6.4.7 *Nematoda (roundworms or eelworms)*

Nematodes are the most common organism in drinking water, although they are too small to be seen without the aid of a microscope. They live off bacteria and so are found in areas of the distribution system which are rich in organic matter, especially dead-ends. These can be effectively controlled by flushing or swabbing.

6.6.4.8 *Smaller Crustacea, e.g.* Cyclops, Chydorus

Another common group are the small crustaceans, which have a thin, hard outer shell (carapace) which they shed as they grow. The most common species is *Cyclops*, which are transparent and only 1.5 mm long. They are only noticed in water when they move with their characteristic darting action. *Chydorus* are even smaller and cannot be seen with the naked eye. However, they do occur in very large numbers, and as they cast their carapaces, which adsorb iron from the water and so become coloured, the water can become severely discoloured. These are usually successfully controlled by flushing and good residual disinfection.

6.6.4.9 *Terrestrial animals, e.g. slugs and earthworms*

Terrestrial animals in the mains usually die rapidly and so do not require control as such. However, the potential points of entry of these animals should be checked, especially service reservoirs. Normal entry is during repair and maintenance work close to the affected consumers' homes, and so are transient problems.

6.6.5 Sampling Water Mains for Animals

Water supply companies usually take samples of water from the distribution systems when flushing to check on the number and type of animals present. This is performed by allowing a standard volume of water up to 2.5 m^3 (2500 l) to be taken from a hydrant. Special nets are used which have an aperture size of 142 μm, and so retain all the animals large enough to cause complaints. A flow gauge is fitted to the standpipe along with a special net (Figure 6.5). The hydrant is then turned on full. Once the required volume of water has passed through the net, it is rolled up with all the animals inside, sealed in a polythene bag and sent to the company biologist for analysis (Department of the Environment, 1985). The abundance of species is important; for example, hundreds of thousands of water fleas per 2500 l sample will give rise to dirty water complaints, whereas hundreds or even several thousands per sample will not cause any complaints. In contrast, more than 50 *Gammarus* or *Asellus* per sample would be a cause for concern.

Houses connected to the end of mains are the most likely to suffer from animals in their water. However, where larger areas are affected this may be due to the area being supplied from a poorly managed service reservoir. Low pressure can also cause a build up of organic material in some sections of the mains with a resultant increase in the animal population.

So, apart from consumers seeing the animals because of their size, or when they are small because of their active movement, animals can also cause other complaints. The presence of animals can cause discoloured water due to the faeces of larger animals or the exoskeletons cast off by small crustaceans. Also, as the animals die and decay, they will result in intermittent taste and odour problems. None of the animals found in water distribution systems in the UK are known to carry disease, although this is not true of tropical regions. In the USA it has been shown that the invertebrates found in distribution mains can harbour pathogenic bacteria, which are protected from disinfection (Table 6.2) (Levy *et al.*, 1986). However, the problem of animals in drinking water is primarily aesthetic.

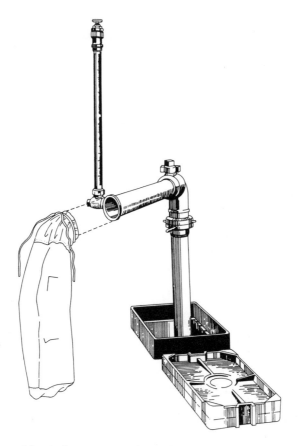

Figure 6.5. Vernon–Morris flow gauge and sampling net used to collect macroinvertebrates from the mains supply. Reproduced from the Department of the Environment (1985) with the permission of the Controller of Her Majesty's Stationery Office

6.7 PATHOGENS IN THE DISTRIBUTION SYSTEM

Water which is bacteriologically pure when it enters the distribution system may undergo deterioration before it reaches the consumers' taps. Contamination by micro-organisms can occur through air valves, hydrants, booster pumps, service reservoirs, cross-connections, back-syphonage or through unsatisfactory repairs to plumbing installations. Further problems can also arise within the domestic plumbing system (Section 7.8).

The main danger associated with drinking water is the possibility of it becoming contaminated during distribution by human or animal faeces. This was the case in the Bristol outbreak of giardiasis in 1985, when contamination occurred through a fractured main. A major outbreak of typhoid fever occurred in 1963 in Switzerland when sewage seeped into the water mains through an undetected leak in the pipe. There

Table 6.2. Bacteria associated with four categories of common invertebrates collected from a drinking water distribution network in the USA. Reproduced from Levy *et al.* (1986). Reprinted from *Journal AWWA*, vol. 78, No. 9 (Sept. 1986), by permission. Copyright © 1986, American Water Works Association

Bacterium	Intertebrate type			
	Amphipod	Copepod	Fly larva	Nematode
Acinetobacter sp.	+			
Achromobacter xylooxidans	+			
Aeromonas hydrophila			+	
Bacillus sp.	+			
Chromobacter violaceum	+			
Flavobacterium meningosepticum		+		
Moraxella sp.	+	+		
Pasteurella sp.	+	+		
Pseudomonas diminuta		+		
Pseudomonas cepacia			+	+
Pseudomonas fluorescens	+		+	
Pseudomonas maltophilia	+			
Pseudomonas paucimobilis	+	+		
Pseudomonas vesicularis	+			
Serratia sp.	+			
Staphyloccocus sp.	+			

are many more examples indicating that the microbial quality of water is potentially at risk while in the distribution system. Uncovered service reservoirs can also be a major source of contamination, especially from birds. Usually the mains is under considerable positive pressure; however, back-syphonage can occur in the distribution system if the water pressure drops and there are faulty connections or fractures in the pipe. In this way contaminants can be sucked into the distribution system. The problems are more likely to occur when water supply pipes are laid alongside sewerage systems. They should be as remote as possible from each other, although in practice this is very difficult.

On passing through the distribution system, the microbiological properties of the water will change. This is due to the growth of micro-organisms on the walls of the pipes and in the bottom sediments and debris. Although not causing any serious health problems these non-pathogenic bacteria can cause serious quality problems, causing discoloration and a deterioration in taste and odour (Section 6.1).

Increased heterotrophic bacteria within the distribution system are due to a number of factors, usually the absence of a residual disinfectant combined with either contamination from outside the distribution network, or more commonly from regrowth. The growth of bacteria in the water and on pipe surfaces is limited by the concentration of essential nutrients in the water. Organic carbon is the limiting nutrient for bacterial growth, with the growth of bacteria directly related to the amount of assimilable organic carbon (AOC) in the water. There are a number of different methods for calculating AOC concentrations in water. These fall into two broad categories: (1) those that measure the concentration of organic carbon that can be converted into microbial biomass using a parameter such as ATP as the determinand (Van der Kooij et al., 1982); and (2) those that measure the potential mineralization of dissolved organic carbon (DOC) by micro-organisms, by assessing the decrease in organic carbon concentration over time. This is commonly referred to as the biodegradable dissolved organic matter (bDOC) (Huck, 1990: Gibbs et al., 1993).

Among the major genera found in distribution systems are *Acinetobacter, Aeromonas, Listeria, Flavobacterium, Mycobacterium, Pseudomonas* and *Plesiomonas*. Some of these organisms can be considered as opportunistic pathogens. The type of micro-organism and the number depend on numerous factors such as the water source, type of treatment, residual disinfectant and nutrient levels in the treated water. The development of slimes, or biofilms as they are known, lead to the survival of other bacteria. Certainly biofilms protect bacteria from disinfectants. *Legionella* in particular is able to survive within biofilms, as are *Pseudomonas* and *Aeromonas* spp. (Stout et al., 1985) (Section 7.8). The development of non-pathogenic coliforms is possible in biofilms, but when detected it is important for the water operator not to assume a non-faecal cause such as regrowth in the distribution system, even in the absence of *Escherichia coli*.

With good treatment and adequate disinfection with chlorine, ensuring a residual disinfection to control regrowth within the distribution system and to deal with any contamination, serious problems should not arise. When supplies have a high chlorine demand due to the presence of organic matter and humic acids, for example, then it is difficult to maintain sufficient residual chlorine in the system. It is also necessary to strike a balance so that those closest to the treatment plant on the distribution network do not receive too large a dose of chlorine in their water and those at the end too little. Research has shown that many bacteria, viruses and protozoan cysts are far more resistant to chlorination than the indicator bacteria used to test disinfection efficiency. New guidelines are therefore being prepared to optimize the disinfection of water supplies. Currently the residual concentration of free chlorine leaving the treatment plant is less than 1.0 mg/l and usually nearer to 0.5 mg/l. Most treated waters continue to exert some chlorine demand, which reduces this free chlorine residual even further. Further chlorine is lost by its interaction with deposits and corrosion products within the distribution system, so that free chlorine residuals do not often persist far into the network, leaving the water and the consumer at risk.

The problem stems from treatment processes being primarily designed to remove particulate matter rather than dissolved organic matter. Chlorination and ozonation lead to increases in the AOC concentration in water, whereas coagulation and sedimentation can remove up to 80% of the AOC present. The AOC is also removed by biological activity during filtration, although residence times are far too short to allow significant reductions. Success has been achieved by chemically oxidizing the AOC into simpler compounds to make them more readily removable by biological filtration. There is evidence that the micro-organisms living on granular activated carbon filters can also significantly reduce AOC concentrations (Huck et al., 1991).

The presence of biofilms is currently causing water microbiologists much concern, especially as pathogenic micro-organisms can be protected from inactivation by disinfectants by the biofilm. Control of these growths is difficult and research in this area is active. I would repeat what I said in the section on chlorination: the risk from chlorination by-products which are known to be carcinogenic at higher concentrations than those found in water must be balanced with the risk of microbial pathogens from inadequate disinfection. The problems of biofilms is reviewed by Van der Wende *et al.* (1989) and Maul *et al.* (1991).

Chapter 7

Problems Arising in the Home Plumbing System

7.1 HOUSEHOLD PLUMBING SYSTEMS

7.1.1 Entry to the Home

Houses are connected to the mains by the service pipe (Section 3.3). The connection from the mains to the boundary stopcock is known as the communication pipe, and this is owned and maintained by the water supply company. The boundary stopcock, as the name suggests, is located just outside the boundary of the property served, with the stopcock being a control valve which turns off the water supply to the house. It is usually buried 1 m below ground, at the bottom of an access chamber known as the guard pipe; which can be a short piece of any suitable pipe, the top access being protected by a small metal cover. The stopcock can be turned on or off using a long-handled key, the exact nature of which depends on the type of handle on the stopcock itself. It is switched off by turning the key in a clockwise direction. From here the service pipe, which is now known as the supply pipe, carries the water to the house. It rises slightly to ensure that all air bubbles escape. From the boundary stopcock onwards all the pipework and the appliances are the householder's responsibility. The supply pipe must be at least 750 mm below the ground at all times to protect it from frost. It usually enters the house at the kitchen and rises up from the floor underneath the kitchen sink. The pipe (the rising main) rises directly either to the cold water tank (often referred to as the cold water cistern because it is operated by a ballcock valve), or to feed the other cold water draw-off points around the house. There should be an indoor stopcock positioned where the pipe enters the house and before the connection from the rising main to the sink to be able to cut off the supply to the house. The kitchen tap should always be connected directly to the rising main. If the supply pipe enters the house at another point, for example through a cloakroom or garage, then the stopcock will be fitted there. To close a stopcock it should be turned clockwise. In some houses, generally newer ones, there may be a drain cock positioned just above the house stopcock to allow the rising main to be drained if necessary (Figure 7.1).

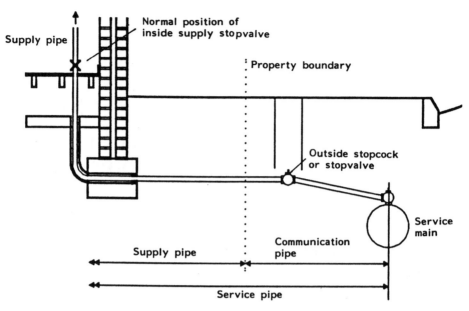

Figure 7.1. Definition and location of service pipe, boundary stop valve and supply pipe. Reproduced from White and Mays (1989) by permission of the WRc Evaluation and Testing Centre

7.1.2 Plumbing Systems

The layout of the water system in the house depends on a number of factors including whether there is non-electrical central heating. Essentially houses have either a high-pressure direct system, or a low-pressure indirect system which requires a cold water storage cistern in the attic or in another elevated position.

High-pressure direct water systems are common in the USA, Canada and most of Europe. The water comes into the building and directly feeds all the cold taps and any appliances which also use cold water. This includes the WC (lavatory), hot water storage cylinder and even the washing machine and dishwasher. These systems are much simpler and cheaper to install than the indirect system and avoid having lots of pipework and a storage cistern for the cold water in the attic, or occasionally in the airing cupboard. In the UK high pressure systems are only found in older properties built before 1939. In many countries such as Sweden, The Netherlands, Germany, Denmark and Belgium storage tanks or cisterns are considered to be unhealthy, and a potential health hazard, and so are prohibited.

The vast majority of houses in the UK and Ireland have indirect or low-pressure systems, with only the cold water tap that serves the kitchen, and possibly the cold water tap to the washing machine, directly connected to the rising main (Figure 7.2). The rising main goes up into the attic to discharge into a cold water storage cistern.

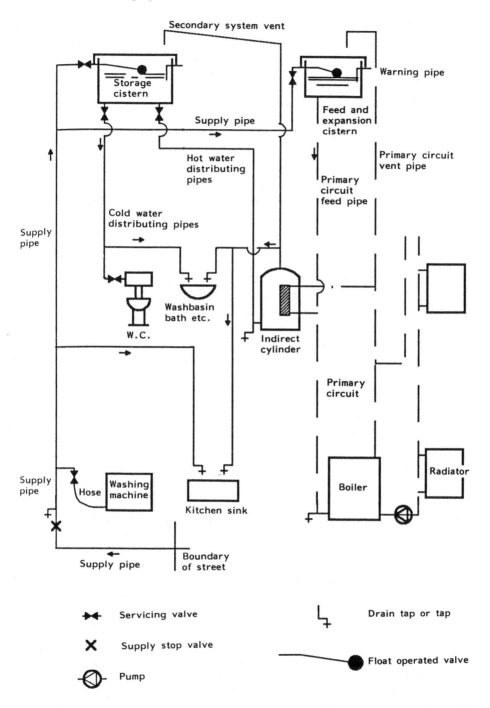

Figure 7.2. Schematic diagram of typical household plumbing system. Reproduced from White and Mays (1989) by permission of the WRc Evaluation and Testing Centre

From this cistern distribution pipes feed all the other cold taps, the lavatory cistern and the hot water system. Most water supply companies insist on the use of a storage cistern to ensure that at times of peak usage all demands are satisfied, which would not be the case if everyone had direct systems. This is why on occasions the water tank may start to fill even though water has not been used recently. This is because, due to peak demand, the pressure has been too low to refill the tank. A major advantage of having a low-pressure system is that there is less noise, especially from lavatory systems. Burst pipes can be awful, but under mains pressure they can be extremely serious. Also, when the water supply is cut off for some reason, for example during maintenance or repair work, a household with an indirect system still has plenty of water stored in the attic for essential use.

The rising main feeds the storage tank or cistern. Older tanks are made of galvanized iron, which is very heavy and liable to corrode. Modern tanks are made of plastic, which is much lighter, and are generally 50 gallons in size (227 l), although different sizes are available.

The rising main will be a 15 mm diameter pipe, although with low pressures a slightly larger pipe (22 mm) may be used. The supply of water enters at the top of the storage tank via a float (ball) valve similar to that in the WC. It is operated by a lever which is opened and shut by a float made of copper or plastic which floats on the surface of the water. Also positioned at the top of the tank is the overflow pipe to carry away the excess water if the ball valve fails for some reason, or the washer needs replacing so that the supply of water from the rising main cannot be fully shut off. This is to prevent the tank overflowing and pouring through the ceiling. A leaking overflow needs fairly quick attention; not only is it a nuisance to neighbours due to the noise it makes, especially late at night, but it is an offence to allow water to be wasted in this way. During cold weather, if the overflow freezes up, the water storage tank will overflow and flood the house. The WC operates in exactly the same way as the water storage tank, using a float valve, so the same applies to the overflow for the WC cistern. The overflow pipes are usually located above the back door of the house, to ensure maximum attention.

The water from the tank is distributed to the rest of the house by two 22 mm diameter pipes fitted about 75 mm from the base of the tank. This is to allow debris to settle so that it is not carried into the plumbing system. One of the pipes will supply the cold taps and the WC. Usually the pipe goes directly to the bath, with the cold taps of the wash basins and the WC cistern taken off this pipe using 15 mm branches. The water flows through these pipes by gravity and so is at a lower pressure than that from the rising main. There is usually another stopcock on the rising main above the connection to the cold water tap in the kitchen. This allows the water supply to the water storage tank, and thus the whole house, to be shut off while retaining the supply for drinking at the cold water tap in the kitchen. The second pipe from the tank feeds into the bottom of a hot water cylinder. This feed pipe also has a stopcock fitted to allow the flow of water to the hot water cylinder to be shut off during maintenance. The cylinder can be drained using the drainage valve at the base of the cylinder. The hot water cylinder is made of copper and can hold about 30 gallons (136 l) of water. The water is heated either by an electric element heater and/or by a heating coil from a boiler. The hot taps are then fed from the hot water cylinder. If the water overheats in the cylinder it expands and the excess escapes into the cold water cistern in the attic (Figure 7.2).

Cold water storage tanks in older houses are often open. This allows access for dust, insects, birds, bats, rodents and, of course, their droppings, and, as is discussed in Section 7.7, fibres from the insulation. It is therefore important that the tank is covered. Water tanks now come with closely fitting lids; however, existing tanks can be effectively covered by placing a thick slab of expanded polystyrene insulation on the top. It may be a good idea to weigh the cover down so that it does not move or fall into the tank due to excessive air currents in the attic, especially if it is cross-ventilated.

Water is heated either directly or indirectly depending whether it comes into contact with a heater. With direct heaters the water is heated within the hot water cylinder by one or more immersion heater elements and the hot water is then supplied to the hot taps. If a boiler is used the water is heated by circulation through the boiler and the heat is transferred to the water in the hot water cylinder and stored there. An immersion heater is often used in the cylinder as well. Neither of these systems can supply central heating radiators. Indirect heating is usually combined with the central heating system. The water in the cylinders is heated indirectly by a separate water circuit which goes through the boiler, whereas the radiators are supplied by a separate hot water circuit. The central heating system is fed by a separate water circuit, supplied by a small feed and expansion cistern which is also supplied by the rising main via a small ball valve (Figure 7.3). It is also located in the roof and is needed only to top up the system due to evaporation losses. If the feed and expansion tank is being refilled by a constant drip, check that the ball valve does not need adjusting. If there is no water in the overflow pipe then the dripping means that there is a leak somewhere in the central heating system and the water is escaping from the enclosed system.

Hot water always rises above cold water because as it heats up it becomes lighter (less dense). In the hot water cylinder the hot water therefore collects at the top, so cold water is fed in at the base to replace any hot water drawn. When a boiler is fitted the heat exchange pipe runs from the top to the bottom of the cylinder, so that the hot water enters at the top ensuring the most efficient exchange of heat. The hot water is taken from the top of the cylinder to feed the hot water taps around the house (Figure 7.3).

A branch from the hot water outlet pipe from the hot water cylinder goes back up to the cold water cistern in the attic. This hangs over the edge of the tank and is not immersed. It allows for expansion and for the escape of air from the cylinder. The pipe from the boiler does not discharge its hot water, it runs through the cylinder in a sealed coil. This allows heat exchange.

7.1.3 Back-syphonage

It is possible, if there is a drop in pressure in the mains due to a burst or perhaps due to the fire brigade using large amounts of water, that, if the plumbing has been installed incorrectly, back-suction or back-syphonage will occur. Water is siphoned back through the plumbing system to the rising main and into the water supply distribution main. For example, hose extensions are commonly fitted to cold water taps and if the

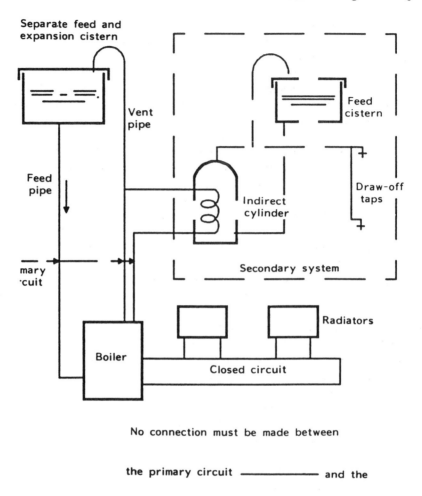

Figure 7.3. Schematic diagram of an indirect heating system. Reproduced from White and Mays (1989) by permission of the WRc Evaluation and Testing Centre

hose is immersed in water and back-syphonage occurs, all the water in the sink is sucked back into the mains until the siphon is broken. The example often quoted is the garden hose left in a bucket of weedkiller or insecticide which is drawn back up the hose and into the public supply. It is an unlikely scenario, but not impossible. If the bucket of weedkiller is replaced by a water butt containing lots of murky water, or a fish pond that is being filled, then the scenario becomes more plausible. Back-syphonage is considered in detail in Section 7.8. Devices are now available to prevent back-syphonage and there is now a law that a back-flow prevention device (or pipe interrupter) should be attached to any tap onto which a hose pipe is connected. These are widely available and simply screw onto the tap, or can be fitted into a section of the hose (Figure 7.4).

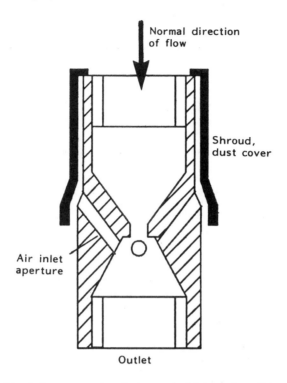

Figure 7.4. Typical back-flow prevention device used with hosepipes. Reproduced from White and Mays (1989) by permission of the WRc Evaluation and Testing Centre

7.1.4 Water Byelaws

The 10 water service companies and the 29 former statutory water companies in England and Wales, the 12 Regional and Islands Councils in Scotland, and the Department of the Environment in Northern Ireland all have the duty to make and enforce byelaws (known as Regulations in Northern Ireland). The byelaws are used to prevent waste, undue consumption, misuse, or the contamination of water supplied by the water undertakers. All byelaws are essentially the same and are based on the Model Water Byelaws 1986 Edition. A summary of the byelaws is given in Table 7.1, and a useful explanatory guide has been prepared by the WRc Evaluation and Testing Centre (White and Mays, 1989). If any changes or alterations are made to a consumer's supply system then the byelaw inspector must be notified so that he or she can inspect it, regardless whether it is carried out by the householder or a plumber. Simple repairs or renewals are exempt. However, water byelaws are legal requirements and if they are not adhered to then prosecution could follow. Current penalties are £400 in respect of each offence and £40 in respect of each continuing offence for each day during which the offence continues after conviction. All new connections to the mains will require the plumbing to be inspected before the connection is made to ensure that it is satisfactory.

Table 7.1. Summary of the main provisions in the water supply byelaws used by all the water undertakers in the UK

Part I General provisions
   Byelaw
1	Interpretations
2,3	General prohibitions
4	Savings for fittings lawfully used and fittings used for temporary non-domestic purpose

Part II Prevention of contamination of water from contact with unsuitable materials or substances
   Byelaw
5	Prohibition of installation of pipes in contact with contaminating material
6	Permeation and deterioration of pipes
7	Materials in contact with water
8	Prohibition of coal tar
9	Prohibition of lead
10	Prohibition of copper in repairs to lead pipes

Part III Prevention of contamination of water by back-syphonage, back-flow or cross-connection
   Byelaw
11	Definitions
12	Taking of supplies
13	Prevention of cross-connections
14	Connections to closed circuits
15	Prevention of cross-connection between cistern-fed primary circuits and secondary hot water systems
16	Draw-off taps to baths, sinks, etc.
17	Shower hose connections
18	Hose connections
19–21	Bidets
22,23	Clothes and dishwashing machines
24	Pipes conveying water to cisterns
25	General requirements for protection at draw-off points
26	Secondary back-flow protection
27	Pipes to be readily distinguishable
28	Separation of pipes in fire-fighting installations from other fittings
29	Accessibility of back-flow prevention devices

Part IV Prevention of waste or contamination of stored water
   Byelaw
30	Cisterns storing water for domestic purposes
31	Placing of storage cisterns
32	Support of storage cisterns
33	Pipes supplying water to storage cistern
34	Fixing and adjustment of float-operated valves
35–37	Prevention of contamination from primary heating circuits and feed cisterns
38–41	Warning and overflow pipes
42	Requirements for float-operated valves
43	Float-operated valves conveying hot water
44	Flow control devices other than float-operated valves
45	Location of storage cisterns storing water for non-domestic purposes

Table 7.1. (*continued*)

46	Animal drinking troughs or bowls
47	Ponds, fountains and pools

Part V   Prevention of waste of water from damage to water fittings from causes other than corrosion
Byelaw

48	Covering of pipes
49	Protection from damage by freezing and other causes
50	Protection of plastic pipes from petroleum products

Part VI   Prevention of waste from or contamination by unsuitable or improperly installed water fittings
Byelaw

51,52	Materials and construction of water fitting, pipes, pipe joints and pipe fittings, cisterns and cylinders
53,54	Laying and jointing of pipes
55	Cleaning pipes after installation or repair
56	Prevention of warming of water in pipes
57	Use of adhesives in jointing metal pipes
58	Accessibility of pipes and pipe fittings
59	Disconnection of water fittings
60	Connecting water fittings of dissimilar metals

Part VII   Stop valves, etc.
Byelaw

61–63	Provision and location of stop valves
64	Requirements for stop valves
65	Draining of supply pipe
66	Requirements for draining taps
67	Location of draining taps
68–70	Provision and location of servicing valves
71	Requirements for servicing valves
72	Watertightness of back-flow prevention devices
73	Accessibility of stop valves and servicing valves

Part VIII   Water closets and urinals
Byelaw

74	Requirements for water closet pans
75–79	Water used in cleaning water closet pans
80,81	Warning pipes and markings of flushing cisterns or troughs
82	Water used in flushing urinals
83	Control of flow to urinal cisterns
84	Pipes discharging to water closet pans and urinals

Part IX   Prevention of waste, misuse and contamination of water from draw-off taps, baths, basins, sinks and other fittings
Byelaw

85	Inlets and outlets of baths, wash basins and sinks
86	Washing troughs
87	Requirements for draw-off taps
88	Water used by clothes and dishwashing machines

continued overleaf

Table 7.1. (*continued*)

Part X   Prevention of waste or contamination of water from any hot water system
Byelaw
89	Secondary system vent pipes
90	Accommodation of expansion water in cistern-fed systems
91	Accommodation of expansion water in systems connected to a supply pipe
92	Capacity of expansion cisterns or vessels, etc.
93	Boilers
94,95	Pressure relief valves, expansion valves, temperature relief valves, etc.

Part XI   Taps for drawing drinking water
Byelaw
96	Taps for drinking water

Part XII   Notices to undertakers
Byelaw
97,98	Notices to undertakers

Part XIII   Penalties
Byelaw
99	Penalties
100	Defence for person charged with an offence

Connection to the mains is not free and can be expensive, as is the cost of replacing a joint supply pipe with a new single supply pipe.

7.2   CORROSION

Problems with iron are most likely to arise at the treatment plant or within the distribution network rather than the home plumbing system. However, corrosion can result in a number of metals or alloys from which pipework or plumbing fittings are made contaminating drinking water. The most important of these are lead, copper, zinc and iron.

Most types of corrosion involve electrochemistry. For corrosion to occur we need the components of an electrochemical cell, i.e. an anode, a cathode, a connection between the anode and cathode (external circuit) and finally a conducting solution (internal circuit), in this instance drinking water. The anode and cathode are sites on the metal pipework (it may be the same metal or different metals) which have a difference in potential between them. When this occurs then oxidation (the removal of electrons) occurs at the cathode, which is negatively charged. In practice metal dissolution occurs at the anode, releasing the metal into solution (Figure 7.5).

Corrosion cells form on the same piece of metal where there are adjacent anodes and cathodes. These are due to the non-uniformity of the surface and can be caused by minute differences in the pipe surface formed during manufacture, or by stress imposed during installation. Any imperfections in the pipe will create tiny areas of metal with different potentials.

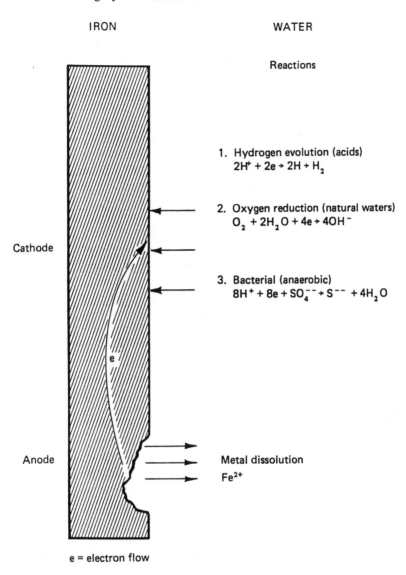

IRON   WATER

Reactions

1. Hydrogen evolution (acids)
   $2H^+ + 2e \rightarrow 2H \rightarrow H_2$

2. Oxygen reduction (natural waters)
   $O_2 + 2H_2O + 4e \rightarrow 4OH^-$

Cathode

3. Bacterial (anaerobic)
   $8H^+ + 8e + SO_4^{--} \rightarrow S^{--} + 4H_2O$

e

Anode

Metal dissolution
Fe^{2+}

e = electron flow

Figure 7.5. Main reaction pathways leading to the corrosion of metal surfaces. Reproduced from Ainsworth *et al.* (1981) by permission of the Water Research Centre

Corrosion is more rapid where two different metals are coupled together. This is known as galvanic corrosion. The most serious situation is where a copper pipe is used to replace a section of lead pipe (Section 7.3). The conducting solution (electrolyte) is the water, the copper pipe is the cathode and the lead pipe is the anode. The lead pipe slowly corrodes, releasing the metal into solution. Metals can be listed in order of their tendency to corrode or go into solution, i.e. the galvanic series. The metals and alloys most commonly used in water treatment and distribution can be placed in electrochemical order—that is, in the decreasing order in which they tend to corrode when connected under ordinary environmental conditions. The order is: manganese, zinc, aluminium, steel or iron, lead, tin, brass, copper and bronze. The further apart two metals are on this scale, the greater the potential difference between them and so the greater the rate of corrosion. Iron pipes will corrode if connected by brass fittings, and of course lead pipes will corrode if connected to copper fittings. In household plumbing it is therefore best to use all the same metal, e.g. copper pipes and copper fittings, or metals very close on the electrochemical scale, e.g. copper pipes and brass or bronze fittings.

Galvanized steel, steel coated with zinc, was widely used for water storage tanks and is still available. Zinc is more likely to corrode than steel so any crack or blemish on the surface coating of zinc will result in a cell being formed with zinc as the anode and zinc being released into the water (Section 7.5).

The rate of corrosion increases rapidly at higher temperatures and more acidic pH values. The corrosion of metals is therefore more pronounced in softer water and in the hot water circuit. The hot water cylinder is usually made from copper, which is a particular target for corrosion. Differences in dissolved oxygen or the hydrogen ion concentration (pH) can also set up a difference in potential within a pipe. This occurs most often under rivets, washers or in crevices where dissolved oxygen may be lower. The area of the metal in contact with the higher dissolved oxygen concentration will become the cathode and the area with the lower dissolved oxygen concentration becomes the anode and so corrodes. The same occurs with differences in hydrogen ion concentration.

Once the surface of the pipework has been corroded, it provides a habitat for micro-organisms. These are able to attach themselves to the pipe and are not flushed away in the water flowing past. As the micro-organisms build up they increase the resistance to flow of water through the pipework. Some of these micro-organisms, in particular iron and sulphur bacteria, can also corrode the pipework (both metal and non-metal). This is mainly a problem in the mains and at the treatment works as it is mainly buried metal pipes which are affected, although concrete pipes can also be attacked.

When corrosion occurs it is in fact a chemical reaction accompanied by the generation of a very small amount of electrical energy. This results in metal ions leaving the site of corrosion (the anode) and entering the water (Figure 7.5). This dissolution of the metal ion leaves behind excess electrons which migrate towards the cathode, hence the electric current. These excess electrons are consumed at the cathode in a balancing reaction. There are in essence three types of balancing reaction. The first, which is most common in drinking water, is the reduction of oxygen

$$O_2 + 2H_2O + 4e \rightarrow 4OH^-$$

where e are electrons. This means that where the corrosion is due to differences in the dissolved oxygen concentration there is a tendency for the oxygen concentration to equilibrate and the corrosion will stop. However, if the cathode is situated where there is plenty of water movement and the anode is in an area where the water is stagnant (e.g. under a rivet), then equilibrium will not occur and corrosion will continue.

Where the water is acidic, and this becomes increasingly important below pH 6, and where oxygen is absent from the surface of the metal such as occurs beneath deposits, then hydrogen ions are released to form hydrogen

$$2H^+ + 2e \rightarrow 2H \rightarrow H_2$$

This is the reaction we see when we charge a car battery, with hydrogen bubbles being formed at the cathode.

The final reaction involves a type of bacteria, which are only found under anaerobic conditions, which reduce sulphate to sulphide

$$8H^+ + 8e + SO_4^{2-} \rightarrow S^{2-} + 4H_2O$$

In practice these reactions ensure the assimilation of electrons at the cathode, thus driving the rate of corrosion at the anode, regardless of whether the water is aerobic or anaerobic (De Rosa and Parkinson, 1986).

Corrosion can be controlled by choosing the correct metal and alloys, or by using special corrosion-resistant metals. Surfaces can be coated or lined with special protective films. Lining is more commonly carried out in water mains to protect iron pipes, but is also possible in smaller pipes. Corrosion can also be prevented by connecting the pipework to another piece of metal which will act as an anode and so will be deliberately corroded instead of the existing pipework. This is known as cathode protection and the metal insert, a sacrificial anode, should be installed in metal storage tanks to protect them from aggressive water, and also in the hot water cylinder, which is particularly at risk. Sacrificial anodes are generally made of magnesium alloy fitted onto a steel rod for use with galvanized steel tanks, whereas copper cylinders are protected using aluminium rods. Although there are many other ways of protecting metal pipework from electrochemical corrosion, the selection of the correct materials and fitting a sacrificial anode are all that is normally required.

Chemical corrosion, simply dissolving the metal, can also occur in soft water areas with lead the most readily corroded. The only effective way to control this is to increase the pH of the water to between 8.0 and 8.5. This is done at the water treatment plant by adding various alkali hydroxides, silicates or borates. Under alkaline conditions these form an impervious layer on the wall of metal pipes, usually calcium carbonate, protecting them from further corrosion. Excessive scaling can reduce the diameter of the pipe significantly, seriously reducing the flow-rate of water through the pipe. The whole question of the formation of scale in pipes to offer protection, and the measurement of the corrosiveness of water, is under debate at present. Because of the widespread use of copper pipework, much attention has been paid to the forms of corrosion of this metal. Corrosion causes pitting of the inner surface of the pipes leading to eventual perforation. Two forms are well documented (Table 7.2), although other forms have been described.

In household plumbing the corrosion of lead, galvanized steel and copper pipework leads to discoloration and occasionally taste problems. Lead is also toxic, and the corrosion of lead and copper is dealt with in the following sections. The corrosion of galvanized steel is slightly different. The zinc layer is there to protect the steel underneath by corroding preferentially, ensuring that any exposed steel becomes the cathode rather than the anode. As the zinc corrodes it forms zinc hydroxycarbonate,

Table 7.2. Summary of types I and II pit corrosion of copper pieces

Feature	Type I corrosion	Type II corrosion
Pit shape	Broad and hemispherical	Steep sided and narrow
Pit contents	Copper(II) oxide and chloride	Crystalline copper(II) oxide
Corrosion product covering pit	Thick hard green layer of calcium and copper carbonates	Small black mound of copper(II) oxide and basic copper sulphate
Cause	Cold hard waters, presence of carbon (graphite) film produced during manufacture	Hot ($>60°C$), soft water (<40 mg $CaCO_3$/l), low pH, hydrogen-carbonate: sulphate ratio >1
Typical time for pipe failure	Three years	15–20 years
Control	Removal of carbon film during manufacture	pH control to $\geqslant 7.4$

which then forms a protective layer on the steel. However, in soft waters where the alkalinity is low (less than 50 mg/l as $CaCO_3$), or in water with a high carbon dioxide concentration, which is common in groundwaters from private wells and boreholes, the corrosion product tends to be zinc carbonate. This does not adhere to the metal surface and so is lost into the water, giving the appearance of sand in the water. It is advisable not to use galvanized tanks or pipework for such aggressive waters. The corrosion of galvanized steel is increased significantly if it is coupled with copper tubing, especially if it is used upstream of the galvanized tank and the water is cuprosolvent (i.e. able to corrode copper pipework). Even a small amount of copper can significantly increase the rate of corrosion. So for domestic purposes, stay clear of galvanized steel.

The water byelaws (Byelaw 60) do not permit dissimilar metals to be connected unless either effective measures are taken to prevent deterioration (i.e. the installation of a sacrificial anode) or where deterioration is unlikely to occur through galvanic action. Metals can be safely mixed as long as the sequence of metals is kept as follows: downstream order of use—galvanized steel, uncoated iron, lead and copper. Plastic pipes can be used in conjunction with any materials fairly safely. A common problem is that even if this sequence has been strictly adhered to, the water return (overflow) pipe from the copper hot water cylinder may be returned to a galvanized tank. If the water supply byelaws are carefully followed then corrosion will be minimal (White and Mays, 1989). A new test method to assess the potential of metallic products to contaminate water supplies has been published by the British Standards Institution (1991).

7.3 LEAD

For centuries, lead has been known to be toxic. There is perhaps more known about the acute and chronic adverse effects of lead on humans than about any other poison. Although the effects of high concentrations of lead on the body are known, research over the past 15 years has begun to highlight the problems caused by much lower levels of exposure which were previously thought to be harmless. The effects of this long-term exposure is neural; it affects the brain, resulting in behavioural changes and deficits in intelligence levels. The problem has been brought to light by a number of studies in Britain and the USA, although it is the specific studies carried out at Glasgow and Edinburgh Universities which have shown fairly conclusively that exposure to low levels of lead has adverse effects on the learning ability of children (Lansdown and Yule, 1986; Thornton and Culbard, 1987).

7.3.1 Sources of Lead

Raw waters, both groundwater and surface waters, rarely contain lead in concentrations in excess of 10 $\mu g/l$. Lead salts are not used in water treatment and there is little chance of lead levels increasing in the mains. So where does it come from?

Houses built before 1964 have some lead piping and fitments present, and lead lined storage tanks are still common in some areas of Scotland, whereas houses built before 1939 will have had extensive lead piping fitted. All water supply companies used lead for service pipe connections, mainly because lead is so malleable and thus will not fracture if the mains or another part of the pipe move slightly. The use of such connections was phased out many years ago, but there are still many millions of houses connected to the mains by lead service pipes, even though there is no lead piping in the house itself. The problem arises where the water is plumbosolvent, which means that it can dissolve lead. The most important factor affecting plumbosolvency is the pH of the water. The rate of leaching increases dramatically below a pH of 8.0, so that soft acidic waters (pH 6.5), which are so common in certain areas, will rapidly dissolve the lead from the pipes into solution. Those who are most at risk are those living in older houses which still have lead pipes, and which have acidic water. On this basis the worst affected areas in the UK include most of Scotland, the north of England, Wales and the south-west of England. Problems are found in some major British cities including Glasgow, Edinburgh, Birmingham, Liverpool, Manchester and Hull. However, the softness of water is not the only factor related to plumbosolvency, and even in hard water areas lead can still be a problem. Other important factors include the water temperature, the contact time between the water and the source of lead and the amount of lead piping, which affects the surface area of lead exposed to the water. These are all considered in the following. There has been an enormous amount of research into the factors affecting plumbosolvency in the hope that a simple solution might be found to this major problem. Among the factors affecting lead leaching into drinking water are vibrations in pipes, scouring by high water velocities, thermal expansion effects, the age and nature of the pipes, the presence of particulate lead deposits and electrochemical reactions caused by mixing metal pipes (Section 7.2). Because of the complexity of all these factors interacting together, there is a significant variation in the lead concentrations in water within individual water supplies over time and between supplies to different houses in the same area. This has led to much research and debate over how to actually take samples for the determination of lead in drinking water.

A number of interesting studies have been carried out. One of the most fascinating is the Renfrew lead pipe survey carried out between 1976 and 1977 in which every house in the Renfrew district was checked for lead piping. Lead piping was not used in Renfrew after 1968, so although the total housing stock in the district is 77 806, only those built before 1968 were inspected, about 55 393 in all. As many of the houses share a communication pipe or supply pipe, it was found that there were only 21 070 communication pipes, of which 96% were lead. Of the 21 000 supply pipes identified, over 80% were lead. Although the average length of the supply pipe was 19 m, some houses receiving private water supplies had supply pipes in excess of 1.5 km. Of the houses inspected 1592 had lead-only pipes from the point of entry into the house to the kitchen cold tap. A further 13 349 homes had the usually worse situation of having a combination of lead and copper pipes. The majority, about 40 452 (73%) had no lead pipes running to the cold water tap, most of these being made of copper. Therefore 27% (14 941) of homes had some lead leading from the mains to the cold water tap in the kitchen. It was important to identify those houses where the kitchen cold water tap was not connected directly to the supply pipe, but was fed from a cold water storage tank. In Renfrew, 875 houses (1.6%) fell into this category. The percentage of these storage tanks that were lead-lined is unknown, but it is estimated to be about 45%. The survey also noted that just over 2% of houses had lead pipework leading to the hot tap and that 6% of bathroom taps were connected with lead pipework. The survey, the first of its kind, was enormously useful. Each household was made aware of the presence of lead pipes and was given a leaflet explaining the problems and the measures that could be taken to avoid excessive lead levels in drinking water. The council determined the extent of the problem and had the opportunity to identify those most at risk (Britton and Richards, 1981).

Lead is ubiquitous in the environment. It is found in food, especially canned foods, air, paint, glazes, dust, fumes and, of course, drinking water. Although the exposure to lead from some of these sources has occurred for decades and in some instances for centuries, there are clear indications that exposure levels are decreasing. Legislation to limit the use of lead in paint and industrial sources of lead has had a significant effect. The introduction of lead-free petrol has contributed enormously to reducing the levels in fumes and dust, especially in our towns, with reduced levels in gardens and garden produce as a result. The main sources of lead are food, water and air, and although the contributions from food and air are decreasing, the contribution from drinking water has remained largely unaltered and in many instances is the major single source of the metal in our diet.

7.3.2 Intake of Lead and Health Effects

The intake of lead from food and water in the UK is between 70 and 150 μg/day for adults and between 70 and 80 μg/day for children between the ages of two and four years. This is only an approximation and may in fact be unhelpful. It is perhaps more pertinent to consider how much lead is being taken in by consumers in their tap water. The Department of the Environment carried out a survey of lead in drinking water during 1975–6. Samples were taken of tap water from 3000 households throughout the

Table 7.3. Lead concentrations in households in the UK during a survey in 1975–6. Reproduced from the Department of the Environment (1977) with the permission of the Controller of Her Majesty's Stationery Office

Lead concentration in daytime sample ($\mu g/l$)	Percentage of households			
	England	Scotland	Wales	Total
0–9	66.0	46.4	70.5	64.4
10–50	26.2	19.2	20.7	25.3
51–100	5.2	13.4	6.5	6.0
101–300	2.2	16.0	1.5	3.4
$\geqslant 301$	0.4	5.0	0.8	0.9
Total	100.0	100.0	100.0	100.0

UK. The results are shown in Table 7.3, and it is clear that although most houses were receiving drinking water with less than 10 $\mu g/l$ lead, significant numbers were receiving drinking water in excess of 50 $\mu g/l$: 7.8% in England, 8.8% in Wales and a staggering 34.4% in Scotland. These samples were taken during the daytime and although the water was not flushed before sampling the water had not been standing in the pipes overnight. These results therefore do not tell the true extent of the problem. Although the contribution of lead from drinking water to the diet of most people will be small, it contributes in excess of 50% when households have lead pipes and receive plumbosolvent water. Of course, this percentage increases dramatically in individuals who drink a lot of water, which children do, especially infants fed with formula milk feeds made up with tap water.

Lead is odourless, tasteless and colourless when in solution, which makes even fairly high levels in drinking water undetectable unless chemically analysed. Another problem is that lead is reasonably soluble, so it is readily absorbed by the body. A considerable proportion of lead in water may be in particulate form rather than in solution and this may affect its uptake (Hulsmann, 1990). Absorption varies widely between individuals, with children absorbing lead far more readily than adults. This is especially true of young children, infants and fetuses, who are all growing rapidly; in fact, they absorb lead five times faster than adults. Lead attacks the nervous system and can result in mental retardation and behavioural abnormalities in the young and unborn. This has been supported by studies carried out by a number of research teams in Scotland and the USA. A study by Edinburgh University in 1987 found that even among children who were not exposed to high levels of lead and who had blood lead levels 50% less than that considered safe were showing signs of learning difficulties. In fact, the study, which looked at the mental aptitude of 500 children, concluded that they performed on average 6% worse. The research team concluded that there is perhaps no safe level of lead exposure for children (Thornton and Culbard, 1987).

The World Health Organization have identified three especially vulnerable groups. These are fetuses, younger children and patients with specific disorders which require an increased intake of water or the need for renal dialysis. Even so, the standards set by the World Health Organization and the EC do not take into account the facts (1) that lead is more readily taken up by children, (2) that it causes them more serious damage in the early years when the brain is developing, or (3) that certain groups are more at risk.

Research by Glasgow University in the city, which is one of the worst affected areas in Europe for increased lead levels in drinking water, produced a number of worrying facts about the risks that unborn children face from lead. For example, they found that infants whose mothers drank water with a high lead concentration during pregnancy were twice as likely to be mentally retarded. The rate of stillbirths also increased within this group. When placentas were examined only 7% of the placentas from normal births had high levels of lead compared with 61% of those from infants who were stillborn or died shortly after birth. In a city where in the late 1970s and early 1980s it was common to find tap water with 10, 15 and even 20 times more lead than the 100 μg/l considered safe at that time, it is perhaps not surprising that more than 10% of all the infants born had lead levels in their blood at the time of birth which would have been considered unsafe for fully grown adults.

7.3.3 Standards

The World Health Organization (1984) has published guidelines for lead in drinking water and recommends that a maximum acceptable limit is 50 μg/l. The EC Drinking Water Directive places lead in section D, which covers toxic substances (Table 1.10). The MAC level is given as 50 μg/l, although this applies to running water when the sample is taken after flushing. If the water is allowed to stand and not flushed before sampling, then 100 μg/l is allowed, although in the Directive it makes it clear that where the upper lead limit is frequently exceeded, then suitable measures must be taken to reduce the exposure of consumers. Until the mid-1980s, 100 μg/l was considered to be a safe maximum in the UK, but public pressure resulted in a new standard being set by the Water Supply (Water Quality) Regulations in England and Wales, and similar values have been set for the rest of the UK of 50 μg/l. The UK Government felt that the provisions in the Directive relating to flushed samples were ambiguous in the absence of definitions of the terms flushing, frequency and appreciable extent, and did not include them in the new Regulations.

The 50 μg/l limit is based on an earlier EC Directive (EC, 1977), which requires that no more than 10% of the population should exceed a blood lead level of 30 μg/100 ml blood. However, it is extremely difficult to relate lead levels in water directly to lead levels in blood, although this has been done for bottle-fed infants in Glasgow (Lacey et al., 1983). The blood lead level of 30 μg/100 ml is fairly arbitrary and there is now considerable evidence that much lower blood lead levels can cause health problems. In the USA, blood lead levels above 15–20 μg/100 ml are considered by the USEPA to cause concern, although there is still much debate over what is a safe blood level, and subsequently a safe lead concentration in drinking water (Mullenix, 1992). The USEPA set the current MCL for lead at 50 μg/l in 1975 with no MCLG value specified. In August 1988 it proposed a MCLG value of zero and a MCL value of just 5 μg/l for water leaving the treatment plant and entering the distribution system. If the lead level exceeds certain threshold values, the so-called 'no-action levels', then certain remedial measures must be taken by the supplier. For example, if the average lead level in the system exceeds 10 $\mu g/l$ or the pH is less than 8.0 in more than 5% of the samples collected at consumers' taps, then corrosion control is required. If 5% of these samples exceed 20 $\mu g/l$ then a public education programme must be instigated (Cox *et al.*, 1992).

The new USEPA standards are in line with the revised World Health Organization guidelines for lead in drinking water. Originally 50 $\mu g/l$, the new guideline is only 10 $\mu g/l$, reflecting the toxicity of lead (World Health Organization, 1993). This will place increasing pressure on the member states of the EC to introduce similar standards.

7.3.4 Levels in Water

It is recognized that soft acidic waters are plumbosolvent, therefore areas where substantial plumbosolvency problems will occur can easily be predicted by carrying out simple water quality measurements. Although it is also known that some hard waters can also give rise to high lead levels, it is not so easy to identify such sources accurately. In 1980 it was estimated that between seven and 10.5 million of the total 18.5 million households in the UK had lead pipes somewhere between the connection with the mains and the kitchen tap. It was thought that of the five million homes in areas where water is significantly plumbosolvent about 50% had lead in the connection pipes and that many more had lead pipework within the house.

The Water Supply (Water Quality) Regulations require companies to treat water supplies if a significant reduction in lead concentration can be achieved by doing so, and if treatment is reasonably practicable. In 1991 the water supply companies took over 58 000 samples of drinking water from 2577 water supply zones in England and Wales and tested them for lead. Three per cent of these samples exceeded the PCV standard, showing 661 (26%) water supply zones to be at risk, which is a significant increase from the previous year. Currently 2174 of the water supply zones are covered by undertakings to comply with EC standards by 1995 (Department of the Environment, 1992).

Some of the worst affected areas outside Scotland, where the situation is extremely serious, include parts of Lancashire and Greater Manchester, which are served by reservoirs located in the Lake District and operated by North West Water. Other areas of concern where the water is potentially very plumbosolvent are Preston, Hull, Blackburn and Blackpool. Two horrendous examples were found by Friends of the Earth (1989) in tap water in Dilmorth near Preston and in Blackburn, which contained 3600 and 7750 $\mu g/l$ of lead, respectively. Such high levels are generally due to old and damaged lead pipes where particulate lead is scoured from the surface of the pipe. Many houses are served by very long lead service pipes (Section 3.3) and this can result in long pipe–water contact times, resulting in high lead levels where the water is soft and plumbosolvent. In a survey of 2000 houses in the UK the lead concentration in tap water was shown to be related to the pH and the length of the service pipe (Table 7.4). Service pipes shorter than 10 m were shown to have significantly lower lead levels (Pocock, 1980). Most of the Midlands and London are free from problems. The situation in other developed countries is similar to the UK (Gendlebien *et al.*, 1992).

Table 7.4. Percentage of houses, based on length of service pipe and water pH, with a first flush of lead in excess of 100 $\mu g/l$. Reproduced from Department of the Environment (1977) with the permission of the Controller of Her Majesty's Stationery Office

Length of lead household and supply pipes (ft)	pH of mains water			
	<6.8	6.8–7.5	7.6–8.3	>8.4
<16	60	22	4	10
16–32	62	16	6	10
32–65	66	28	15	4
≥65	74	30	20	7

7.3.5 Factors Affecting Lead Levels

Many factors affect the concentration of lead in water and these have been studied extensively. The solubility of lead is temperature dependent, so much so that marked differences are seen over the year, with summer levels often double those recorded in the winter. As a guideline, an increase of 10°C will double the rate of leaching through oxidation. This is also seen when samples are taken from hot and cold taps, with levels significantly higher in hot water drawn from taps. This may cause problems in bathrooms, where the hot tap is often used to rinse teeth. Hot water taps should never be used for drinking. In Scotland, a survey found that it was widespread practice to fill the kettle from the hot water tap to make it boil faster. These results are shown in Table 7.5. It is important to wash out the kettle each time it is used if a lead problem is suspected. This is because boiling water increases the lead concentration. Table 7.5 shows the difference between first draw samples, where the water has been stored in the pipe overnight and so the lead has had plenty of time to solubilize, and samples taken during the day, which show the normal background level expected from the plumbing. Levels of lead taken during the day from the hot water tap show a three to 10-fold increase in lead concentration. The range of lead levels in the kettle from 14 households had maximum concentrations over 200 times higher than the EC limit. In a survey conducted in 1979 on water usage, it was discovered that only one-third of the population emptied their kettle before refilling it each time (Hopkins and Ellis, 1980). It is important for consumers never to top up kettles, but to always empty them before refilling.

Apart from the contact period, temperature and pH, there are many more secondary factors which contribute to the overall lead concentration. The velocity of water flowing through lead pipes has an important influence. At very low flows there is an increased contact time, so that the lead levels are high (Figure 7.6). This is why the water first drawn from taps after it has been standing in the pipes overnight is rich in lead and why it is always best to flush this lead-rich water from the pipework before using the water. The first water drawn from the kitchen tap in the morning will invariably be to fill the kettle, so this, combined with the concentration effect of boiling, will mean that the first cup of tea of the day will contain the highest concentrations of

Table 7.5. Lead concentrations from a number of different households in the cold water taken first thing in the morning (first draw) and during the day, from the hot tap during the day and in the kettle. Reproduced from Britton and Richards (1981) by permission of the *Journal of the Institution of Water Engineers and Scientists*

Household No.	Lead concentration in samples (mg/l)			
	Cold water first draw	Cold water random daytime	Hot water (random)	Kettle (range)
1	0.19	0.04	0.41	0.03–1.03
2*	0.06	0.04	0.28	0.26–0.37
3*	0.03	0.02	0.43	0.04–0.53
4	0.28	0.15	0.88	0.18–0.30
5*	0.05	0.03	0.71	0.03–1.05
6*	0.55	0.07	0.39	0.16–0.65
7*	0.48	0.15	0.47	0.23–0.61
8	0.04	<0.01	0.46	0.03–0.13
9*	0.12	0.05	0.43	0.12–0.32
10	0.68	0.10	0.31	0.38–1.09
11	0.21	0.04	0.15	0.34–0.51
12*	0.16	0.09	0.37	0.06–0.39
13*	0.50	0.26	0.34	0.23–0.45
14*	0.07	0.07	0.55	0.07–0.56

*Households where hot water is known to be drawn regularly or occasionally when filling kettle.

lead of the day as well! The rule for consumers is to always flush taps, if they have lead pipes, before drinking the water. The period of flushing depends largely on the length of the lead service pipe to the house, which if it is a joint supply pipe or is quite a distance from the boundary stoptap (i.e. more than 30 m) will take much longer. This is considered in more detail at the end of this section.

As expected, as the velocity of flow in the pipe increases, the lead concentration begins to decrease in the water. However, at high flow velocities the lead concentration can, in older pipes, rapidly increase again due to lead deposits being physically scoured from the pipe. This may be a problem with joint supply pipes made from lead which serve a number of households and where the water usage is high at certain times of the day or week. The lowest concentrations of lead are found at intermediate flow-rates. What is important is not to drink water from the tap without flushing first, if the water has been standing in the tap for a long period. As can be seen in Figure 7.6, 30 minutes standing in the pipe can double the lead concentration in water and in some instances it can treble in 60 minutes. This is, of course, dependent on the plumbosolvency of the water, but illustrates the point that consumers must be careful.

The age of lead pipes is also critical in terms of the amount of lead leached. As they age the plumbosolvent action increases due to a greater surface area being exposed, with localized corrosion accelerating at sites where there are casting faults and impurities. In Glasgow the average life span of a communication pipe is 58 years. Britton and Richards (1981) have calculated the life span of such pipes to be between 43

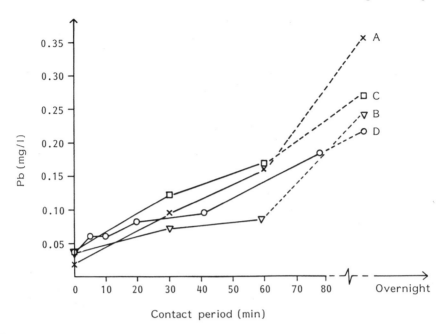

Figure 7.6. The longer the contact time between the water and the lead pipes, the greater the concentration of lead in drinking water. This is shown for four homes. Reproduced from Britton and Richards (1981) by permission of the *Journal of the Institution of Water Engineers and Scientists*

and 55 years, although with effective water treatment to reduce plumbosolvency this may be increased to between 135 and 170 years.

Lead can be a short-term problem in modern buildings. Tin–lead soldered joints are widely used to connect copper pipework. There is always a danger when different metals are used in plumbing that a galvanic interaction will occur, in this instance between the lead solder and the copper pipe, resulting in electrochemical corrosion (Section 7.2). The rate of corrosion is much higher under these circumstances than that which occurs naturally, it is enhanced by the flow of water and will be much higher in hot water pipes. This source of lead does not usually result in serious increases in the lead intake as it is generally only the internal pipework that is affected, and not that to the cold water tap (Oliphant, 1983). High lead concentrations have been reported in newly constructed hospitals and schools with pipework made entirely from copper pipes. Table 7.6 shows such an example for a school on Arran, Scotland. These samples were taken three years after the lead plumbing had been replaced by new copper pipes. The over-zealous use of lead solder should clearly be avoided by plumbers. The corrosion of tin–lead solder is reviewed by Gregory (1990). In 1986 the US Congress banned the use of solder containing lead in potable water supply and plumbing systems. It did this by classifying it as a hazardous substance and requiring warning labels to be used stating that it was prohibited for use with potable water systems. Only lead-free materials can now be used. Solder is considered to be lead-free if it contains no more than 0.2% lead, whereas pipes and fittings must not exceed 8.0% lead (USEPA,

Table 7.6. Illustration of the problem with plumbosolvency after the removal of lead plumbing at a school. Reproduced from Britton and Richards (1981) by permission of the *Journal of the Institution of Water Engineers and Scientists*

Sampling point*	pH	Copper concentration (mg/l)	Lead concentration (mg/l)
Water supply to school	6.9	0.01	0.01
Cold tap: sink unit (single) (classroom unit)	6.4	0.55	0.12
Cold tap: standpipe (old building)	6.3	0.78	0.01
Cold tap: wash-hand basin (boys' toilets)	5.8	0.99	0.09
Cold tap: wash-hand basin (girls' toilets)	6.3	1.57	0.12
Cold tap: sink unit (primary 7 class)	6.6	1.44	0.28
Servery: drinking tap	6.5	2.20	0.26
Cold tap: sink unit (servery)	6.6	1.89	0.14
Cold tap: sterilizer (servery)	6.7	1.70	0.08
Cold tap: wash-hand basin (foyer of servery) right hand	6.6	1.44	0.60
Cold tap: wash-hand basin (foyer of servery) left hand	6.6	1.78	0.17
Servery: drinking tap	6.8	0.56	0.70
Servery: drinking tap	6.3	1.73	0.23
Servery: drinking tap	6.3	1.58	0.11

*All are random daytime samples taken in May and June 1980. All lead replaced with copper in 1977–8

1988). Water coolers (drinking fountains) are commonly used in schools in the USA and were found to have lead-lined water reservoirs and lead-soldered parts. This was leading to increased lead levels in the water. Legislation, in the form of the Lead Contamination Control Act 1988, amended the Safe Drinking Water Act to protect children from this exposure by banning the sale or manufacture of coolers that are not completely lead-free.

Lead pipe, because of its age, is far more likely to burst and need repair. Often a section of copper pipe is inserted. This causes a significant increase in the water lead level in this section by direct electrochemical corrosion of the adjacent lead surfaces. The problem is extended by copper being taken up in the water and deposited onto the lead pipe as the water flows through the rest of the pipework. At each site where copper is deposited, a new galvanic cell is produced causing further corrosion. In the Renfrew study it was shown that much higher lead levels were recorded after copper inserts had been made either in the service pipe connection with the main or in the household plumbing. If property owners need to replace a section of lead pipe then they should, if possible, replace the whole pipe. Otherwise they should use plastic pipe, and special connections. In the past they would have been recommended to replace sections with new lead pipe as a last resort, rather than to use a section of copper tubing if they were unable to replace all the pipework. Since the introduction of the water byelaws in the UK it is now illegal to install any lead pipework, fittings, storage tank or a tank lined with lead, or to use lead in any repair in the home plumbing system (Byelaw 9). Copper pipes are no longer permitted to be used for repairs to lead pipework unless galvanic corrosion is prevented (Byelaw 10). In practice this is almost impossible and so plastic pipe should be used to repair lead pipes using special fittings. Remember that removing a section of metal pipe and replacing it with plastic pipe may affect the earthing of any electrical system.

7.3.6 Remedial Measures

There are a number of possible remedial measures that can be taken to reduce lead levels in drinking water. These are: (1) to reduce the level of lead in the resource, although this is rarely a problem; (2) to alert the consumer of the possibility of high lead levels and to identify who is at risk and what action should be taken to reduce exposure; (3) to reduce the corrosivity of water by neutralization; (4) to line lead service pipes; or (5) to replace lead service pipes.

An education programme is only an interim solution. All consumers receiving increased lead levels should flush their tap before using the water for drinking or preparing food. For most people about 40 l/day would have to be flushed to be effective, which is about 14.3 m^3/year. As much as 100 l may have to be flushed for those with joint supply pipes or longer supply pipes, which should be effective in 95% of cases. This is equivalent to 36.5 m^3/year. The total water usage each year is given to give some idea of the annual cost involved if supplies are metered. However, it is still a cheap option for consumers, although it is a waste of water. Water companies would clearly be unhappy about widespread flushing in the long term because of the extra water demand and the extra waste water generated which will require treatment at the sewage treatment plant. Many people now install a household treatment system. These are examined in detail in Section 8.2, but in reality none are 100% effective, although those using reverse osmosis or ion exchange are capable of removing 90–95% of the lead present under optimum conditions. It is undesirable to rely on a household system to remove a toxic substance such as lead in the long-term.

The usual remedial action is to treat all the water supplied to consumers at the water treatment plant to reduce its plumbosolvency, but this is expensive and causes secondary problems. This often involves relatively simple changes in water treatment practice to ensure that the water reaches the consumer at a pH of between 8.0 and 8.5. However, in practice operational problems can make this difficult. The required pH is achieved by adding alkalis such as lime, caustic soda or soda ash (Section 3.2). Lime is the cheapest, but is more difficult to handle resulting in excessive costs in terms of capital equipment and labour at smaller treatment plants. Some small plants use a filter containing semi-calcined (slaked) dolomite, which is simple to operate. Orthophosphate is also known to help control lead concentrations in waters with high alkalinity, although there may be long-term operational, quality and environmental problems associated with its use (Lee *et al.*, 1989; Colling *et al.*, 1992).

An example of water neutralization is the water supply to Glasgow from Loch Katrine, which had a pH of 6.3. This resulted in random daytime water samples containing on average more than 100 μg/l lead. From April 1978 to April 1980, the pH was increased to 7.8 so that by the time it reached the consumer the water had a pH of between 6.9 and 7.6. This resulted in 80% of the random daytime samples being below the 100 μg/l limit. Since April 1980 the mains have been dosed to a pH of 9, so that samples at the tap now have a pH in excess of pH 8.5, which has resulted in 95% of the homes treated now having lead levels below 100 μg/l, the remainder having lead-lined tanks or long service pipes.

Lead pipes can be lined with a plastic coating to prevent corrosion. This is cheaper than replacing the lead pipes, but so far no details have been made available about (1) its effectiveness, (2) the safety and suitability of the lining materials and (3) their durability. At the present time this option should only be considered in consultation with the water supply company. The best option, and the most successful and reliable, is to replace the lead pipes with copper and plastic. It is desirable, but not necessary, to replace all the household plumbing. Just the service pipe from the mains to your drinking water tap which is normally in the kitchen needs to be replaced. The new regulations which implement the EC Drinking Water Directive in the UK require the water undertakers (either a company or a council) to replace their part of the service pipe when requested to do so in writing by a consumer who has replaced his or her part of the lead service pipe in water supply zones considered to be at risk. It is difficult to estimate a cost for replacing a supply pipe as this depends on exactly where it runs. If it is currently under a driveway it may be possible to disconnect it entirely and run a new supply pipe to the house by a different route. If the lead pipework in the rest of the house is to be retained, it may be best to use plastic for the new supply pipe; certainly do not use copper upstream of any lead pipe for the reasons already outlined (Section 7.2).

If the cost of replacing lead pipes is compared with the cost of adjusting the pH of the finished water at the treatment plant, then pipe replacement is far more expensive in terms of capital costs. For the water supply companies, replacing all the lead service pipes, which they are in fact legally required to do if requested, would probably cost the same as identifying and treating all suspect supplies. If they did replace all the communication pipes then they would certainly save in the long term, as the extra operating costs of treating the water would be very high indeed, possibly in excess of £12 million each year, and this would be a continuing annual cost. The cost of replacing the household supply pipes and associated plumbing would cost almost double what it would cost the water supply companies to replace their lead communication pipes. The household costs would vary, possibly averaging out at about £350, although it is hard to be exact. There are still between 30 000 and 60 000 lead-lined tanks supplying water for all domestic purposes, including drinking, in the affected households. Their use is restricted to dwellings, mainly tenement properties, in the Edinburgh and Glasgow areas. These are often exposed to the atmosphere and although the water serving Glasgow from Loch Katrine, for example, is treated to keep its pH well above 8, the pH in these tanks falls to less than 8 on storage and results in the leaching of lead.

All the EC countries have indicated that if the current MAC value for lead is reduced to 10 μg/l, in line with the World Health Organization revised guidelines, there will be considerable difficulties in compliance, especially with pipe replacement seen as the only long-term solution. In France it is estimated that it would take 30 years to replace all the lead pipes currently in use. In Belgium the problem is particularly acute in the eastern region where soft moorland waters are used for supply. Also, in some areas the raw water contains natural concentrations of lead and cadmium which have to be removed by treatment at high pH. It is estimated that to conform with the new standard would require 50 000 pipe replacements, which could take up to 20 years to complete. In most European countries lead is now banned. However, in Finland, where lead has been banned for over 100 years, there are no known problems with lead in the water supply and any new lower limit will be readily adopted.

7.3.7 Babies and Lead

People planning to try for a family, or indeed who have a family, should be advised to have their water tested for lead. If the water contains high levels, and during the vital time of pregnancy anything over 10 μg/l is too high, the mother should drink an alternative supply of water. The EC says that 50 μg/l is safe, whereas the World Health Organization specifies 10 μg/l. However, based on more recent studies the EC value is clearly too high for young children and especially for fetuses. In my opinion I would prefer women not to drink water containing more than 10 μg/l while they are pregnant. Personally, I feel that 10 μg/l is about the right permissible level for very young children and pregnant women. Ideally, the lead service pipe should be replaced or, alternatively, bottled water should be used. If the mother is going to breast-feed, she should continue to drink bottled water. If the baby is to be bottle-fed and there are increased lead levels in the kitchen tap water, then bottled water should be recommended for making up the feeds, but *ONLY IF THE FOLLOWING ADVICE IS CAREFULLY ADHERED TO*. (1) *DO NOT* use sparkling mineral water, use only still water. (2) Check that the bottled water has a low nitrate and sulphate concentration, and has as low a mineral content as possible. (3) If it does not give details of the composition of the water on the label, then it should not be used. (4) If it tastes salty or has a discernible mineral taste, then it should not be used. (5) Before use, empty the kettle fully and then *BOIL* the bottled water before making up the feed.

All bottled waters are fine for older children, except for the very salty or strong mineral waters (Section 8.1). For children below the age of five years use a non-sparkling, low nitrate and low sulphate water. As you can see, I am not too happy about recommending the use of bottled water to feed babies. This is because there are real risks involved, as I explain in Section 8.1. Only a few bottled waters are suitable for bottle-feeding babies and infants, and some may pose a serious threat to health if used for making up formula feeds. Extreme care must therefore be taken.

7.4 COPPER

Copper is an essential element for human health. The average western diet will contain between 2 and 5 mg of copper each day; it is also widely used in drugs with concentrations up to 20 mg/l common. Although it can be toxic at high concentrations, it rarely causes problems in genetically normal subjects. This is because the body can excrete excessive copper, unlike lead which is accumulated. Because copper is essential to the body it is bound up in protein complexes, which reduces the concentration of free copper ions throughout the body, thus avoiding toxicity. It has been noted that copper lowers the toxicity of lead, and that lead induces copper deficiency symptoms. Tissue concentrations of copper remain fairly constant, except for changes due to age, regardless of exposure. This was shown by a study on copper miners in South America who were exposed to an atmosphere containing 2% copper ore dust for up to 20 years. Liver and blood analysis showed their copper levels to be normal.

Concentrations in drinking water are generally low, with an astringent taste caused by copper at concentrations in excess of 3 mg/l (3000 μg/l). Copper is naturally present in waters in metalliferous areas, but more often its presence in water is due to the attack of copper plumbing. Where copper sulphate has been used as an algicide in reservoirs, increased copper levels have occasionally been reported in drinking water, although this is rare. If acidic drinks or food are stored in copper vessels for long periods, then high concentrations can result, which will lead to mild toxicity, the symptoms of which are nausea, vomiting, diarrhoea and general malaise.

Copper pipes are widely used in houses for the hot and cold water systems. There are in excess of 6.5 million copper service pipes in the UK, which represents about 42% of all service pipe connections. The corrosion of copper pipework is known as cuprosolvency and results in copper being leached into solution, giving rise to blue-coloured water and occasionally taste problems. These problems occur with soft acidic waters or with private supplies from boreholes where free carbon dioxide in the water produces a low pH. Apart from pH, other important factors in cuprosolvency are hardness, temperature, the age of the pipe and oxygen availability. The inside of the pipe is protected by a layer of oxides, but water with a pH less than 6.5 and a low hardness less than 60 mg/l as $CaCO_3$ will result in corrosion. Hot water accelerates the corrosion rate, although oxygen must be present. If oxygen is absent, corrosion will be negligible regardless of the temperature. The corrosion of pipes can be severe causing bursts in the pipework, usually the hot water pipes, the central heating system or the hot water cylinder. Under such conditions some form of corrosion control is recommended. The corrosion of copper pipework is considered in more detail in Section 7.2.

Where the water is cuprosolvent an astringent taste may be produced and copper levels may increase enough to cause nausea. This is why water should not be drunk from hot water taps, as the water heated in the copper cylinder will have been in contact with copper for some time at increased temperatures, allowing maximum corrosion to occur. Copper is a relatively soft metal and corrosion due to turbulent or excessive flows can occur. Where the water is also soft with a low pH and warm, this will seriously aggravate the situation. Pitting of copper pipes is associated with hard to moderately hard waters, often occurring in the cold water pipes where carbon dioxide exceeds 5 mg/l and dissolved oxygen is high. This also gives rise to copper in the water. Raising the pH of the water to 8.0–8.5 overcomes most corrosion problems.

The EC has set two guide (G) values, but no mandatory values, for copper. A G value of 100 μg/l applies to outlets of pumping and/or treatment plants and their substations, whereas a level of 3000 μg/l (3 mg/l) applies after the water has been standing in the pipes for 12 hours at the point where the water is made available to the consumer (i.e. the kitchen tap). The EC notes that above 3000 μg/l taste problems, discoloration and corrosion will occur. The UK Government has set a single standard for copper of 3000

μg/l. The Irish national limit is 500 μg/l at the outlets of pumping and/or treatment plants, and 3000 μg/l from taps after 12 hours within the pipes. In my experience levels in excess of 2000 μg/l will cause taste problems, so the PCV set by the Water Supply (Water Quality) Regulations appears to be at variance with its definition of wholesomeness. The National Primary Drinking Water Regulations in the USA have a threshold value of 1 mg/l for copper. In accordance with the 1986 amendment to the Safe Drinking Water Act, the USEPA have proposed guide (MCLG) and mandatory (MCL) levels of 1.3 mg/l entering the distribution system. Corrosion control is required if copper levels at the kitchen tap exceed 1.3 mg/l in 5% of the samples tested, or, as with lead, if the pH is less than 8.0 in more than 5% of the samples tested. In their new guidelines, the World Health Organization has set a maximum value for copper of 1 mg/l. This concentration is lower than the health-based guide value of 2 mg/l, as at concentrations above 1 mg/l staining of laundry and sanitaryware occurs.

Copper rarely causes health problems, although householders can have problems with taste, especially if they are particularly sensitive. The most common problem relating to copper will be blue water and the staining of sanitaryware. A particular problem associated with new plumbing is increased concentrations of copper. In these circumstances children should not be allowed to drink from bathroom taps or drink their bath water, as the levels of copper can be high enough to make them feel nauseous. If the water is to be used to make a drink the cold water taps should be flushed for a few minutes in the morning or if the water has been standing in the pipes for over an hour. Unless the water is acidic the problem should quickly resolve itself as the pipework becomes coated with a layer of oxides. Children should be taught only to drink water from the cold water tap in the kitchen. Copper begins to impart a blue colour to water at concentrations above 4–5 mg/l. Such water should not be drunk or used in cooking or used for washing hair which has been highlighted or bleached.

7.5 ZINC

Zinc, like copper, is essential to human health but if ingested in gross amounts from water it will have an emetic effect. The inhalation of zinc-containing fumes is much more serious, but in water fairly high levels are permissible, the limiting factor being its taste threshold. The EC has taken a similar approach to zinc and copper and has not set a mandatory (MAC) value, only a guideline (G) values. For zinc these are 100 μg/l at the outlets of pumping and/or treatment plants and their substations, and 5000 μg/l regardless of whether the water has been in the pipe for 12 seconds or 12 days. The Irish standards are 1000 and 5000 μg/l respectively (Table 1.10), whereas the UK PCV of 5000 μg/l applies to samples collected at the tap only. The World Health Organization (1993) has set a guide value of 3000 μg/l based on aesthetic reasons only.

Zinc can arise from metalliferous areas where zinc ore has been mined. For example, in County Wicklow, Ireland, which is an old mining area, the groundwater is so rich in zinc, copper and lead, that the water has to be supplied from a nearby aquifer which is free from contamination. The prime source of zinc in most homes will be from the corrosion of galvanized steel tanks and pipework. This is caused by low pH, so increasing the pH will help. When copper and galvanized zinc are used together then corrosion of galvanized steel occurs due to galvanic action. This is particularly important if the water is cuprosolvent (acidic) and the copper piping is located upstream of the galvanized steel. Just a small amount of copper (0.1 mg/l) is enough to increase the corrosion rate of the galvanized steel considerably. The water byelaws do not permit different metals to be connected where there is a chance of corrosion occurring through galvanic action. If copper is used downstream of galvanized steel, no problem will occur. The corrosion of galvanized steel is discussed fully in Section 7.2, and is the source of the sandy water often noticed when running water in the bath.

The use of galvanized steel is not recommended, especially with other metals such as copper. New plastic water tanks are more effective than the old galvanized tanks, are considerably lighter and are less likely to burst.

7.6 ODOUR AND TASTE

Household plumbing systems are in many ways ideal places for microbial growth to occur. The many connections leave areas of stagnant water where sediment can accumulate and micro-organisms can survive. The system may contain organic-based pipes, tanks and even washers where micro-organisms can live and obtain nutrients. The problem is acute in plumbing systems which have a low turnover of water and where water is warmed by running cold water pipes close to those carrying hot water. Warming also occurs in large blocks of flats where the cold water tap may be fed from a storage tank which in turn is fed by an auxiliary pump because the pressure in the water mains is insufficient to reach the top of the building (Section 3.2). In large buildings and offices it is important that the pipes carrying drinking water are insulated by lagging and that they are distributed through the building via separate ducts away from central heating systems and hot water pipes. Under warm conditions actinomycetes in particular thrive, so that musty and earthy odours will be produced. In large buildings the central storage tanks should be examined periodically for microbial growth, and if necessary, cleaned out. The materials and fittings used in plumbing systems should be those tested and certified as not supporting microbial growth.

An astringent taste in water can be caused by the corrosion of pipes and fittings containing copper or zinc. Where corrosion is severe due to water acidity, the water may become undrinkable. Standard taste threshold concentrations for copper vary from 3 mg/l for the most sensitive 5% of the population to 7 mg/l, which is the mean taste threshold concentration. This compares with 5 and 20 mg/l, respectively, for zinc. These figures may in practice be too high, especially in soft water areas.

In Section 3.3 it was pointed out that polyethylene, especially low density polyethylene, is permeable to a wide range of compounds, especially hydrocarbons and phenols. Although convenient for use with cold water in household plumbing systems, polyethylene can occasionally be the unexpected source of an odour problem. Careful consideration must be given to where these pipes are to be laid. For example, contamination could occur if the pipework is in contact with bituminous materials or the solvents which are now so widely used in the construction and building industries. The diffusion of materials from gas pipes can also lead to problems. An interesting example is a polyethylene pipe which ran along the walls of a cellar. The walls had been painted with bitumen paint to dampproof the cellar and the gas supply entered the house through the cellar and ran parallel against the water pipe to the meter. As a result of the permeability of the polyethylene pipe the water occasionally had a strong tarry taste.

Further information on odour and taste is given in Section 4.4.

7.7 FIBRES, INCLUDING ASBESTOS

Most of the fibres found in drinking water are asbestos. These originate either from the resource, or more likely in the UK from asbestos cement pipes used in the water distribution network. Asbestos also can originate in the home plumbing system from the water storage tank. This is only a problem where open-topped tanks are located in the roof, which can receive asbestos fibres from atmospheric fallout. This may be significantly higher in cities and industrial areas, so open-air tanks in particular need to be sealed. In Ireland many isolated cottages and dwellings which do not have a natural water supply close at hand collect rain-water from their roofs for use in the home. This is especially common on farms, where roof water is usually collected for watering the animals. Where this water drains off asbestos sheeting or asbestos roof tiles, then increased levels of asbestos fibres are found in drinking water. Water collected from roofs is fine for most household purposes, but cannot be recommended for drinking without treatment. In practice it usually tastes awful, as organic matter from the roof quickly decays anaerobically in the bottom of the tank producing sulphides which smell and taste dreadful. Iron levels can also be high if the water is collected from an old corrugated roof and may stain clothes during washing. If rain-water has to be used, then consumers must be advised to regularly clean out any debris from the storage tank, to keep the tank covered to prevent algal growth and colonization by dipteran larvae and other insects and, if possible, prevent any organic debris, especially leaves, from entering the tank. Rain-water is naturally acidic and so acts aggressively on the roof surface and the tank. If the roof is made of asbestos based material, galvanized steel or has been bitumen-sealed, then it should not be used for collecting drinking water.

Glass, carbon and other mineral fibres may also be carcinogenic to humans. Animal experiments have confirmed that they can produce tumours. Most attics are now insulated, and indeed the water tank is also insulated, often with glass fibre. Attics have an air flow which can at times be fairly turbulent, so fibres are constantly being blown around the attic space. If the water storage tank is open these fibres will fall into and collect in the tank. It is therefore advisable to cover it to prevent fibres, and other material, from falling into the tank (Section 7.1). Asbestos fibres can accumulate in the tank and if the house is situated at the end of a cul de sac or if the distribution mains is a spur, then the tank should be checked periodically to see if there is any accumulated debris. If there is a deep deposit (greater then 10 mm) then the tank should be drained and all the accumulated debris carefully cleaned out. This should only be necessary at very long intervals.

If sediment in the tank is a serious problem then the distribution system should be suspected and the water supply company contacted to flush out the mains. After flushing has been completed by the company the consumer should shut off the supply to the storage tank, and then run the kitchen tap, which is connected directly to the supply pipe, until the water is absolutely clear. The consumer will usually only need to run the tap for 10 minutes or so, but where there is a long service pipe it may take a long time, perhaps an hour or more, before all the debris is finally flushed out.

7.8 MICRO-ORGANISMS IN PLUMBING SYSTEMS

We have seen that the contamination of drinking water by pathogenic and non-pathogenic micro-organisms occurs mainly at source (Section 4.7), although contamination can also occur during treatment or within the distribution systems (Sections 5.3 and 6.7). The contamination of otherwise potable water can also occur within the consumer's premises. This is generally due to the type of plumbing system installed, a lack of basic maintenance and the careless use of appliances.

The householder is responsible for the maintenance and repair of the supply pipe, which runs from the water supply company's connection pipe at the boundary stoptap to the house. All the water entering the house passes through this pipe so any fracture will allow possible contamination to enter. Sewerage pipes should be located well away from the supply pipe, preferably on the other side of the building.

Back-syphonage in plumbing systems is more of a problem in older buildings as modern building regulations and water byelaws incorporate measures to prevent it. It generally occurs when a rising main supplying more than one floor suffers a loss of pressure at a low point in the system, causing a partial vacuum in the rising main. Atmospheric pressure on the surface of, for example, a bath full of water on an upper floor in which a hose extension or a shower attachment from an open tap has been left, will push the contents of the bath back up through the hose and tap into the plumbing system to fill the partial vacuum. Plumbing systems suffer from constant changes in pressure, so care should always be taken when hoses are left to run water into containers. Farmers, who often need to fill large pesticide sprayers with water, need to take great care to ensure that the end of the hosepipe is never allowed to fall below the liquid surface in case of back-syphonage. This is extremely important if private supplies are pumped by a submersible pumping system from an underground storage tank. Once the pump is switched off any water left in the pipework connected directly to the pump will drain back into the tank. If the pump has been used to fill a watering can or a water trough, for example, and if the end of the hose is not removed before the pump is switched off, then the contents of the container will be drained by back-syphonage into the storage tank.

Mechanical back-flow prevention devices are available to prevent water flowing back into the household plumbing system or even the mains distribution system from the consumer's premises. Such devices are widespread in the USA and other European countries but are only now becoming widely used in the UK. A list of devices permitted by the UK byelaws (Byelaw 11) can be obtained from water supply companies. There are a number of back-flow devices, both mechanical and non-mechanical, available in the UK at present, including pipe interrupters which are now mandatory for all hosepipes (Figure 7.4). These are fully explained by White and Mays (1989) and more technically by Halford and Bond (1986).

Water storage tanks must be covered (Section 7.1). When left open they become a breeding ground for a wide variety of micro-organisms. They are also fouled by birds and vermin. Many houses have mice in the attic, a favourite place for overwintering, under or in the six inches of insulation. The only source of water in the attic is the cold water storage tank, so fouling by vermin is fairly common. In 1984 there was an outbreak of *Shigella sonnei* at a boarding school in County Dublin, which caused widespread gastrointestinal illness among the pupils. The source was identified as the water storage tank which was supplying all their drinking water and which had been fouled by pigeons.

Contamination by rats can also lead to leptospirosis or Weil's syndrome, although entry of the bacteria into the body is normally via small cuts or the mucous membranes from where they enter the bloodstream and affect the kidneys, liver and central nervous system. Nearly all those affected work in high-risk occupations, e.g. handling animals, meat processing, sewage workers, fish-farmers, river inspectors and water scientists. Infection of people in low-risk categories comes almost exclusively from swimming in contaminated water, although all those engaged in water sports on both inland and coastal waters are at risk. The rapid increase in reported cases in the UK since 1986 has been linked with the increasing popularity of water sports and of windsurfing in particular. The chance of infection from drinking water is slight, but tanks should be covered to be on the safe side.

7.8.1 Legionnaires Disease

Legionnaires disease was first reported in 1976 when there was a major epidemic among the residents of a hotel in Philadelphia. The American Legion was holding its annual conference at this hotel, hence the name of the disease. Since then numerous outbreaks of the disease have been reported, including many in the UK. In 1985, 75 000 cases were reported in the USA with 11 250 deaths. The disease is a severe form of pneumonia. Although there are over 20 strains of *Legionella pneumophila* known, only one serotype is thought to be a serious threat to health. The bacterium has been associated with domestic water systems, especially hot water which is stored between 20 and 50°C. It appears that a long retention time and the presence of key nutrients such as iron provides ideal conditions for the bacterium to develop. Therefore iron storage cisterns subject to corrosion will be susceptible. In addition, water pipes which are installed alongside hot water pipes or other sources of heat may also allow the bacteria to develop. The bacterium is widespread in natural waters so any water supply can become contaminated (States *et al.*, 1990). The main mode of transmission appears to be from bathing, especially showers.

7.8.2 Other Sources

Non-pathogenic bacteria can develop in rubber extension pipes on taps and in mixer taps, leading to taste and odour problems. Long rubber extension pipes which extend into contaminated water (during dish or clothes washing at the kitchen sink) will quickly develop thick microbial contamination inside the pipe. Such pipes should be avoided whenever possible because they can cause a significant reduction in water quality, and can also result in back-syphonage. If consumers need to use these pipes, and many elderly or disabled people find them especially useful, then they should be removed and cleaned regularly.

People using home treatment systems should ensure that they are regularly cleaned or the cartridges replaced as stated in the instructions. Micro-organisms will grow on activated carbon granules held in cartridges and may, ironically, cause taste and odour problems. If consumers are not changing the cartridges regularly, then it is best not use them at all (Section 8.2).

Chapter 8
_____ Alternatives to Tap Water

8.1 BOTTLED WATER

The increase in the consumption of bottled water in the UK and Ireland is phenomenal, with the UK market now worth over £200 million annually. In 1991 between 200 and 250 million litres of bottled water were drunk in the UK and this is increasing by 30% each year. In 1990 British and Irish people drank on average three litres of bottled water each. This is still low compared with about 50 litres per person each year in other European countries (France, 75; Belgium, Germany and Italy 60; and Spain and Portugal 25 litres). By 1992 this UK and Ireland figure of three litres per person was estimated to have doubled and to be rapidly increasing.

With so many new companies springing up, and with so much supermarket shelf space given over to water, bottled water is clearly very big business indeed. There are numerous mineral water associations throughout Europe such as those in France, Italy and a new one recently formed in Britain, the British Natural Mineral Water Association. There are currently over 40 bottled water companies in the UK; not all companies survive, however, and a number have disappeared in what is an extremely competitive market—for example, Bronte and Kristal. Bottled water is expensive because it is heavy to transport and overseas companies such as Perrier, Badoit, Evian and Volvic from France, Spa from Belgium, San Pellegrino and Fiuggi from Italy, Ramolosa from Sweden, and Apollinaris and Gerolsteiner from Germany are at a disadvantage compared with local bottled water suppliers. However, as the cost margin is small, the major companies have started buying up or developing new companies in target countries rather than exporting water direct. For example, Perrier has acquired Buxton Spring, and Spa has launched Brecon. With smaller overheads combined with experience and support from the parent company, these new labels are generally successful.

Bottled water is a generic term which describes all water sold in containers. There are other terms such as spring, table or natural mineral water, which are specific terms defined by legislation. Mineral waters are waters from recognized and licensed sources. They have no restrictions in mineral content and are exempt from the normal guidelines and limits laid down by the EC Drinking Water Directive. They are controlled by their own Directive. Spring, table or any other descriptive term other than mineral are not exempt from the Drinking Water Directive, and so must conform in every way, including the limits on minerals (e.g. calcium, sodium, magnesium, sulphate, hydrogencarbonate, chloride). Bottled waters are therefore of two types: natural mineral waters and other types of bottled water. The latter are just as pure, if not purer, than the mineral waters and unlike the mineral waters conform to the higher standards as laid down for drinking water. The exemption for mineral waters was to allow for personal taste differences. For example, some people like the very strong mineral waters such as Vittal; others prefer the medium mineral content found in brands such as Perrier.

8.1.1 EC Directive

During the mid-1970s the EC identified a number of concerns relating to bottled and natural mineral waters. These were: (1) the national requirements of individual Member States constituted barriers to free trade; (2) doubtful claims were being made about the medical or health benefits of such waters; (3) microbiologically unfit products were on the market; (4) some sources were contaminated or were at serious risk from contamination; and (5) the standard of bottling practice left much to be desired. This led to the development and finally the introduction of the Natural Mineral Waters Directive in August 1980 (80/777/EEC). The Directive only applies to waters described as natural mineral waters which are sold within the EC. It put into place a scheme for the recognition and exploitation of natural sources of mineral waters implemented by national authorities. The Directive stresses that only the purest of groundwaters with no organoleptic defect may be bottled and marketed as natural mineral waters. Only those products that have been recognized by the national authority can be called a natural mineral water, even though their purity and composition may be excellent. In the Directive there is a long list of requirements which have to be investigated and fulfilled before an application for natural mineral water status is even considered.

Apart from microbiological standards there are no limit values laid down in the Directive. Most importantly, there is no maximum permissible concentration for individual anions (e.g. chloride, sulphate, nitrate) or for total dissolved solids. Many natural mineral waters therefore exceed the MAC values for a range of parameters listed in the Drinking Water Directive (Table 8.1). Natural mineral waters can be classified into three groups: (1) natural mineral water, which is still; (2) carbonated mineral water, which is still water with added carbon dioxide; and (3) naturally carbonated natural mineral water, which is water that is carbonated in its natural form.

Among the restrictions in the Natural Mineral Water Directive is the prohibition of transport of water to bottling plants, so that the water must be bottled at source. Prohibition on treatment of the water is included, although a number of treatments are allowed particularly those relating to the bacteriological quality (e.g. simple filtration and carbonation), as long as no change is made to the essential characteristics of the water. Bottles must be securely closed with tamper-proof seals and there are restrictions on the claimed health benefits. Labelling requirements include the name and location of the source. The microbiological criteria laid down refer to the quality of the source and the quality of the water in the bottle during marketing. The source must

Table 8.1. Chemical constituents of some major bottled waters marketed in the British Isles. The list is by no means comprehensive and the selection of waters included is purely arbitrary. Waters are listed according to the country of origin

Country	Trade name	Ca^{2+}	Mg^{2+}	K^+	Na^+	HCO_3^-	SO_4^{2-}	NO_3^-	Cl^-	pH	Dry solids
Belgium	Bru	23	22.6	1.8	10.0	209	—	1.0	4.0	—	—
	Spa Reine	3	1.3	0.5	2.5	11	5	1.9	2.7	—	—
France	Evian	78	24.0	1.0	5.0	357	10	3.8	4.5	7.2	309
	Perrier	140	3.5	1.0	14.0	348	51	—	30.9	—	500
	Vichy	100	9.0	60	1200	3000	130	—	220	—	—
	Vittel	505	110	4.0	14.0	403	1479	0.7	11.0	7.0	2580
	Volvic	10	6.0	5.4	8.0	64	7	4.0	7.5	7.0	110
Ireland	Ballygowan	117	18.0	3.0	17.0	400	15	2.0	28.0	6.9	450
	Carlow Castle	117	15.4	5.3	13.1	355	61	8.6	10.2	7.4	560
	Glenpatrick	112	15.0	1.1	12.2	400	19	—	26.0	7.5	—
	Kerry Spring	76	9.6	—	24.0	244	8	2.1	46.0	6.9	—
	Slievenamon	112	15.0	1.1	12.2	400	19	—	20.0	7.5	—
	Tipperary	37	23.0	17.0	25.0	282	10	2.3	19.0	7.7	272
Italy	Aqua Fabia	129	4.0	0.9	11.2	384	—	—	19.5	—	—
	Clavdia	96	30.0	88.0	66.0	561	49	4.5	58.0	—	—
	Crodo lisiel	53	6.7	2.8	4.7	103	79	3.2	2.5	—	—
	C. Valle D'oro	519	49.0	5.0	1.8	69	1398	—	1.8	—	—
	San Pellegrino	208	56.4	3.0	41.1	226	539	1.0	71.0	7.1	11 204
	S. Antonio	61	8.9	0.7	3.9	196	10	18.7	9.0	7.7	218
	Sorgente Panna	31	6.1	0.9	6.5	104	19	1.8	8.9	7.4	136
Sweden	Ramlosa	2	0.5	1.8	220	535	14	1.9	24.0	8.7	515
UK	Ashbourne	102	24	3.0	10.0	—	60	6.0	30.0	7.4	420
	Buxton Spring	56	20	1.0	24	123	7	0.01	38.0	6.3	—
	Cerist	3	1.5	0.4	3.0	—	—	0.8	8.1	—	19
	Highland Spring	45	20	—	12.5	181	10	0.2	16.0	7.1	—
	Malvern Spring*	76	32	3.0	9.0	—	41	6.0	16.0	8.2	360
	Pentre Nant Spring	35	12	1.9	50	—	28	—	34.0	—	—
	Prysg	11	4.3	0.1	8.2	40	4	—	—	—	—
	Snowdonia Spring	7	6.1	1.0	5.5	6	6	—	17.7	—	—
	Strathglen Spring	24	5.0	0.3	7.5	95	10	1.8	11.0	7.3	119
	Stretton Hills	41	7.2	1.6	10.8	106	33	12.6	20.0	—	—

*The Malvern Springs Pure Water Company Ltd bottled at Aston Manor Brewery in Birmingham. There is another Malvern brand produced by Schweppes Ltd, which is a mineral water.

Table 8.2. Parameters required by the EC to be monitored to show the general composition and quality of mineral waters, and those substances regarded as toxic or undesirable. Values in parentheses are maximum allowable concentrations in $\mu g/l$

Parameters required to be measured		
pH	Calcium	Magnesium
Sodium	Potassium	Iron
Manganese	Hydrogencarbonate	Carbonate
Chloride	Sulphate	Ammonium
Nitrate	Nitrite	Total organic carbon
Electrical conductivity	Temperature	Phosphate
Free carbon dioxide	Fluoride	Silica
Toxic and dangerous substances		
Lead(10)	Arsenic(50)	Cyanide(50)
Cadmium(5)	Chromium(50)	Selenium(10)
Nickel(50)	Antimony(10)	Mercury(1)
Undesirable substances		
Copper	Zinc	Aluminium
Cobalt	Strontium	Lithium
Barium	Iodine	Bromine
Sulphide	Molybdenum	Silver

be free from pathogenic micro-organisms, with maximum total bacteria content less than 100/ml at 20–22°C and less than 20/ml at 37°C. Table 8.2 lists the parameters required to be measured to show the constancy of mineral composition and general organic quality.

Although the Natural Mineral Water Directive is operative throughout the European Community, different regulations apply to other bottled waters. There is separate national legislation specifically covering spring and table waters in France, The Netherlands, Germany and Italy. In the UK, mineral waters must comply with the Natural Mineral Water Regulations 1985, whereas all other bottled waters must conform with the Drinking Water Directive.

Bottled waters are covered by specific and general Directives, and so should be of excellent quality. However, there is little evidence that they are better than most tap waters; in fact, there has been much disquiet in recent years that some parameters are rarely, or even never, tested in natural mineral waters. Complex organic compounds can be found at high concentrations in some of the groundwaters from where these mineral waters are abstracted and the chemical quality can be variable. Some companies have over-pumped their groundwater reserves in an attempt to satisfy the growing demand and have found that the quality has altered. Problems at bottling plants have led to numerous breaches of bacteriological quality standards. Factors such as shelf life (water is like any other food and so once opened will begin to deteriorate even when kept in a refrigerator) all make the quality of bottled water every bit as suspect as tap water.

8.1.2 Mineral Content

The mineral content of bottled water is dependent solely on the rocks that the water comes into contact with and the length of time it has been in contact with them. This is explained more fully in Section 2.4. Some highly mineralized waters have very complicated hydrogeological origins, and some are naturally carbonated due to the upwelling of carbon dioxide. The sparkle in sparkling water, which is so much enjoyed in the British Isles and Germany, although it makes up less than 20% of mineral water sales in France, is caused by carbon dioxide. It can occur naturally, but more often than not it is added later. Perrier, for example, has found itself under criticism for labelling its waters as naturally carbonated. In the past the carbonated water was bottled directly, but problems were encountered with fluctuating carbon dioxide concentrations. It now extracts the water and gas separately and mixes them under controlled conditions to ensure a consistent product.

Europeans traditionally drank mineral waters as an aid to health, a tradition which is well known and still preserved at many British spa towns. Highly mineralized waters such as Contrexeville (France), Fiuggi (Italy) and Radenska (the former Yugoslavia) are renowned for curing urinary disorders and helping to break up kidney stones. Other waters with a high hydrogencarbonate concentration are used to ease indigestion. Evian was supposed to soothe skin diseases, whereas Apollinaris helped bronchial complaints. These were the traditional reasons for drinking mineral waters.

Some idea of the contents of a range of bottled waters widely available in the British Isles is given in Table 8.1. Most bottled waters declare a basic chemical breakdown on the side of the bottle. In many other European countries, such as Italy, for example, a more comprehensive breakdown of the contents of the water is given, including the name of the laboratory and the date the analysis was carried out. Full details of the resource are given, unlike Irish waters which have tended to concentrate on the historical and romantic aspects of the source. Italian companies also give a declaration about the bacteriological quality by the laboratory which carried out the analysis, as well as the date of bottling. However, if the labels from a bottle of Italian water, for example San Pellegrino, sold in Italy and in Ireland are compared, then some interesting differences emerge. There is no chemical or bacteriological analysis on the label of the bottle sold in Ireland and the information has reverted back to romantic descriptions of the source instead of hard facts. The declaration 'composition in accordance with the results of the officially recognized analysis of 27th September, 1990', which is on the export label of San Pellegrino, is unhelpful to anyone wanting to make an informed purchase. This declaration is common on many imported waters from outside the British Isles, and the problem is that even when the customer makes an effort to write to the company asking for a copy of the officially recognized analysis, they do not always oblige. The Italians treat bottled water more as a food, as indeed do the Americans, which is why it is so difficult to export mineral water to the USA. Clearly a full breakdown on the label of the contents of the water, including the date of bottling, is a prerequisite for consumers to make an informed choice. Many companies do this already, although the information is sometimes limited, often omitting certain key parameters such as nitrate.

The key parameters to look for on the label of a bottled water are discussed in the following.

8.1.2.1 *Calcium (Ca)*

Calcium concentrations range from less than 10 to over 520 mg/l (e.g. Crodo Valle D'oro contains 519 mg Ca/l). Most bottled waters contain around 100–120 mg/l calcium. There is no danger to health at these levels as calcium is an essential element for building bones and teeth, especially in children. Calcium contributes to water hardness, which is known to reduce cardiovascular disease (Section 4.6), and high levels of calcium make the water palatable.

8.1.2.2 *Magnesium (Mg)*

Along with calcium, magnesium is a major constituent of hardness and a major dietary requirement for humans. Levels are generally low in bottled water, although there are exceptions. Vittel contains 110 mg/l and Apollinaris 122 mg/l magnesium. When present as magnesium sulphate it makes an effective laxative, so avoid high concentrations unless you need help in this direction. It is best not to give bottled waters with high magnesium concentrations to young children (less than seven years of age).

8.1.2.3 *Potassium (K)*

Although it is an essential element, the body finds it difficult to deal with excess potassium, resulting in kidney stress and possible kidney failure. Although potassium is not considered to be toxic, long-term exposure to high potassium concentrations should be avoided. It is advisable not to select any bottled water with more than 12 mg/l potassium for regular drinking. Check waters carefully as this is one element that is often exceeded. Typical low potassium waters are Buxton Spring, Glenpatrick, Perrier, Evian, Aqua Fabia and Spa Reine.

8.1.2.4 *Sodium (Na)*

This is a key element in bottled waters and is essential for health. In healthy adults excess sodium is excreted, but in sensitive adults such as those with hypertension and heart weakness, infants with immature kidneys and the elderly, high sodium levels may cause problems and so a low sodium diet is often advised. The EC MAC for sodium is 150 mg/l, and many mineral waters exceed this. For those identified as being at risk, all high sodium waters should be avoided and those with low sodium levels such as Evian, Volvic and Cardo used instead. High sodium waters such as Vichy, Vichy Calalan, San Narciso and the Russian Borjomi, which contains nearly 1500 mg/l of sodium, could be injurious to people at risk. For routine use by the family look for a water with less than 150 mg/l, and preferably less than 50 mg/l, such as Glenpatrick, Buxton Spring, Volvic or Evian. The others should be kept for the occasional glass of something interesting and refreshing, not drunk regularly. Sodium can act as a laxative when present as sodium sulphate (Glauber salts).

Deaths of babies have occurred from drinking formula feeds made up with high sodium mineral waters, so use the water with the lowest sodium content you can get. Remember that other minerals which may be found in excess (apart from sodium) can also cause potentially serious problems when used to make up babies' feeds. The use of mineral waters for bottle-feeding babies is considered at the end of this section.

8.1.2.5 Sulphate (SO_4)

Magnesium and sodium sulphate are both strong laxatives, so although the human body can adapt to moderately high levels of sulphate in water, sudden increases will lead to a purgative effect. This could be extremely serious in young infants and sensitive adults, so mineral waters with high sulphate levels should be avoided by families with young children. The EC MAC is 250 mg/l, with many mineral waters containing much more than this. Avoid any mineral waters with a sulphate level in excess of 30 mg/l if you want to use them on a regular basis. For very young children about 10 mg/l seems a sensible maximum.

8.1.2.6 Nitrate (NO_3) and nitrite (NO_2)

Many aquifers are showing increasing nitrate levels and there is a trend in many countries for bottled water companies not to give details of nitrate. If they do not, then avoid them. Excessive nitrate can give rise to two problems: infantile methaemoglobinaemia and the formation of nitrosamines (Section 4.2). The nitrate itself is not a direct toxicant, but it is the conversion of nitrate to nitrite which causes the problems. This can occur readily in the digestive tract under certain conditions, so low nitrate levels are desirable. The EC set a MAC of 50 mg/l nitrate, but there is no real safe limit; it is a case of the lower the better. Of those companies that do give information on nitrate, all have low concentrations, with Buxton Spring having less than 0.01 mg/l. Any bottled water containing more than 0.05 mg/l nitrite should be avoided and only nitrite-free bottled water should be used to make up formulae feeds for babies.

8.1.2.7 Chloride (Cl)

Chloride is not dangerous at the concentrations found in bottled waters. It has a taste threshold of about 200 mg/l, but levels are generally much lower than this in mineral waters. If bottled water is to be used for making up drinks such as tea, coffee and fruit juices, then the lower the chloride concentration the better. Natural surface waters contain between 15 and 35 mg/l chloride, whereas seawater contains 35 000 mg/l. Concentrations up to 1500 mg/l are probably drinkable without any adverse effects.

8.1.2.8 Fluoride (F)

Fluoride levels should be given on the side of bottles. In general they are not, but when they are, they are generally less than 1 mg/l. Many bottled waters contain more than 1.0

mg/l fluoride; for example, Ramlosa contains 2.2 mg/l. If consumers are going to use a bottled water as a complete replacement for tap water, for both drinking and cooking, then the fluoride level must be 1.0 mg/l or less to avoid dental fluorosis and other associated problems (Section 5.6).

8.1.2.9 *Hydrogen ion concentration (pH)*

In still mineral waters the pH ranges from 6.0 to 8.0, which is fine. In carbonated waters, however, the pH can decrease in some brands to 5.0 or even 4.5. Although this keeps the water bacteriologically pure, it does make the water aggressive and can lead to problems such as the leaching of minerals from teeth if drunk regularly. Bear in mind, however, that many canned carbonated soft drinks are equally aggressive.

8.1.2.10 *Total dissolved solids (TDS)*

This is a gross measure of all the solid material left after the water has been evaporated at 180°C. This will include mainly dissolved minerals and so can be related to the water's salty taste. The higher the total dissolved solids, the more salty it will taste. The EC set an MAC of 1500 mg/l. Generally water with more than about 500 mg/l total dissolved solids has a distinctive mineral taste, and above 1000 mg/l it has a strong mineral taste. Above 1500 mg/l it will taste salty, although this will vary according to personal palates.

8.1.3 Contamination

Bottled drinking water should be microbiologically pure and free from other contaminants. However, as Section 2.4 shows, groundwaters are at risk from a number of contaminants, especially pesticides and industrial solvents. These do not appear to be widely monitored at present, but every effort should be made to ensure that sources of bottled water are protected to ensure that they remain as pure as possible.

Perrier, who have held the major share of the market for mineral water for many years, had an operational problem which came to light in February 1990. They found that benzene had managed to contaminate some of their water. This did not affect the source, but was solely an operational problem at the bottling plant, caused by a filter on a gas line which had not been changed when necessary. This led to contaminated bottles being found in a number of countries, including Britain and the USA. Perrier, who acted most correctly, ordered that all stocks should be withdrawn immediately from shops, apparently world-wide. In the USA 72 million bottles were withdrawn and in Britain over 10 million bottles were cleared off the shelves, with an estimated 160 million bottles world-wide being withdrawn at a cost of US$35 million. Perrier, with its distinctive green bottle which has been the market leader for many years and in so many countries, took a severe knock, and allowed new companies to steal a share of the market. However, the marketing carried out largely by Perrier has put bottled water on a firm basis in Britain and Ireland. Perrier has become almost a generic term for mineral water, just like Hoover for vacuum cleaners, whereas in Ireland the brand name Ballygowan has become such a generic term.

Each country is now dominated by its own brands. In Ireland Ballygowan, Tipperary and Glenpatrick are firmly the market leaders. In Britain the market is far more open, with Buxton Spring and Highland Spring being two popular and excellent brands.

8.1.4 Long-term Use

The sparkling (carbonated) mineral waters are fun, and can possibly do us good. As a substitute for drinking water, however, care must be taken. Identify and consider any member of the family who might be at risk, especially the young and the very old. If bottled water is required to supplement or substitute piped drinking water, then identify why. It will be cheaper to correct the problem with the tap water by correcting plumbing problems, or at worst by having to install a home treatment system. If, however, the consumer needs to use bottled water to supplement the tap water, then non-carbonated water should be used. Use water with as little in it as possible and select a water which is low in sodium, fluoride, potassium, sulphate and nitrate. Mineral-rich water should not be consumed continuously over a long period of time.

If consumers wish to use bottled water for making up formula milk for a baby, then they must ensure that it is very low in sodium and sulphate (< 10 mg/l), with low nitrate (< 5 mg/l as nitrate) and no nitrite. CARBONATED WATER SHOULD NOT BE USED, only still waters. Suitable waters are Evian and Volvic. THE WATER MUST BE BOILED before making up the feed. Remember, bottled water has not been disinfected, and even though it is normally free from pathogenic micro-organisms at source, there are other normally harmless bacteria present. Contamination can occur, so it must be boiled to be safe. Finally, opened bottles of water should not be stored for longer than 36 hours if they are being used to feed an infant, unless they are in a special sterilized bulk container with a tap. Even so, this should be stored in a refrigerator (Section 4.2).

8.2 WATER TREATMENT IN THE HOME

Over the past 15 years the number of manufacturers and distributors of water treatment systems has mushroomed. The diversity of systems available has also expanded. The systems range from small simple jug filters which produce about two litres at a time and just remove sediment and fine solids from the water, to in-line systems producing such pure water that it can be used in a car battery instead of distilled water. Most of the systems are based on physical filtration, ion exchange, activated carbon, reverse osmosis or ultraviolet radiation. Often a combination of these processes are used (Table 8.3). The basis of all these processes has already been described in relation to full-scale water treatment plants. The same technology has been scaled down for use in the home.

Table 8.3. Most effective home treatment methods for the removal of selected chemical and physical problems

Problem material or substance	Fibre filter	Activated carbon filter	Reverse osmosis	Deionization	Distillation	Ultraviolet radiation
Sodium	−	−	+ +	+ +	+ +	−
Arsenic	−	−	+ +	+ +	+ +	−
Lead	−	−	+ +	+ +	+ +	−
Cadmium	−	−	+ +	+ +	+ +	−
Potassium	−	−	+ +	+ +	+ +	−
Sulphate	−	−	+ +	+ +	+ +	−
Hardness (Ca)	−	−	+ +	+ +	+ +	−
Hardness (Mg)	−	−	+ +	+ +	+ +	−
Nitrate	−	−	+ +	+ +	+ +	−
Chloride	−	−	+ +	+ +	+ +	−
Faecal bacteria	−	−	+ +	−	+ +	+ +
Viruses	−	−	+ +	−	+ +	+ +
Protozoan cysts*	+ +	−	+ +	−	+ +	+
Organics	−	+ +	+ +	−	+	−
THM, TCE†	−	+ +	+ +	−	+	−
Dioxins	−	+	+ +	−	+	−
Chlorine	−	+ +	+ +	−	+	−
Pesticides	−	+ +	+ +	−	+ +	−
Sediment	+ +	+/−	+ +	−	+ +	−
Taste/odour	−	+ +	+ +	−	−	−
Asbestos*	+ +	−	+ +	−	+ +	−

Key: −, Non-effective; +/−, some reduction; +, good to moderate reduction; + +, excellent reduction.
*Requires a filter with a pore size of <1 μm to be effective.
†THM = Trihalomethanes; TCE = trichloroethylene.

8.2.1 Physical Filtration and Activated Carbon

Water treatment starts with water filters. The simplest systems are physical filters which just remove sediment such as sand, silt and rust particles; anything which is larger than the holes in the filter. There are two types of system. In-line cartridge filters are fitted into the plumbing system, usually under the kitchen sink so that all the water to the kitchen tap is treated. Alternatively, jug filters are small containers of about two litres capacity fitted with a filter reservoir which is also used as the lid. Water is poured into the reservoir and passes through the filter into the jug.

Cartridge filters are usually plumbed into the mains connection to the cold water tap (Figure 8.1). This is very simple and can be done by any DIY enthusiast. Of course, not all the water from the cold water tap needs to be treated to this new high level and so an alternative system is to take a branch connection from the mains pipe and then have a separate tap at the sink (Figure 8.2). In fact, a number of manufacturers have produced easy to install systems using a simple self-piercing connection, and all that is required is

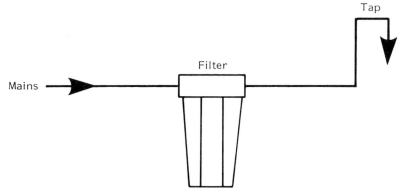

Figure 8.1. Typical plumbing arrangement for a water filter fitted to treat all the water to the kitchen cold tap

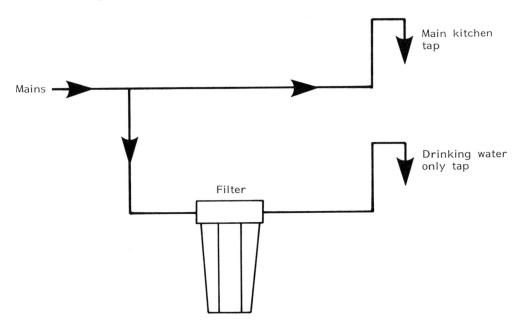

Figure 8.2. Alternative method of plumbing a water filter into the cold water supply in the kitchen. This only treats a proportion of the water to the sink and so requires a separate tap. The smaller volume passing through the filter unit, however, will prolong the life of the cartridge

to tighten it onto the incoming mains water pipe. They supply a simple to fit small bore tap and all the other pipe connections are push-fit. The plastic housing for the filter cartridge is simply screwed into the wall (Figure 8.3). The water from the mains enters the plastic housing and then passes under pressure through the cartridge filter, the purified water passing up the centre of the housing flowing onwards to the tap. This is the most common type of system and there is a wide variety of cartridge filters that can be used.

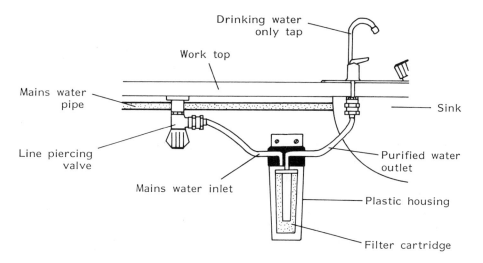

Figure 8.3. Typical layout of a DIY filter kit. Redrawn and reproduced with permission of Doulton Water Care

Sediment and turbidity are removed by a physical cartridge filter. These can be made of corrugated paper, which removes particles down to 5 μm diameter but not very fine particles. They are only useful for gross sediment. Finer particles are removed by cartridges made from fibres. These are either made out of single strands of yarn which are wound spirally to form the filter, or consist of pads containing felted or bound fibres. Spirally wound filters are made from cotton, wool, rayon, fibre glass, polypropylene, acrylic, nylon and other polymers. The yarns are brushed to raise the fibres to create a filtering medium. These cartridges come in three general grades depending on the type of yarn used. Coarse woven filter cartridges are used to pretreat household waters to remove particles larger than 50 μm before further finer filtration or treatment by another process. Medium woven filter cartridges remove particles down to 20 μm in diameter. These last longer than the finer filters and are usually adequate for drinking waters. Fine woven filter cartridges are generally used in laboratories as they remove all particles down to 5 μm or less. They provide very good quality drinking water and may be needed if the turbidity in the water is made up of very fine particles. The larger the pore size, the longer the life of the filter cartridge. Therefore, when fine filters are used, prefiltering using a coarse grade of filter will significantly increase the life of the second filter.

Apart from the spirally wound filters, cartridges can also be packed with loose felted fibres. Cellulose, wool, glass fibre and polypropylene are all used. These can remove particles as small as 0.5 μm, but will quickly become blocked if the water is visibly dirty. Other materials, such as ceramics, can be used.

As filtration continues the pores become progressively clogged with retained material, and so the average pore size gradually becomes smaller. All particles will be retained, including any substances adsorbed onto the particles. Even a physical filter will remove aluminium or pesticide residuals that are adsorbed onto particulate matter. If the pore size is small enough, even bacteria may be removed. Protozoan cysts of *Giardia* and *Cryptosporidium* are effectively removed by fibre cartridge filters with a pore size of 1 μm. Physical filters are able to remove turbidity and red water caused by iron particles and rust. These sediment filters are designed to treat all the water coming into the house and so should be fitted on the rising main before the junction feeding the kitchen tap. If a water softener is installed it must be installed after the filter. If the water is seriously cloudy or rusty, a coarse filter cartridge must be used before a finer filter cartridge (in series); the cartridges should be changed frequently.

Taste and odours caused by chlorine, decaying organic matter and metals can also be removed by using simple cartridge systems. These consist of a hollow cartridge filled with granular activated carbon instead of fibres. Activated carbon is also excellent at removing organic chemicals such as chlorinated hydrocarbons including trihalomethanes and trichloroethylene, pesticides, industrial solvents and naturally occurring humic substances. These humic substances also cause colour, which can also be removed by activated carbon. They operate under normal pipe pressures, as do fibre cartridges, and although the fibre cartridge becomes discoloured indicating that it needs to be replaced, it is not possible to tell when activated carbon cartridges need to be replaced. Their effective life span depends on the concentration of the organic material and residual chlorine in the water, so advice must be sought from the supplier about how often they should be replaced. Certainly twice yearly replacement will be required under normal use. Prefiltration using a fibre filter is usually necessary unless the water is exceptionally clear and free from solids. The activated carbon cartridge system needs only to be placed on the supply to the kitchen cold water tap. Special units are available which only treat the water required for drinking via a special tap (Figure 8.3). This considerably prolongs the life of the filter cartridge.

When using cartridge filters the system should always be flushed through for at least 10 seconds before using the water for drinking or cooking. This is especially important if the tap has not been used for 24 hours or longer. The cartridges should be replaced if the water pressure drops or if taste and odours begin to reappear. Fibre filters and activated carbon cartridges should be changed fairly regularly to maintain maximum performance. Certainly do not use any cartridge filter for longer than six months because foul tastes and odours can originate from old cartridges. As with full-scale granular activated carbon filters, heterotrophic bacteria can colonize the surface of activated carbon, leading not only to taste and odour problems but also to a risk of colonization by opportunistic pathogens (Reasoner *et al.*, 1987).

8.2.2 Reverse Osmosis

Reverse osmosis removes dissolved impurities such as manganese and sulphate. This is achieved by using a semipermeable membrane. The water is forced through the membrane under pressure while dissolved and particulate materials are left behind. This ensures that there is a constant concentration gradient with the liquid passing through the membrane from the concentrated side. This means that there is a production of waste water in which all the various contaminants are concentrated and

Table 8.4. Nominal retention characteristics of a reverse osmosis membrane. Reproduced with the permission of the Bruner Corporation

Chemical	Symbol	Rejection (%)
Sodium	Na^+	87–93
Calcium	Ca^{2+}	94–97
Magnesium	Mg^{2+}	96–98
Potassium	K^+	87–94
Iron	Fe^{2+}	95–98
Manganese	Mn^{2+}	95–98
Aluminium	Al^{3+}	98–99
Ammonium	NH_4^+	86–92
Copper	Cu^{2+}	98–99
Nickel	Ni^{2+}	98–99
Zinc	Zn^{2+}	98–99
Strontium	Sr^{2+}	96–98
Cadmium	Cd^{2+}	96–98
Silver	Ag^+	93–96
Mercury	Hg^{2+}	96–98
Barium	Ba^{2+}	96–98
Chromium	Cr^{3+}	96–98
Lead	Pb^{2+}	96–98
Chloride	Cl^-	87–93
Hydrogencarbonate	HCO_3^-	90–95
Nitrate	NO_3^-	60–75
Fluoride	F^-	87–93
Silicate	SiO_2^{2-}	85–90
Phosphate	PO_4^{3-}	98–99
Chromate	CrO_4^{2-}	86–92
Cyanide	CN^{4-}	86–92
Sulphite	SO_3^{2-}	96–98
Thiosulphate	$S_2O_3^{2-}$	98–99
Ferrocyanide	$Fe(CN)_6^{3-}$	98–99
Bromide	Br^-	87–93
Borate	$B_4O_2^{2-}$	30–50
Sulphate	SO_4^{2-}	98–99
Arsenic	As	94–96
Selenium	Se^{2-}	94–96

Also removes 85–90% of all organics, i.e.: THM's, PCB's, pesticides, herbicides, benzene, etc.

which has to run to the drain. Typical removal (or rejection as it is referred to in reverse osmosis) characteristics for a reverse osmosis system are given in Table 8.4, with on average between 94 and 98% of the total dissolved solids removed. The most critical factor in home units is the type of membrane used. Different membranes are used depending on whether the water has been chlorinated or not.

The performance of the membrane depends very much on the quality of the raw water. For example, its life is prolonged if chlorine is removed by activated carbon pretreatment and if iron is removed using a catalysed manganese(II) dioxide filter, which oxidizes the iron to an insoluble form which can then be removed by a simple fibre filter. The membrane may need to be flushed occasionally and bacterial development can also become a problem. Reverse osmosis is ideal for removing aluminium, copper, nickel, zinc and lead. It can also remove 85–90% of all organic compounds including trihalomethanes, polychlorinated biphenyls, pesticides and benzene. When used downstream of an activated carbon system, organic compounds are essentially eliminated. Nitrate is not completely removed, however, and so to remove essentially all the nitrate present (99%) an ion-exchange system should be used upstream of the reverse osmosis unit. The installation of ion-exchange systems will increase the capacity of the reverse osmosis unit ten-fold, also making it much more efficient. Some problems have been reported due to excessive water pressure or microbial degradation of the membrane material, causing membrane fracture and thus treatment failure.

8.2.3 Ion Exchange

Ion-exchange systems use a resin onto which certain undesirable ions are adsorbed and replaced by different ions. The usual ion-exchange reaction exchanges calcium and magnesium ions for sodium ions, which reduces hardness. Most domestic water softeners work on this principle. Softening does not reduce the alkalinity or total dissolved solids, but scale formation caused by calcium, magnesium, carbonate and hydrogencarbonate is eliminated.

Nitrate can be removed by using a strongly basic anion resin for the preferential removal of nitrate and sulphate anions. These are replaced by chloride ions thus:

$$CaNO_3 + RCl \rightarrow RNO_3 + CaCl$$
$$MgNO_3 + RCl \rightarrow RNO_3 + MgCl$$

where R is the resin within the unit.

Exchange only continues as long as there are sufficient ions in the resin to be exchanged. The resin is regenerated by flushing with concentrated brine as explained in Section 3.2. Ion-exchange systems are used to treat the water for the entire home, whereas reverse osmosis is used for the kitchen tap only.

Ion exchange does not remove bacteria or viruses, sediment, taste, odours or organisms. Apart from softening water, such systems need to be used with fibre and activated carbon filters for full treatment (Table 8.3). There is concern that there is a higher rate of cardiovascular disease in soft water areas; also, the corrosion of certain metals and materials are increased in soft waters. This option should therefore be considered carefully. If a consumer is receiving a sodium-free diet, then remember that base exchange removes calcium and magnesium at the expense of increasing the sodium concentration.

8.2.4 Disinfection

One of the most effective ways of ensuring that your water is microbially safe is to install your own disinfection system. Ultraviolet radiation is extremely effective and is able to kill almost all known micro-organisms found in water. The systems available are all generally similar. An ultraviolet lamp is housed in a transparent quartz sleeve. The water enters one end of the unit and flows around the protected lamp. This ensures adequate exposure. The greater the volume of water required, the larger the lamp that is used to ensure an adequate exposure time. These units are plumbed into the incoming water main and are generally designed to treat all the water, although smaller units are available to fit just the supply to the kitchen cold water tap. If all the water is supplied from a storage tank, the ultraviolet sterilizer can be fitted directly to the incoming water main in the attic. The lamp is housed in a metal box with a warning light and also an audible warning signal, if required, to indicate lamp failure. The lamp indicator and control switch can be remote from the actual unit if the sterilizer itself is in a cellar, attic, or otherwise not visible. The lamp runs for six to nine months on a continuous basis using either 18 or 35 W. In terms of running costs it is equivalent to having a 100 W light bulb on for five to eight hours each day. Even at high flows removal of 99.7% of faecal coliforms has been achieved in trials and so at the lower flows expected in the average household even higher levels can be achieved. Bacteria may survive, if they are protected by particulate matter, and the efficiency of ultraviolet sterilizers can be increased by installing a fibre filter upstream of the ultraviolet lamp. A word of caution: the units come with full instructions and usually special safety devices to ensure that the lamp is switched off before the protective housing is removed. Ultraviolet light can severely damage your eyesight so under NO, ABSOLUTELY NO, circumstances should you allow the light to be activated if the protective cover is not in place (Section 3.2).

An alternative method of disinfection is to install a chlorinator. Most piped water supplies are already chlorinated and so it is not advisable to chlorinate the water a second time. However, those served by private supplies, especially surface streams, springs or shallow wells, should consider either an ultraviolet system or a chlorinator. Chlorine tablets (calcium hypochlorite) are held in a tall tube, the base of which is perforated to allow the water to flow past the tablets, slowly dissolving them and releasing chlorine. The dosing chamber is a simple box which is plumbed into the pipework, although it is best to have the chlorinator as far upstream of your system as possible to reduce chlorinous tastes and odours. In terms of aesthetic quality, a fibre filter followed by an ultraviolet sterilizer is preferable. However, where there is no power, or the power is intermittent, passive chlorination is the only answer.

8.2.5 Other Systems

There are a number of more specific treatment systems also available. Acid water neutralizers reduce the corrosion of copper and lead pipes. They neutralize the water by passing it through a special medium which increases the pH. The water supply flows upwards through the unit and does not require any electricity, back-washing or drainage. Other systems add soda ash or lime to the water. There are also a number of iron removal systems available, including one used for private groundwater supplies which are rich in iron. The water is aerated to bring the iron from its soluble iron(II) state into the insoluble iron(III) state, which is then filtered out.

All these water treatment systems require regular maintenance and occasional replacement of the active units (cartridges, etc.). Systems not looked after properly may result in much poorer quality water than before. Also, apart from the initial cost, service arrangements, replacement cartridges and other parts can be expensive.

Currently there are no British Standards covering home treatment systems, and no government guidelines. Neither are they covered by the current water byelaws. This looks as though it may change, and over the next few years we should begin to see some control over the standards of these units. The Drinking Water Inspectorate highlighted in its 1990 annual report that extravagant claims for home treatment systems were often made by suppliers and some manufacturers, and that some devices could even lead to the deterioration of water quality by, for example, encouraging microbial growths. The Inspectorate stated that it did not see a need on health grounds for these devices to be used on water supplies drawn directly from the public mains. However, it recognized the right of consumers who may object to the taste or appearance of their water, or the presence of a particular substance in the supply, to use such a device. It stresses that the installation of these devices must comply with the water byelaws. For consumers with private supplies, the use of such devices may be the only way to obtain water of reasonable quality. The Inspectorate has set up an expert group to devise national test procedures for home treatment systems, which they prefer to call 'point of use devices'. This is going to be helpful in providing potential users with accurate information on the capability of the available systems.

Attempts have also been made in the USA to regulate home treatment systems. In Iowa, such systems are required to be (1) registered with the Department of Health detailing treatment claims; (2) independently performance tested using a specified protocol; and (3) details of the performance analysis, along with details of the law and effectiveness of water treatment units prepared by the Department of Health, must be supplied to customers before the delivery of the unit (Chorquette and Gergely, 1992). Further details of the effectiveness of home treatment systems are given by Bell et al. (1984) and Geldreich and Reasoner (1990).

Chapter 9
The Water We Drink

9.1 PUBLIC PERCEPTION OF DRINKING WATER QUALITY

Over the past five to ten years increased public awareness of environmental issues has caused consumers to be concerned about the quality of drinking water. This is reflected by a large increase in the sales of bottled water and home treatment systems (Chapter 8). Owing to a lack of knowledge, consumers will generally interpret their risk and safety in an emotional rather than a scientific way. This has led to considerable pressure being put on the water undertakers to implement often unnecessarily stringent treatment and control measures to reduce the concentrations of particular parameters. Most water companies have an action plan of remedial measures to improve water quality based on appropriate risk-benefit and cost-benefit analysis. However, pressure from consumers, politicians and, of course, the media can result in this structured approach being set aside and inappropriate action being taken, often at considerable cost.

One of the major challenges facing the water industry is understanding how consumers perceive water quality, why they are so unhappy about drinking tap water and, finally, educating the consumer about water quality and the regulatory functions of water undertakers. To answer some of these questions the Office of Water Services (Ofwat) commissioned a major survey of consumer attitudes (MORI, 1992).

Consumers were asked if they thought that their tap water was safe to drink. Over two-thirds were satisfied, whereas 20% expressed concern. Surprisingly, only 16% responded as being very satisfied, so most consumers have some doubts about the safety of their drinking water. Asked about the reason for their dissatisfaction, 49% referred to the taste or smell and the presence of chemicals, 38% indicated that their worries were related to stories in the media, whereas 29% referred to the visual appearance of the water. The results showed that the consumers' view of safety was based heavily on the smell of the water.

Seventy per cent of consumers found the taste of their tap water to be acceptable, whereas 20% found it unacceptable. This was due to chlorine (25%), unspecified chemicals (17%), a general dislike of tap water (15%) or a metallic taste (9%). The taste of water was the strongest motivation in the purchase of bottled water. The detection of chlorine in drinking water appears to be a major source of complaint with consumers. Ninety per cent thought it either essential or very important that drinking water should be clear and colourless. Only 18% of consumers were dissatisfied with their water in this respect, reporting occasional cloudiness, 'bits' or colour.

A good indicator of true consumer dissatisfaction is the use of alternative water supplies or the use of home treatment systems. The survey showed that 39% of consumers did take some personal action to improve quality. Fourteen per cent of consumers boiled their water before drinking, 21% purchased bottled water and 11% used home treatment devices (9% jug filters and 2% in-line systems). The reasons for boiling water or the use of home treatment systems were primarily concerns over the safety of their supplies, although an improvement in taste was also an important factor. The reasons for purchasing bottled water were more complex. Most used bottled water due to taste considerations (74%), although 37% stressed safety reasons. The survey identified a strong belief that bottled water was purer and safer to drink, which, as shown in Section 8.1, is not always true.

Consumers were asked if they had contacted their water supply company over the past three years, apart from paying bills. Of the fifth who replied that they had, only 20% of these queries had been in relation to water quality. There had been surprisingly few requests for information about water quality (4%), and these were mainly in relation to complaints. However, 44% of those questioned during the survey wanted to know more about their drinking water quality, where the water came from and how it was treated, and especially how well their water conformed to current standards. Improving tap water quality appeared to be high on consumers' list of priorities for future investment by the water industry, although the survey showed that those who were least satisfied with their water quality were least prepared to pay more to improve it. This supports the general view that good water quality is a right, rather than something consumers should pay for directly.

The Ofwat survey was unique in that it showed that the public's perception of drinking water was largely based on the physical characteristics of their own water supply, rather than its quality in relation to the prescribed standards. The lack of knowledge about water supply and quality was highlighted, as was the desire to know more. The media is also a significant influence on attitudes towards water and the assessment of risk. With up to 60% of quality problems in some areas related to householders' own plumbing systems, it is clear that companies will have to look carefully at the education of consumers if they are to overcome some of the basic aesthetic-based problems highlighted by the survey.

9.2 CUSTOMER–SUPPLIER INTERACTION

Problems can arise at source through natural or man-made contamination, from chemicals added during water purification, during distribution and, of course, in the consumer's own home. For example, discoloured or dirty water could be caused by corrosion of the iron mains, poor or insufficient treatment, microbiological growths in the mains or even replacement of the natural supply with one of a different composition. It may even be a combination of problems.

Wholesomeness as defined in the Water Industry Act 1991 implies that water is not only safe to drink but that it is aesthetically acceptable to consumers in terms of taste, odour, colour and clarity. In fact, consumers tend to assess quality subjectively using these four parameters, and although water may be absolutely safe to drink, if it is cloudy or turbid the consumer will assume that it is in fact unfit to drink. A good example is milky water. The water is overaerated as it leaves the tap, usually due to high pressure in the mains, and takes on a milky appearance due to the presence of millions of tiny bubbles. These gradually rise to the surface and the water becomes totally clear and wholesome.

Consumer complaints fall into a number of categories: discoloured water, with or without particulate matter; particulate matter such as rust, grit or sediment; staining of laundry; staining of bathroom fittings; cloudy, milky or chalky water; tastes and odours; the presence of animals such as worms or insects; scale deposits within domestic plumbing; alleged illness; noise; corrosion; and low pressure. Water supply companies are now getting more specific complaints—for example, about aluminium, lead, pesticides and nitrate. However, these are more concerns, as the aesthetic quality of the water is generally unaffected. This supports the findings in the Ofwat survey that such complaints are often the result of media attention.

The water companies believe that if consumers are unhappy about the quality of their water then they should complain. Only in this way is the water supply company aware that problems actually exist and so can try to rectify them, especially as now in the UK it is a criminal offence to supply water which is unfit for human consumption. As water supply companies carry a legal liability to fulfil their obligations to consumers, if consumers are unhappy, they want to know about it.

The Water Supply and Sewerage Services (Customer Service Standards) Regulations 1989 came into force at the same time as the new companies were formed: these regulations were meant to protect the public (HMSO, 1989d). The regulations have given rise to customer guarantee schemes as well as codes of practice. The guarantee scheme gives undertakings relating to keeping appointments, written responses about accounts, responses about complaints and interruptions and restoration of supplies. A failure to meet with any of these guarantees results in penalties against the companies. The companies are also obliged to keep a register of complaints, including the date of the original complaint, the date received, the name and address of complainant, the address where the complaint arose, the type of complaint (quality, pressure, flow) and the action taken. These regulations were amended by the Water Supply and Sewerage Services (Customer Service Standards) (Amendment) Regulations 1993 (HMSO, 1993).

Wessex Water, like all the water supply companies in England and Wales, publishes a code of practice for customers. It really is a user's handbook describing the company, the areas covered by the company, and the services provided; in fact, just about everything the customer needs to know. The customer's code of practice gives a general synopsis of the Customer Service Standards Regulations, and Regulation 8 requires that customers are to be reminded of their rights at least every 12 months. Importantly, it explains how and for what charges are made, who to telephone in emergencies and how to complain, including details of the Customer Service Committee (CSC). The production of such a code of practice is a requirement under the licence of appointment, and covers a number of specified areas including customer relations with the CSCs, disconnections and leakage. There is also, for each company, details of the guaranteed standards scheme, which entitles customers to a payment or credit when the company fails to maintain certain minimum standards of service to domestic premises. They pay £5 each day or for each failure of these standards. Wessex's guarantee scheme is similar to all the others, but the code of practice issued to customers is clear and comprehensive. Wessex Water operate a customer service unit where all enquiries, whether by phone or letter, are dealt with. If these enquiries, including complaints, cannot be dealt with by the unit, then they are given a priority rating (i.e. immediate (same day), urgent (within one day), three days, one week, routine (14 days) or by appointment). Customer relations are clearly important to Wessex Water, as reflected by the low number of disconnections (along with Thames Water). Wessex Water received a commendation from the British Quality Association in 1989 for its customer services. The Three Valleys Water Services plc operates a similar system to Wessex and is in the process of compiling a care register of special customers. The register is intended for disabled, elderly or disadvantaged customers who might require extra help in their dealings with the company.

The levels of service standards have been set by Ofwat, who have also implemented procedures to ensure that these are being met. It is mandatory for the water supply companies to remind its customers at least once a year of the standards of service that they should expect and receive during their dealings with companies. Customers must also be informed of the address to which complaints should be sent. The standards specify the performance required in the areas of water quality and adequacy of supply, keeping agreed appointments, notifying customers of planned interruptions to supply and response times to written complaints and requests. Each company must publish 'Level of Service Information' and 'Service Target' Reports, which are sent to Ofwat each year and from which new targets are set. Performance data for all the companies are published annually by Ofwat, but companies are also required to supply this information to customers on request.

All the water supply companies have detailed action plans to deal with complaints. An initial reply must be dispatched to the customer no more than 10 working days from the receipt of the complaint. A substantive reply must be dispatched to the customer not more than 10 working days from the receipt of the complaint unless an initial reply has been sent, in which case the substantive response must be dispatched not more than 20 working days from the receipt of the complaint. The Director General of Water Services monitors the performance of the guarantee schemes to make sure that the companies meet customer service targets, which are published annually.

Codes of practice for domestic consumers are also required to be published by the individual water supply companies. At present there are a number of such codes. The code of practice for customers gives details of the services available to customers, how the company works, how and where to get help and advice and also how to make a complaint and details of the guarantee scheme. The code of practice for leaks from the supply pipe of metered supplies safeguards customers against paying for leakages from private pipes when the meter is first installed. The code of practice and procedure on disconnection for domestic customers explains what happens when bills are not paid and how the company can help. The code of practice on sewage flooding covers damage from overflows from sewers. The code of practice on the environment, access and conservation outlines the measures taken to protect the environment, allow access to land and water owned by the companies and explains conservation programmes.

The Director General of Water Services is also charged with protecting customers' interests and to ensure this CSCs have been set up in each of the water service company areas. If the consumer has a complaint, the first stage is to complain to the company directly. If the customer is still unhappy they should then contact the CSC.

The companies must also keep a register which contains details of the quality of drinking water by water supply zone. Anyone can obtain a copy of any part of the register by either writing or telephoning the water supply company. The registers are open for inspection during normal office hours. The companies must publish a copy of the water quality report annually, giving a full breakdown of the results and a detailed explanation. Other details must also be given; for example, incidents or accidents leading to a loss in quality, or any relaxations of EC standards which may apply. These reports form the basis of the annual report on water quality in England and Wales published by the Drinking Water Inspectorate.

Apart from the statutory reports on service and quality and the codes of practice, the water supply companies also publish an impressive array of literature explaining the operation and management of the industry. Open days are common and there is a real commitment from the companies to educate consumers. As the head of customer services from Severn and Trent explained to me: 'If they [the customer] understand what we are doing and how it is done then they will be happier with the service that we are giving them'. To this end, early in 1993, Severn and Trent sent a video explaining their services to every one of their customers.

If the customer complains, he or she is certain to obtain a quick response. However, the answer received may not always be satisfactory. It is then that the customer may need to go back to the company again, and if the problem cannot be resolved then the company will refer him or her to the regional CSC, who can refer quality problems to the Drinking Water Inspectorate if necessary (Section 1.2). It does seem that on the whole most complaints are dealt with satisfactorily by the companies themselves, with remarkably few complaints in 1990–1; only 4627 complaints from 50 million consumers.

It is not only the large water supply companies which are able to excel in the area of customer relations; most of the former statutory water companies also do an excellent job. North Surrey Water Ltd have an impressive system for dealing with water quality complaints. They use a standardized step by step approach including at each stage standard forms filled in by the inspector, sampler and analyst. It is clear to me that the water supply companies take complaints very seriously indeed.

The water supply companies do other things as well. For example, Southern Water plc have moved into the bottled water market, producing a new sparkling mineral water called Hazeley Down. It comes from deep within a chalk aquifer and is claimed to be of outstanding quality. They have formed a separate company to produce the new product, Hazeley Down Mineral Water Co. Ltd, which illustrates the flexibility of the water holding company structure. Another subsidiary of Southern Water plc is Southern Watercare Ltd. This is a contract plumbing service which provides all the normal services including an emergency 24 hour, 365 day call-out service. They have begun work on a £1.3 million contract with Southampton City Council to replace lead service pipes to council houses. Several other water supply companies also operate plumbing subsidiaries, including service agreements where the company inspects the consumers' plumbing annually and provides a complete maintenance and repair service for a fixed annual service charge. Remember that many water quality problems arise in the home plumbing system, so it is clearly a logical step for the water supply companies to want to move into this area as well.

9.3   FINAL ANALYSIS

Is my drinking water safe to drink? This is the question that anyone involved in the water supply industry is constantly asked. You should now be able to make up your own mind based on the facts. Accidents can and do happen; for example, the Drinking Water Inspectorate reported 79 such incidents in 1990, of which only 54 actually resulted in a reduction in quality. Thirty-four of these incidents involved microbial contamination due to operational problems with the disinfectant process or due to contamination at the service reservoir within the distribution system. These were swiftly dealt with by the water supply companies and consumers were asked to boil their water while the problem was being resolved. The remaining 20 incidents involved contamination by pesticides, nitrates and chlorinated solvents, and increased aluminium levels due to operational problems at treatment plants. Where these occurred, alternative supplies were made available to the public as they were resolved. In most of these cases, although standards were broken, the water was still safe to drink. Incidents on the scale of Camelford are rare. Problems also occur with the quality of bottled water, and accidents such as that at the Perrier factory are also thankfully rare. At the end of the day, however, the regulations and standards relating to piped water ensure that the water is safe to drink. In the UK there are independent channels to help and advise consumers, to follow up complaints and to check out worries and concerns. Water companies maintain registers of water quality analyses and these are open for public scrutiny. In reality, the public has much less information about many of the suppliers of bottled water, and fewer channels of complaint.

The improvement programme currently being carried out by the water supply companies will bring almost all piped drinking water supplies in England and Wales up to EC standards by the year 1995. These will include standards for aluminium, nitrate and lead. To achieve this the water service companies are planning to spend £5985 million on improvements to distribution, £3060 million on improved treatment and £1095 million on resource development over the period 31 March 1990 to 31 March 2000. This is part of their £24 585 million investment programme for this decade. In 1990–1 the industry invested £2295.5 million in its infrastructure, of which over 37% was spent on water supply. The water-only companies (the former statutory water companies) invested £196 million in 1990–1 as part of their £2110 million investment programme for the decade. Investment is being made in Scotland and Northern Ireland as well, with £128.5 million and £27.7 million, respectively, invested in 1989–90. However, Ofwat has warned that this will mean large increases in water charges over the next few years.

A major concern must be the level of organic compounds and pesticides in particular in drinking water. There are areas in the UK where 100% of the water supplied by some companies exceeds the limit for total or individual pesticides at some time. Non-compliance occurs in excess of 70% of the time in some areas. This is totally unacceptable, especially where the technology for full-scale treatment is not yet sufficiently developed. Effective solutions must be found and research and development in this area must be a priority for the last decade of this century. Sophisticated treatment technologies such as activated carbon and membrane filtration can only provide the last line of defence. We must, however, look beyond the water industry for solutions, and look towards those who use such chemicals at source, especially the non-agricultural herbicides atrazine and simazine. Water resources are an inseparable part of the environment, so good water quality may only be achievable through the sensitive management and protection of the resources themselves. Better control, application and use of chemicals is often more cost effective than trying to remove them at the water treatment plant. However, for some areas of south-east England where water is coming from chalk aquifers, the levels of pesticides in groundwater are likely to continue to increase for decades to come, so treatment in these areas is clearly a priority.

The importance of closer links between suppliers and consumers is self-evident. It is paramount that consumers should be readily able to obtain information about water quality, and help where necessary. The industry needs to invest more in risk-benefit analysis (Gilbert and Calabrese, 1992; Keller and Wilson, 1992) to assess the dangers of certain pollutants to consumers. This is highlighted by the problems surrounding nitrate and aluminium, where the perceived threat is clearly exaggerated and is taking valuable investment capital away from dealing with the more important quality problems.

It is time for the consumers themselves to take a share of the huge responsibility involved in achieving high quality water, and this can only be achieved by openness, education and, above all, a willingness on the behalf of the professionals to interact more closely with consumers. Privatization has to some degree achieved this, and more importantly has put the issue of water quality on the agenda for public debate. Even in countries where the water supply is still in the control of the public sector the consumer should be perceived as the customer, and be treated as such. In the words of the chairman of the Water Service Association in the UK, 'If the consumer is unhappy then we have failed.'

Although I am sure that new issues and problems will come to light in the future, we will continue to see an improvement in water quality throughout Europe. Safeguarding drinking water quality is a shared responsibility, shared between those who use and dispose of chemicals, who treat and supply water, and all of us who use it. A greater understanding of the problems, and an acceptance of that responsibility by all, will be needed if we are going to preserve one of our greatest assets for future generations: clean, safe drinking water on tap.

MAJOR PESTICIDES AND THEIR DEGRADATION (BREAKDOWN) PRODUCTS WITH THEIR RELATIVE TOXICITY LIMITS IN DRINKING WATER.

Category A pesticides should not exceed 1 μg/l, category B 3 μg/l and category C 10 μg/l. These interim limits were set by the Federal Health Authority in the former Federal Republic of Germany (Miller *et al.*, 1990).

Active ingredient	Category*	Degradation products[†]	Category[‡]
Alachlor	A	2,6-Diethylaniline	—
Aldicarb	B	Aldicarbsulphone (= Aldoxycarb)	B
		aldicarbsulphoxide	B
		Total concentration of aldicarb and main decomposition products	B
Alloxydim	C		
Amitrole	—		
Anilazine	C	2-Chloroaniline	—
		Dichloro-*s*-triazine	C
Asulam	C	*p*-Aminobenzine sulphonic acid	C
Atrazine	B	Desethylatrazine	B
		2-Chloro-4-Ethylamino-6-amino-1,3,5-triazine	B
		Total concentration of atrazine and main decomposition products	B
Azinphos-ethyl	C		
Benalaxyl	C	2,6-Dimethylaniline	—
Benazolin	C		
Bendiocarb	C		
Bentazone	C		
Bromacil	C		
Carbetamide	C	Aniline	A
Carbofuran	C		

Active ingredient	Category*	Degradation products[†]	Category[‡]
Carbosulfan	C		
Chloramben	C		
Chloridazon	C		
Chlorfenvinphos	C		
Chlorthiamid	C	Dichlorobenzamide	B
		Dichlobenil	C
		Total concentration of chlorthiamid and main decomposition product	C
Chlortoluron	C	5-Chloro-*p*-toluidine	—
Clopyralid	C		
Cyanazine	C		
2,4-D	C	2,4-Dichlorophenol	A
Dazomet	A		
Diazinon	A		
Dicamba	C	3,6-Dichlorophenol	A
		3,6-Dichlorosalicylic acid	B
Dichlobenil	C	2,6-Dichlorobenzamide	B
		Total concentration of dichlobenil and main decomposition product	C
Dichlorprop	C	2,4-Dichlorophenol	A
1,2-Dichloropropane	C		
1,3-Dichloropropene	—		
Dikegulac	C		
Dimefuron	C	3-Chloroaniline	—
Dimethoate	C		
Dinoseb	B	Aromatic amines[§] and nitroaromatics[§]	A[‡]
Dinoseb-acetate	B	Aromatic amines[§] and nitroaromatics[§]	A[‡]
Dinoterb	C	Aromatic amines[§] and nitroaromatics[§]	A[‡]
Diuron	C	3,4-Dichloroaniline	—
DNOC	C	Aromatic amines[§] and nitroaromatics[§]	A[‡]
Endosulfan	B	Chlorinated cyclic compounds[§]	A[‡]
Ethidimuron	C		
Ethiofencarb	C	Ethiofencarbsulphone	

Active ingredient	Category*	Degradation products†	Category‡
		Ethiofencarbsulphoxide	C
		Total concentration of ethiofencarb and main decomposition products	C
Ethoprophos	A		
Etrimfos	C		
Fenpropimorph	C		
Flamprop-methyl	B	3-Chloro-4-fluoroaniline	—
Fluazifop	C		
Fluroxypyr	C		
Haloxyfop	A		
Hexazinone	C		
Isocarbamid	C		
Isoproturon	C	p-Isopropylaniline	—
Karbutilate	C		
Lindane	B	Chlorinated cyclohexene§	A‡
Linuron	C	3,4-Dichloroaniline	—
Maleic hydrazide	C		
MCPA	A	p-Chlorophenol	A
Mecoprop (=MCPP)	C	p-Chlorophenol	A
Mefluidide	C		
Metalaxyl	C	2,6-Dimethylaniline	—
Metham sodium	C		
Metazachlor	C	2,6-Dimethylaniline	—
Methabenzthiazuron	C		
Methamidophos	B		
Methomyl	B		
Methyl bromide	—		
Methyl isothiocyanate	B		
Metobromuron	B	p-Bromoaniline	—
Metolachlor	B	2-Methyl-6-ethylaniline	—
Metoxuron	C	3-Chloro-4-methoxyaniline	
		1-Chloro-p-aminophenol	—
Metribuzin	C		
Monuron	C	p-Chloroaniline	—
Nitrothalisopropyl	C	Nitro-aromatics§	A‡
Oxadixyl	C	2,6-Dimethylaniline	
Oxamyl	C		
Oxycarboxin	C		
Parathion	C		
Pendimethalin	C	Aromatic amines§ and nitroaromatics§	A‡

Active ingredient	Category*	Degradation products[†]	Category[‡]
Pichoram	C		
Pirimicarb	C		
Pirimiphos-methyl	C		
Propachlor	C	N-Isopropylaniline	A[‡]
		Aniline	A[‡]
Propazine	C	Desethylatrazine	B
		Total concentration of propazine and main decomposition product	C
Propoxur	C		
Pyridate	C	3-Phenyl-6-hydroxy-6-chloropyridazine	C
		Total concentration of pyridate and main decomposition product	C
S 421	A	Chlorinated unsaturated aliphatic compounds	A[‡]
Sebuthylazine	C	Desethylsebuthylazine	C
		2-Chloro-4-ethylmino-6-1,3,5-triazine	B
		Total concentration of sebuthylazine and main decomposition products	C
Sethoxydim	C		
Simazine	C	2-Chloro-4-ethylamino-1,3,5-triazine	B
		Total concentration of simazine and main decomposition products	C
TCA	C		
Tebuthiuron	C		
Terbacil	C		
Terbumeton	C	Desethylterbumeton	C
		2-Methoxy-4-ethylamino-6-amino-1,3,5-triazine	C
		Total concentration of terbumeton and main decomposition products	C
Terbuthylazine	C	Desethylterbuthylazine	C
		2-Chlor-4-Ethylamino-6-amino-1,3,5-triazine	B

Active ingredient	Category*	Degradation products[†]	Category[‡]
		Total concentration of terbuthylazine and main composition products	C
Thiofanox	A	Thiofanoxsulphone	A
		Thiofanoxsulphoxide	A
		Total concentration of thiofanox and main decomposition products	A
Triclopyr	C		
Trifluralin	C	Aromatic amines[§] and nitroaromatics[§]	A[‡]

*Groups of substances are identified by[§]. Subdegradation products and reaction products are not listed. Their existence must be verified in each individual case.

[†]The substances are classified according to the level of knowledge about chronic toxicity (category A to C) or genotoxic potential (limit values must not be exceeded). When determining the acceptable concentration of chlorophenol it is important that the taste of the drinking water is not tainted.

[‡]If decomposition products of these groups of substances exist they must be differentiated and identified. Degradation products with genotoxic potential must not exceed their limit values.

_____ Appendix 2

THE PERCENTAGE OF WATER SUPPLY ZONES IN EACH WATER COMPANY REGION IN ENGLAND AND WALES WHICH DO NOT MEET THE PCV OR RELAXED PCV FOR TOTAL AND INDIVIDUAL PESTICIDES, WITH THE PERCENTAGE OF DETERMINATIONS EXCEEDING THE LIMITS GIVEN IN PARENTHESES, FOR ALL THE WATER SUPPLY COMPANIES DURING 1990

Pesticides	Water supply company (number of water supply zones)*									
	1(200)	2(6)	3(56)	4(12)	5(6)	6(1)	7(23)	8(16)	9(27)	10(25)
Total pesticides	22(5.7)	0(0.0)	25(11.6)	17(1.4)	0(0)	—†	91(35.7)	19(5.0)	0(0)	0(0)
Atrazine	37(34.0)	33(8.3)	54(33.2)	17(5.1)			100(74.4)	50(14.5)	7(6.2)	80(38.1)
Simazine	20(19.9)		23(7.9)	17(2.0)			96(70.1)	50(10.0)	4(0.9)	52(8.7)
Chlortoluron	8(0.8)						61(11.7)	6(1.7)	11(2.8)	24(6.3)
Dichlorprop	13(1.0)									
Dichlorvos	0.5(<0.1)									
Isoproturon	27(13.9)						48(12.1)		7(1.9)	8(1.6)
Ioxynil	0.5(0.1)									
Linuron	1(0.1)									
MCPB	2(0.1)									
Mecoprop	33(32.6)		3.6(0.9)				9(0.9)		11(3.4)	20(10.9)
Propyzamide	13(1.3)									
Pirimicarb				50(5.0)						
Carbetamide							4(0.5)			
Fenpropimorph							4(4.2)			
MCPA									4(0.8)	
2-4,D										
TCA									4(0.9)	
Flutriafol										
β-HCH										
γ-HCH										
Difenzoquat										
Bromoxynil										
Propazine										

Other pesticides	0(0)	0(0)	0(0)	0(0)	0(0)	0(0)	0(0)	0(0)	0(0)	0(0)
Permethrin-*cis*										
Permethrin-*trans*										
Dicamba										
Prometryn										
2,4,5-TCPA										
Pentachlorophenol										
Diuron										
MCPP										

Pesticides	Water supply company (number of water supply zones)*									
	11(13)	12(36)	13(12)	14(3)	15(53)	16(36)	17(42)	18(26)	19(66)	20(13)
Total pesticides	0(0)	100(54.8)	0(0)	0(0)	0(0)	—†	29(8.7)	39(15.5)	0(0)	—†(6.6)
Atrazine	—‡	100(66.1)			40(30.1)		55(35.7)	46(20.8)		—†(46.3)
Simazine		100(60.3)	8(4.4)		38(19.5)	22(18.0)	26(4.8)	39(21.3)		—†(9.2)
Chlortoluron		72(9.5)			15(7.6)		38(7.4)			—†(10.4)
Dichlorprop										
Dichlorvos										
Isoproturon		100(41.6)			6(3.5)		21(5.4)	23(12.5)		—†(11.6)
Ioxynil										
Linuron										
MCPB										
Mecoprop		44(7.1)			9(14.3)			35(13.3)		—†(6.5)
Propyzamide		3(0.2)								
Pirimicarb										
Carbetamide										
Fenpropimorph										
MCPA										
2-4,D		33(2.8)						4(1.4)		
TCA										
Flutriafol								4(1.2)		
β-HCH										
γ-HCH										
Difenzoquat										
Bromoxynil										
Propazine										

Other pesticides	0(0)	0(0)	0(0)	0(0)	0(0)	0(0)	0(0)	—†	—†(0)
Permethrin-*cis*									
Permethrin-*trans*									
Dicamba									
Prometryn									
2,4,5-TCPA									
Pentachlorophenol									
Diuron									
MCPP									

Pesticides	Water supply company (number of water supply zones)*									
	21(79)	22(311)	23(21)	24(24)	25(316)	26(134)	27(38)	28(72)	29(15)	30(7)
Total pesticides	0(0)	1(<0.1)	0(0)	13(1.0)	—†	13(4.2)	16(4.2)	0(0)	0(0)	0(0)
Atrazine	1(0.3)	1(0.1)	14(11.2)	42(17.2)	28(6.2)	27(8.8)	50(12.7)			100(47.7)
Simazine				17(5.2)	29(3.3)	13(8.9)	50(12.7)			71(3.8)
Chlortoluron				25(3.6)						
Dichlorprop						3(0.9)				
Dichlorvos										
Isoproturon				4(10.0)		2(0.5)				
Ioxynil					4(0.1)					
Linuron										
MCPB										
Mecoprop					25(2.8)	7(1.7)				
Propyzamide										
Pirimicarb										
Carbetamide										
Fenpropimorph										
MCPA					10(0.4)	2(0.7)				
2-4,D					3(0.1)	3(0.7)				
TCA										
Flutriafol	1(0.3)									
β-HCH	1(0.3)									
γ-HCH	1(0.3)									
Difenzoquat					20(0.8)	0.7(0.3)				
Bromoxynil					13(0.7)	0.7(0.3)				
Propazine										

Other pesticides	0(0)	0(0)	0(0)	0(0)	0(0)	0(0)	0(0)	0(0)	0(0)
Permethrin-*cis*					0.3(<0.1)			
Permethrin-*trans*					0.3(<0.1)			
Dicamba					4(0.2)			
Prometryn					2(0.1)			
2,4,5-TCPA					4(0.1)			
Pentachlorophenol					4(0.1)			
Diuron								
MCPP								

Pesticides	Water supply company (number of water supply zones)*								
	31(4)	32(224)	33(267)	34(114)	35(6)	36(5)	37(15)	38(4)	39(208)
Total pesticides	0(0)	68(50.3)	0.4(<0.1)	4(0.7)	0(0)	20(3.1)	0(0)	0(0)	—†
Atrazine	50(18.5)	82(71.2)	1.1(0.5)	17(8.2)		20(6.3)		100(20.0)	3.4(1.3)
Simazine	25(14.8)	63(34.1)	0.4(<0.1)	5(1.4)		20(6.3)		100(6.7)	5.3(2.2)
Chlortoluron		55(26.6)		3(0.6)					
Dichlorprop									
Dichlorvos									
Isopropturon		63(38.5)							1.0(1.4)
Ioxynil									
Linuron									
MCPB		2(0.4)							
Mecoprop				0.9(0.2)					
Propyzamide		3(0.4)							
Pirimicarb								100(11.1)	
Carbetamide									
Fenpropimorph									
MCPA		0.5(12.5)							
2-4,D		7(0.5)							
TCA									
Flutriafol									
β-HCH									
γ-HCH									
Difenzoquat									
Bromoxynil									
Propazine									

Other pesticides	0(0)	0(0)	0(0)	0(0)	0(0)	0(0)	0(0)
Permethrin-*cis*							
Permethrin-*trans*							
Dicamba		0.4(<0.1) 0.9(0.2)			7(1.8)		
Prometryn		0.9(0.2)					
2,4,5-TCPA							
Pentachlorophenol							
Diuron	58(6.6)						
MCPP							0.5(0.9)

*Key to water supply companies: 1 = Anglian Water Services Ltd; 2 = Bournmouth and District Water Co.; 3 = Bristol Water Works Co.; 4 = Cambridge Water Company; 5 = Chester Water Works Co.; 6 = Cholderton and District Waterworks Co.; 7 = Colne Valley Water Co.; 8 = East Anglian Water Co.; 9 = Eastbourne Water Co.; 10 = East Surrey Water plc; 11 = East Worcestershire Water Works Co.; 12 = Essex Water Co.; 13 = Folkestone and District Water Co.; 14 = Hartlepools Water Co.; 15 = Lee Valley Water Co.; 16 = Mid-Kent Water Co.; 17 = Mid-Southern Water Co.; 18 = Mid-Sussex Water Co.; 19 = Newcastle and Gateshead Water Co.; 20 = North Surrey Water Co.; 21 = Northumbrian Water Co.; 22 = North West Water Ltd; 23 = Portsmouth Water Co.; 24 = Rickmansworth Water Co.; 25 = Severn and Trent Water Ltd; 26 = Southern Water Services Ltd; 27 = South Staffordshire Water Works Co.; 28 = South West Water Services Ltd; 29 = Sunderland and South Shields Water Co.; 30 = Sutton District Water Co.; 31 = Tendring Hundred Water Works Co.; 32 = Thames Water Utilities; 33 = Welsh Water plc; 34 = Wessex Water Services Ltd; 35 = West Hampshire Water Co.; 36 = West Kent Water Co.; 37 = Wrexham and East Denbighshire Water Co.; 38 = York Water Works Co.; and 39 = Yorkshire Water Services Ltd.

†No information available.

‡Individual pesticides not measured

References

Addiscott, T.M. (1988) Long-term leakage of nitrate from bare unmanured soil. *Soil Use and Management*, **4**, 91.

Ainsworth, R.G., Calcutt, T., Elvidge, A.F., Evins, C., Johnson, D., Lack, T.J., Parkinson, R.W. and Ridgway, J.W. (1981) A guide to solving water quality problems in distribution systems. *Technical Report 167*, Water Research Centre, Medmenham.

Akbar, M. and Johari, B.H. (1990) Treatment techniques for the removal of organic pollutants—the SG. Linggi experience. *Water Supply*, **8** (3–4), 664.

Alvarez, A.J., Hernandez-Delgado, E.A. and Toranzos, G.A. (1993) Advantages and disadvantages of traditional and molecular techniques applied to the detection of pathogens in waters. *Water Science and Technology*, **27** (3–4), 253.

Amoore, J.E. (1986) The chemistry and physiology of odour sensitivity. *Journal of the American Water Works Association*, **78** (3), 70.

Archibald, G. (1986) Demand forecasting in the water industry. In *Water Demand Forecasting* (Eds V. Gardiner and P. Herrington). Proceedings of a workshop sponsored by the Economic and Social Research Council, University of Leicester.

ASTM (1968) Manual on sensory testing methods. *ASTM STP 434*, American Society for Testing and Materials, Washington.

ASTM (1984) Standard test method for odor in water. D 1292-80. In: *Annual Book of ASTM Standards, Section 11: Water and Environmental Technology*, 11.01, 218.

AWWA (1979) Organic removal by coagulation: a review and research needs. *Journal of the American Water Works Association*, **71** (10), 588.

Baier, J.H., Lykins, B.W., Fronk, C.A. and Kramer, S.J. (1987) Using reverse osmosis to remove agricultural chemicals from water. *Journal of the American Water Works Association*, **79** (8), 55.

Bailey, J., Jolly, P.K. and Lacey, R.F. (1986) Domestic water use patterns. *Technical Report 225*, Water Research Centre, Medmenham.

Baker, R.A. (1963) Threshold odors of organic chemicals. *Journal of the American Water Works Association*, **55** (7), 913.

Bartels, J.H.M., Burlingame, G.A. and Suffet, I.H. (1986) Flavour profile analysis: taste and odour control of the future. *Journal of the American Water Works Association*, **78** (3), 50.

Baxter, K.M. and Clark, L. (1984) Effluent recharge. *Technical Report 199*, Water Research Centre, Stevenage.

Bays, L.R., Burman, J.P. and Lewis, W.M. (1970) Taste and odour in water supplies in Great Britain: a survey of the present position and problems for the future. *Water Treatment and Examination*, **19**, 136.

Bear, J. (1979) *Hydraulics of Groundwater*, McGraw-Hill, New York.

Becker, D.L. and Wilson, S.C. (1978) The use of activated carbon for the treatment of pesticides and pesticide wastes. In: *Carbon Adsorption Handbook* (Eds P. Cheremisinoff and F. Ellerbusch), Ann Arbor Science, Ann Arbor, pp. 167–213.

Becker, F.F., Janowsky, U., Overath, H. and Stetter, D. (1989) Die Wirksamkeit von Umehrosmosemembranen bei der Entfernung von Pestiziden. *Wasser/Abwasser*, **130** (9), 425.

Bell, F.A., Perry, D.L., Smith, J.K. and Lynch, S.C. (1984) Studies on home treatment systems. *Journal of the American Water Works Association*, **76** (4), 126.

Beran, M.A. and Rodier, J.A. (1985) *Hydrological Aspects of Drought. Studies and Reports in Hydrology 39*, UNESCO, Paris.

Beresford, S.A. (1980) Water reuse and health in the London area. *Technical Report 138*, Water Research Centre, Stevenage.

Beresford, S.A. (1981) The relationship between water quality and health in the London area. *International Journal of Epidemiology*, **10**, 103.

Beresford, S.A., Carpenter, L.M. and Powell, P. (1984) Epidemiological studies of water reuse and type of water supply. *Technical Report 216*, Water Research Centre, Stevenage.

Berger, P.S. (1992) Revised coliform rule. In: *Regulating drinking water quality* (Eds C.E. Gilbert and E.J. Calabrese), Lewis, Bocta Raton, pp. 161–166.

Blackburn, A.M. (1978) Management strategies dealing with drought. *Journal of the American Water Works Association*, **70** (2), 51.

Blewett, D.A., Wright, S.E., Casemore, D.P., Booth, N.E. and Jones, C.E. (1993) Infective dose size studies on *Cryptosporidium parvum* using gnotobiotic lambs. *Water Science and Technology*, **27** (3–4), 61.

Bowen, R. (1982) *Surface Water*, Applied Science, London.

Breach, R.A. (1989) The EC Directive on drinking water (EEC 80/778). *Water and Environmental Management*, **3**, 323.

British Standards Institution (1991) Specification of requirements for suitability of metallic materials for use in contact with water intended for human consumption with regard to their effects on the quality of the water. *BS DD201:1991*, BSI, London.

Britton, A. and Richards, W.N. (1981) Factors influencing plumbosolvency in Scotland. *Journal of the Institution of Water Engineers and Scientists*, **35**, 349.

Brown, R.H., Konoplyantsev, A.A., Ineson, J. and Kovalvsky, V.S. (1983) *Groundwater Studies. Studies and Reports in Hydrology 7*, UNESCO, Paris.

Bryant, E.A., Fulton, G.P. and Budd, G.C. (1992) *Disinfection Alternatives for Safe Drinking Water*, Van Nostrand Reinhold, New York.

Building Research Establishment (1987) Low water-use washdown WC's. *BRE Report*, Garston, Watford.

Bull, R.J. and Kopfler, F.C. (1991) *Health Effects of Disinfectants and Disinfection By-products*, AWWA Research Foundation and the American Water Works Association, Denver.

Cable, C.J. and Fielding, M. (1992) Review of transformation products in water sources/supplies. *Report FR 0286*, Foundation of Water Research, Marlow.

Cairns, M.A. and Garton, R.R. (1982) Use of fish ventilation frequency to estimate chronically safe toxicant concentration. *Transactions of the American Fisheries Society*, **111**, 70.

Camper, A.K., Le Chevallier, M.W., Broadway, S.C. and McFeters, G.A. (1985) Growth and persistence of pathogens on granular activated carbon. *Applied Environmental Microbiology*, **50**, 1378.

Charley, R.J. (Ed.) (1969) *Introduction to Geographical Hydrology*, Methuen, London.

Chatfield, E.J. and Dillon, M.J. (1979) A national survey for asbestos fibre in Canadian drinking water supplies. *Report 79-EHD-34*, National Health and Welfare, Ottowa, Ontario.

Chilvers, C. and Conway, D. (1985) Cancer mortality in England in relation to levels of naturally occurring fluoride in water supplies. *Journal of Epidemiology and Community Health*, **39**, 44.

Chorquette, K.C. and Gergely, R.M. (1992) Residential water treatment systems: an Iowa consumer protection law. In: *Regulating Drinking Water Quality* (Eds C.E. Gilbert and E.J. Calabrese), Lewis, Boca Raton, pp. 229–310.

Clark, J.A. (1990) The presence-absence test for monitoring drinking water quality. In: *Drinking Water Microbiology* (Ed. G.A. McFeters), Springer-Verlag, New York, pp. 399–411.

Clayton, B. (1991) Water pollution at Lowermoor, North Cornwall. *Second Report of the Lowermoor Incident Health Advisory Group*. Department of Health, HMSO, London.

Clayton, R.C. and Hall, T. (1990) A review of the requirements for treatment works performance specifications. *Report FR 0089*, Foundation for Water Research, Marlow.

Clement International Corporation (1990) *Toxicological Profile for Polycyclic Aromatic Hydrocarbons*. Agency for Toxic Substances and Disease Registry, Public Health Service, US Department of Health and Human Services, Atlanta.

Clesceri, L.S., Greenberg, A.E., Trussell, R.H. and Franson, R.R. (1989a) Taste 2160. In: *Standard Methods for the Examination of Water and Wastewater*, 17th edn, APHA, AWWA, WPCF, 2.23.

Clesceri, L.S., Greenberg, A.E., Trussell, R.H. and Franson, R.R. (1989b) Odor 2150. In: *Standard Methods for the Examination of Water and Wastewater*, 17th edn, APHA, AWWA, WPCF, 2.16.

Codd, A.G., Brooks, W.P., Lawton, L.A. and Beattie, K.A. (1989) Cyanobacterial toxins in European waters: occurrence, properties, problems and requirements. In: *Watershed 89. The Future of Water Quality in Europe* (Eds D. Wheeler, M.L. Richardson and J. Bridges). *Proceedings of the International Association of Water Pollution Research and Control*, 17–20 April, 1989, Pergamon Press, London, pp. 211–220.

Cohen, J.M., Kamphake, L.J., Harris, E.K. and Woodward, R.L. (1960) Taste threshold concentrations of metals in drinking water. *Journal of the American Water Works Association*, **52** (5), 660.

Colling, J.K., Croll, B.T., Whincup, P.A.E. and Harwood, C. (1992) Plumbosolvency effects and control in hard waters. *Journal of the Institution of Water and Environmental Management*, **6**, 259.

Collins, C.H., Lyne, P.M. and Grange, J.M. (1989) *Microbiological Methods*, 6th edn, Butterworths, London.

Commins, B.T. (1979) Asbestos in drinking water: a review. *Technical Report 100*, Water Research Centre, Medmenham.

Commins, B.T. (1988) Asbestos fibres in drinking water. *Scientific and Technical Report 1*, Commins Associates, Maidenhead.

Conrad, J.J. (1990) *Nitrate Pollution and Politics*, Avebury Technical Press, Aldershot.

Conway, D.M. and Lacey, R.F. (1984) Asbestos in drinking water. *Technical Report 202*, Water Research Centre, Medmenham.

Cooper, P., Wickham, C., Lacey, R.F. and Barker, D.J.P. (1990) Water fluoride concentration and fracture of the proximal femur. *Journal of Epidemiology and Community Health*, **44**, 17.

Coughlan, A. (1991) Fresh water from the sea. *New Scientist*, **131**, 37.

Cox, M., Finkel, L. and Cohen, J. (1992) Regulating lead in drinking water: a status report (June, 1990). In *Regulating drinking Water quality* (Ed. C.E. Gilbert and E.J.) Calabrese, Lewis, Boca Raton, pp. 3–14.

Craun, G.F. (1986) *Waterborne Diseases in the United States*, CRC Press, Boca Raton.

Craun, G.F. (1991) Cause of waterborne outbreaks in the United States. *Water Science and Technology*, **24**, 17.

Criddle, J. (1992) The toxicity of manganese in drinking water. *Report FR 0313*, Foundation for Water Research, Marlow.

Croll, B.T. (1985) The effects of the agricultural use of herbicides in freshwater. In: *Water Research Conference Effects of Land-use on Freshwater*, Stirling, 18th June 1983.

Croll, B.T., Chadwick, B. and Knight, B. (1991) The removal of atrazine and other herbicides from water using granular activated carbon. *Water Supply*, **10**, 111.

Cross, F.T., Harley, N.H. and Hofmann, W. (1985) Health effects and risks from [222]Rn in drinking water. *Health Physics*, **48**, 649.

Cubitt, D.W. (1991) A review of the epidemiology and diagnosis of waterborne viral infections. *Water Science and Technology*, **24** (2), 197.

Dean, R.B. and Lund, E. (1981) *Water Reuse: Problems and Solutions*, Academic Press, London.

DeGraeve, G.M. (1982) Avoidance response of rainbow trout to phenol. *Progress in Fish Culture*, **44**, 82.

Denny, S. (1991) Microbiological efficiency of water treatment. *Report FR 0219*, Foundation for Water Research, Marlow.

Denny, S., Broberg, P. and Whitemore, T. (1990) Microbiological hazards in water supplies. *Report FR 0114*, Foundation for Water Research, Marlow.

Department of the Environment (1977) Lead in drinking water: a survey in Great Britain 1975–76. *Pollution Paper 12*, HMSO, London.

Department of the Environment (1980) *Odour and Taste in Raw and Potable Waters 1980. Methods for the Examination of Waters and Associated Materials*, HMSO, London.

Department of the Environment (1985) *Methods of Biological Sampling: Sampling of Macro-invertebrates in Water Supply Systems 1983. Methods for the Examination of Waters and Associated Materials*, HMSO, London.

Department of the Environment (1988a) *Assessment of Ground Water Quality in England and Wales*, HMSO, London.

Department of the Environment (1988b) *The Nitrate Issue*, HMSO, London.

Department of the Environment (1990a) *Isolation and Identification of* Giardia *Cysts,* Cryptosporidium *Oocysts and Free Living Pathogenic Amoebae in Water etc. 1989. Methods for the Examination of Waters and Associated Materials*, HMSO, London.

Department of the Environment (1990b) *Digest of Environmental Protection and Water Statistics No.12*, HMSO, London.

Department of the Environment (1991) *Drinking Water 1990. A Report by the Chief Inspector*, Drinking Water Inspectorate, HMSO, London.

Department of the Environment (1992) *Drinking Water 1991. A Report by the Chief Inspector*, Drinking Water Inspectorate, HMSO, London.

Department of the Environment (1993) *Digest of Environmental Protection and Water Statistics No. 15*, 1992, HMSO, London.

DeZuane, J. (1990) *Handbook of Drinking Water Quality: Standards and Control*, Van Nostrand Reinhold, New York.

DHSS (1980) *Lead and Health*. The report of the DHSS working party on lead in the environment, Department of Health and Social Security, HMSO, London.

Duncan, A. (1988) The ecology of slow sand filters. In: *Slow Sand Filtration: Recent Developments in Water Treatment Technology* (Ed. N.J.D. Graham), Ellis Horwood, Chichester, pp. 163–180.

Dupuy, C.J., Healy, D., Thomas, M., Brown, D., Siniscalchi, A. and Dembek, Z. (1992) A survey of naturally occurring radionuclides in ground water in selected bedrock aquifers in Connecticut and implications for public health policy. In: *Regulating Drinking Water Quality* (Ed. C.E. Gilbert and E.J. Calabrese), Lewis, Boca Raton, pp. 95–119.

EC (1975) Council Directive concerning the quality required of surface water intended for the abstraction of drinking water in the member states (75/440/EEC). *Off. J. Eur. Commun.*, **L194** (25 July 1975), 26–31.

EC (1977) Council Directive concerning the biological screening of the population for lead. (77/312/EEC) *Off. J. Eur. Commun.*, **20**, L105, 10–17, 28 Apr. 1977.

EC (1980) Council Directive relating to the quality of water intended for human consumption (80/778/EEC). *Off. J. Eur. Commun.*, **L229** (30 August 1980), 11–29.

EC (1991) Council Directive concerning the protection of waters against pollution caused by nitrates from agricultural sources (91/676/EEC). *Off. J. Eur. Commun.*, **L375** (31 December 1991), 1–8.

ECETOC (1988) *Nitrates and Drinking Water*. ECETOC Technical Report 27, European Chemical Industry Ecology and Toxicology Centre, Brussels.

Edmunds, W.M., Bath, A.H. and Miles, D.L. (1982) Hydrochemical evolution of the East Midlands Triassic sandstone aquifer, England. *Geochimica et Cosmochimica Acta*, **46**, 2069.

Eriksson, E. (1985) *Principles and Applications of Hydrochemistry*, Chapman and Hall, London.

Falconer, I.R. (1989) Effects on human health of some toxic cyanobacteria (blue–green algae) in reservoirs, lakes and rivers. *Toxicity Assessment*, **4**, 1175.

Falconer, I.R. (1991) Tumour promotion and liver damage caused by oral consumption of cyanobacteria. *Environmental Toxicology and Water Quality*, **6**, 177.

Fawell, J.K., Fielding, M. and Ridgeway, J.W. (1987) Health risks of chlorination: is there a problem? *Water and Environmental Management*, **1**, 61.

Fayer, R. and Ungar, B.L.P. (1986) *Cryptosporidium* species and cryptosporidiosis. *Microbiological Reviews*, **50**, 458.

Ferguson, D.W., Gramith, J.T. and McGuire, M.J. (1991) Applying ozone for organics control and disinfection: a utility perspective. *Journal of the American Water Works Association*, **83** (5), 32.

Fewkes, A. and Ferris, F.A. (1982) The recycling of domestic waste water: factors influencing storage capacity. *Building and Environment*, **17**, 209.

Fielding, M. and Packham, R.F. (1990) Human exposure to water contaminants. *Report FR 0085*, Foundation for Water Research Centre, Marlow.

Fielding, M., Gibson, T.M., James, H.A., McLoughlin, K. and Steel, C.P. (1981) Organic micro-pollutants in drinking water. *Technical Report 159*, Water Research Centre, Medmenham.

Flanagan, P.J. (1988) *Parameters of Water Quality, Interpretation and Standards*, Environmental Research Unit, Dublin.

Forbes, W.F. and MacAiney, C.A. (1992) Aluminium and dementia. *The Lancet*, **340**, 668.

Forman, D., Al-Dabbagh, S. and Doll, R. (1985) Nitrates, nitrites and gastric cancer in Great Britain. *Nature*, **313**, 620.

Foster, D.M., Rachwal, A.J. and White, S.J. (1991) New treatment processes for pesticides and chlorinated organics control in drinking water. *Journal of the Institution of Water and Environmental Management*, **5** (4), 466.

Foster, S.S.D., Bridge, L.R., Geahe, A.K., Lawrence, A.R. and Parker, J.M. (1986) The groundwater nitrate problem. A summary of research on the impact of agricultural land use practices on groundwater quality between 1976 and 1985. *Hydrogeological Report: 86/2*, British Geological Survey, Wallingford.

Foundation for Water Research (1993) *The Service Pipe Manual*. UK Water Industry Engineering and Operations Committee, Foundation for Water Research, Marlowe.

Franks, F. (1987) The hydrologic cycle: turnover, distribution and utilization of water. In: *Handbook of Water Purification* (Ed. W. Lorch), Ellis Horwood, Chichester, pp. 30–49.

Friends of the Earth (1989) Poison on tap. *Observer Magazine*, 6 August, 15–24.

Galbraith, N.S., Barrett, N. and Stanwell-Smith, R. (1987) Water disease after Croydon: a review of water-borne and water associated disease in the UK 1937–86. *Water and Environmental Management*, **1**, 7.

Geldreich, E.E. and Reasoner, D.J. (1990) Home treatment devices and water quality. In: *Drinking Water Microbiology* (Ed. G.A. McFeters), Springer-Verlag, New York, pp. 147–167.

Gendlebien, A., Jakson, P. and Agg, R. (1992) Lead in drinking water: the impact of existing and future standards on other EC member states. *Report FR 0343*, Foundation for Water Research, Marlow.

Gerba, C.P. and Rose, J.B. (1990) Viruses in source and drinking water. In: *Drinking Water Microbiology* (Ed. G.A. McFeters), Springer-Verlag, New York, pp. 380–396.

German, J.C. (Ed.) (1989) *Management Systems to Reduce Impact of Nitrates*, Elsevier, London.

Gibbs, R.A., Scutt, J.E. and Croll, B.T. (1993) Assimilable organic carbon concentrations and bacterial numbers in a water distribution system. *Water Science and Technology*, **27** (3–4), 159.

Gibson, M.T., Welch, I.M., Barrett, P.R.F. and Ridge, I. (1990) Barley straw as an indicator of algal growth. II: Laboratory studies. *Journal of Applied Phycology*, **2**, 241.

Gilbert, C.E. and Calabrese, E.J. (1992) *Regulating Drinking Water Quality*, Lewis, Boca Raton.

Godley, A. and Wilcox, P. (1992) Plastic pipe location and leak detection on plastic pipes. *Report FR 0257*, Foundation for Water Research, Marlow.

Gould, D.J. (1977) Gull droppings and their effects on water quality. *Technical Report 37*, Water Research Centre, Stevenage.

Graham, N.J.D. (Ed.) (1988) *Slow Sand Filtration: Recent Developments in Water Treatment Technology*, Ellis Horwood, Chichester.

Gray, N.F. (1992) *Biology of Wastewater Treatment*, Oxford University Press, Oxford.

Gregory, R. (1990) Galvanic corrosion of lead solder in copper pipework. *Journal of the Institution of Water and Environmental Management*, **4** (2), 112.

Halford, J. and Bond, K. (1986) Devices with moving parts for the prevention of backflow in water installations. *Technical Report 245*, Water Research Centre, Swindon.

Hardy, J., Highsmith, V.R., Costa, D.L. and Krewer, J.A. (1992) Indoor asbestos concentrations associated with the use of asbestos contaminated tap water in portable home humidifiers. *Environmental Science and Technology*, **26**, 680.

Hart, J., Scott, P. and Carlie, P.R. (1992) Algal toxin removal from water. *Report FR 0303*, Foundation for Water Research, Marlow.

Helms, G. and Rydell, S. (1992) Regulation of radon in drinking water. In: *Regulating Drinking Water Quality* (Ed. C.E. Gilbert and E.J. Calabrese), Lewis, Boca Raton, pp. 77–82.

Hibler, C.P. and Hancock, C.M. (1990) Waterborne giardiasis. In: *Drinking Water Microbiology* (Ed. G.A. McFeters), Springer-Verlag, New York, pp. 271–293.

Hill, R. and Lorch, W. (1987) Water purification. In: *Handbook of Water Purification* (Ed. W. Lorch), Ellis Horwood, Chichester, pp. 226–302.

Himberg, K., Keijola, A.M., Hiisvirta, L., Pyyasalo, H. and Sivonen, K. (1989) The effects of water treatment processes on the removal of hepatotoxins from *Microcystis* and *Oscillatoria cyanobacteria*: a laboratory study. *Water Research*, **23**, 979.

HMSO (1982) *The Bacteriological Examination of Drinking Water Supplies*, HMSO, London.

HMSO (1989a) *The Water Supply (Water Quality) Regulations, 1989. Statutory Instrument 1989/1147*, HMSO, London.

HMSO (1989b) *Nitrate in water. Select Committee on the European Communities, 18 July 1989*, HMSO, London.

HMSO (1989c) *Guidance on Safeguarding the Quality of Public Water Supplies*, Department of the Environment, HMSO, London.

HMSO (1989d) *The Water Supply and Sewerage Services (Customer Service Standards) Regulations 1989. Statutory Instrument 1989/1159*, HMSO, London.

HMSO (1990a) *The Water Supply (Water Quality) (Scotland) Regulations, 1990. Statutory Instrument 1990/119 (S11)*, HMSO, London.

HMSO (1990b) *This Common Inheritance: Britain's Environmental Strategy. Command 1200*, HMSO, London.

HMSO (1991) *Private Water Supplies Regulations 1991. Statutory Instrument 1991/2790*, HMSO, London.

HMSO (1993) *The Water Supply and Sewerage Services (Customer Service Standards) (Amendment) Regulations, 1993. Statutory Instrument 1993/500*, HMSO, London.

Hopkins, S.M. and Ellis, J.C. (1980) Drinking water consumption in Great Britain. *Technical Report 137*, Water Research Centre, Medmenham.

Horth, H., Fawell, J.K., James, C.P. and Young, W. (1991) The fate of the chlorinated-derived mutagen MX in vivo. *Report FR 0068*, Foundation for Water Research, Marlow.

Huck, P.M. (1990) Measurement of biodegradable organic matter and bacterial growth potential in drinking water. *Journal of the American Water Works Association*, **82** (7), 78.

Huck, P.M., Fedorak, P.M. and Anderson, W.B. (1991) Formation and removal of assimilable organic carbon during biological treatment. *Journal of the American Water Works Association*, **83** (12), 69.

Hulsmann, A.D. (1990) Particulate lead in water supplies. *Journal of the Institution of Water and Environmental Management*, **4**, 19.

Hunt, S. and Fawell, J. (1990) Review of the toxicity of aluminium with special reference to drinking water. *Report FR 0068*, Foundation for Water Research, Marlow.

Hunter, P.R. (1991) An introduction to the biology, ecology and potential public health significance of the blue–green algae. *PHLS Microbiology Digest*, **8**, 13.

Hydes, O.D., Hill, C.Y., Marsden, P.K. and Waite, W.M. (1992) Nitrate, pesticides and lead, 1989 and 1990. Drinking Water Inspectorate, HMSO.

Jacangelo, J.G., Patania, N.L., Reagan, K.M., Aieta, E.M., Krasner, S.W. and McGuire, M.J.

(1989) Ozonation: assessing its role in the formation and control of disinfection by-products. *Journal of the American Water Works Association*, **81** (8), 74.

Jones, K. and Telford, D. (1991) On the trail of the seasonal microbe. *New Scientist*, **130**, 36.

Keller, A.Z. and Wilson, H.C. (1992) *Hazards to Drinking Water Supplies*, Springer-Verlag, London.

Kendrick, M.A.P., Clark, L., Baxter, K.M., Fleet, M., James, H.A., Gibson, T.M. and Turrell, M.B. (1985) Trace organics in British aquifers: a baseline study. *Technical Report 223*, Water Research Centre, Medmenham.

Kinner, N.E., Malley, J.P. and Clement, J.A. (1990) Radon removal using point of entry water treatment techniques. *EPA/600/2-90/047*, USEPA, Ohio.

Kirby, C. (1979) *Water in Great Britain*, Penguin, London.

Knight, M.S. and Tuckwell, J.B. (1988) Controlling nitrate leaching in water supply catchments. *Water and Environmental Management*, **2**, 248.

Korich, D.G., Mead, J.R., Madore, M.S., Sinclair, M.A. and Sterling, C.R. (1990) Effects of ozone, chlorine dioxide, chlorine and monochloramine on *Cryptosporidium parvum* oocyst viability. *Applied and Environmental Microbiology*, **56** (5), 1423.

Krasner, S.W., McGuire, M.J., Jacangelo, J.G., Patania, N.L., Reagan, K.M. and Aieta, E.M. (1989) The occurrence of disinfection by-products in US drinking water. *Journal of the American Water Works Association*, **81** (8), 41.

Kruithof, J.C., Puijker, L.M. and Janssen, H.M. (1989) Presence and removal of pesticides. H_2O, **22** (17), 526.

Kruithof, J.C., Van Eekeren, M.W.M. and Schippers, J.C. (1991) Membraanfiltratie, geavanceerde oxydatie en UV-desinfectie in de processes, advanced oxidation and UV- USA en Canada. H_2O, **4** (19), 537.

Lacey, R.F. (1981) Changes in water hardness and cardiovascular death-rates. *Technical Report 171*, Water Research Centre, Medmenham.

Lansdown, R. and Yule, W. (1986) *The Lead Debate: Environment, Toxicology and Child Health*, Croom Helm, London.

Latham, B. (1990) *Water Distribution*, Institution of Water and Environmental Management, London.

Lawrence, A.R. and Foster, S.S.D. (1987) The pollution threat from agricultural pesticides and industrial solvents. *Hydrological Report: 87/2*, British Geological Survey, Wallingford.

LeChevallier, M.W. (1990) Coliform regrowth in drinking water: a review. *Journal of the American Water Works Association*, **82** (11), 74.

LeChevallier, M.W. and McFeters, G.A. (1990) Microbiology of activated carbon. In: *Drinking Water Microbiology* (Ed. G.A. McFeters), Springer-Verlag, New York, pp. 104–119.

Lee, R.G., Becker, W.C. and Collins, D.W. (1989) Lead at the tap: sources and control. *Journal of the American Water Works Association*, **81** (7), 52.

Lerner, D.N. and Tellam, J.H. (1993) The protection of urban groundwater from pollution. In: *Water and the Environment* (Eds J.C. Currie and A.T. Pepper), Ellis Horwood, Chichester, pp. 322–37.

Levi, Y. and Jestin, J.M. (1988) Offensive tastes and odors occurring after chlorine addition in water treatment processes. *Water Science and Technology*, **20** (8–9), 269.

Levy, R.V. (1990) Invertebrates and associated bacteria in drinking water distribution lines. In: *Drinking Water Microbiology* (Ed. G.A. McFeters), Springer-Verlag, New York, pp. 225–248.

Levy, R.V., Hart, F.L. and Cheetham, R.D. (1986) Occurrence and public health significance of invertebrates in drinking water systems. *Journal of the American Water Works Association*, **78** (9), 105.

Lillard, D.A. and Powers, J.J. (1975) Aqueous odour thresholds of organic pollutants in industrial effluents. *EPA-660/4-75-002, Environmental Monitoring Series*, US Environmental Protection Agency.

Littlejohn, J.W. and Melvin, M.A.L. (1991) Sheep dips as a source of pollution of freshwaters: a study in Grampian Region. *Water and Environmental Management*, **5**, 21.

Loch, J.P.G., van Dijk-Looyaard, A. and Zoeteman, B.C.J. (1989) Organics in groundwater. In:

Watershed 89, the future of Water Quality in Europe (Eds D. Wheeler, M.L. Richardson and J. Bridges), Proceedings of the International Association of Water Pollution Research and Control, 17–20th April 1989, Pergamon Press, London, pp. 39–55.

Logsdon, G.S. (1990) Microbiology and drinking water filtration. In: *Drinking Water Microbiology* (Ed. G.A. McFeters), Springer-Verlag, New York, pp. 120–146.

Longtin, J.P. (1988) Occurrence of radon, radium, and uranium in groundwater. *Journal of the American Water Works Association*, **80** (7), 84.

Lorch, W. (Ed.) (1987) *Handbook of Water Purification*, Ellis Horwood, Chichester.

Lykins, B.W. and Griese, M.H. (1986) Using chloride dioxide for trihalomethane control. *Journal of the American Water Works Association*, **78** (6), 88.

Lykins, B.W., Clark, R.M. and Adams, J.A. (1988) Granular activated carbon for controlling THMs. *Journal of the American Water Works Association*, **80** (5), 85.

Macan, T.T. (1970) *A Guide to Freshwater Invertebrate Animals*, Longman, London.

MacDonald, A.J., Powlson, D.S., Poulton, P.R. and Jenkinson, D.S. (1989) Unused fertilizer nitrogen in arable soils—its contribution to nitrate leaching. *Journal of the Science of Food and Agriculture*, **46**, 407.

MacLachan, D.R., Fraser, P.E. and Dalton, A.J. (1992) Aluminium and the pathogenesis of Alzheimer's disease: a summary of evidence. In: *Aluminium in Biology and Medicine, Ciba Foundation Symposium No. 169*, Wiley, Chichester, pp. 87–108.

Macrory, R. (1989) *The Water Act 1989: Text and Commentary*, Sweet and Maxwell, London.

Mallevialle, J. and Suffet, I.H. (Eds) (1987) *Identification and Treatment of Tastes and Odors in Drinking Water*, American Water Works Association Research Foundation and Lyonnaise des Eaux, Denver.

Mallevialle, J. and Suffet, I.H. (1992) *Influence and Removal of Organics in Drinking Water*, Lewis, Boca Raton.

Mara, D.D. (1974) *Bacteriology for Sanitary Engineers*, Churchill Livingstone, Edinburgh.

Martyn, C.N., Osmond, C., Edwardson, J.A., Barker, D.J.P., Harris, E.E. and Lacey, R.F. (1989) Geographical relationship between Alzheimer's disease and aluminium in drinking water. *Lancet*, **i**, 14 January, 59.

Masschelein, W.J. (1982) *Ozonization Manual for Water and Wastewater Treatment*, Wiley, Chichester.

Maul, A., Vagost, D. and Block, J.C. (1991) Microbiology of distribution networks for drinking water supplies. In: *Microbiological Analysis in Water Distribution Networks*, Ellis Horwood, Chichester, pp. 11–31.

McDonald, A. and Kay, D. (1988) *Water Resources: Issues and Strategies*, Longman, London.

McFeters, G.A. (Ed.) (1990a) *Drinking Water Microbiology*, Springer-Verlag, New York.

McFeters, G.A. (1990b) Enumeration, occurrence, and significance of injured indicator bacteria in drinking water. In: *Drinking Water Microbiology* (Ed. G.A. McFeters), Springer-Verlag, New York, pp. 478–492.

McFeters, G.A. and Singh, A. (1991) Effects of aquatic environmental stress on enteric bacterial pathogens. *Journal of Applied Bacteriology, Symposium Supplement*, **70**, 1155.

McKinney, J.D., Maurer, R.R., Haas, J.R. and Thomas, R.O. (1976) Possible factors in the drinking water of laboratory animals causing reproductive failure. In: *Identification and Analysis of Organic Pollutants in Water*, Vol. 5 (Ed. L.H. Keith), Ann Arbor Science, Ann Arbor, pp. 417–432.

Miller, D.G., Zabel, T.F. and Newman, P.J. (1990) Summary report on environmental developments June 1989–March, 1990. *Report FR 0088*, Foundation for Water Research, Marlow.

Millette, J.R., Clark, P.J. and Pansing, M.F. (1979) Exposure to asbestos from drinking water in the United States. *EPA-600-1-79-028*, US Environmental Protection Agency, Cincinnati.

Montgomery, J.M. (1975) *Water Treatment Principles and Design*, Wiley, New York.

MORI (1992) A quantitative survey. *Report prepared for the Office of Water Services*, MORI, London.

Morris, R.W., Grabow, W.O.K. and Dufour, A.P. (Eds) (1993) *Health-related Water Microbiology 1992*, Pergamon Press, Oxford.

Mullenix, P.J. (1992) Can safe lead levels in drinking water be deduced from current scientific evidence? In: *Regulating Drinking Water Quality* (Eds C.E. Gilbert and E.J. Calabrese), Lewis, Boca Raton, pp. 37–46.

Musial, C.E., Arrowood, M.J., Sterling, C.R. and Gerba, C.P. (1987) Detection of *Cryptosporidium* in water by using polypropylene cartridge filters. *Applied and Environmental Microbiology*, **53**, 687.

Najm, I.N., Snoeyink, V.L., Lykins, B.W. and Adams, J.Q. (1991) Using powered activated carbon: a critical review. *Journal of the American Water Works Association*, **83** (1), 65.

National Rivers Authority (1990) *Toxic Blue-green Algae*. Water Quality Series 12, Anglian Region, NRA, Peterborough.

National Rivers Authority (1991) *Demand and Resources of Water Undertakers in England and Wales—1991*, NRA, London.

National Water Council (1982) Components of household water demand. *Occasional Technical Paper: 6*, National Water Council, London.

Nature Conservancy Council (1989) *Fish Farming and the Safeguard of the Natural Marine Environment of Scotland*. NCC, Edinburgh.

NEHA (1989) Ensuring safe drinking water. *Environmental Health Trends Report 3*, National Environmental Health Association, Denver.

Neri, L.C. and Hewitt, D. (1991) Aluminium, Alzheimer's disease, and drinking water. *The Lancet*, **338**, 390.

OECD (1989) *Water Resource Management: Integrated Policies*, Organization for Economic Co-operation and Development, Paris.

Ofwat (1991) *Paying for Water: OPCS Omnibus Survey*, Office of Water Services, London.

Oliphant, R.J. (1983) The contamination of potable water by lead from soldered joints. *External Report 125E*, Water Research Centre, Medmenham.

Open University (1974) *The Earth's Physical Resources: 5. Water Resources*, S26, Open University Press, Milton Keynes.

Open University (1975) *Water Distribution, Drainage, Discharge and Disposal*. PT 272–7, Open University Press, Milton Keynes.

Owen, M. (1993) Groundwater abstraction and river flows. In: *Water and the Environment* (Eds J.C. Currie and A.T. Pepper), Ellis Horwood, Chichester, pp. 302–311.

Parekh, B.S. (Ed.) (1988) *Reverse Osmosis Technology*, Marcel Dekker, New York.

Parr, W. and Clarke, S. (1992) A review of potential methods for controlling phytoplankton, with particular reference to cyanobacteria, and sampling guidelines for the water industry. *Report FR 0248*, Foundation for Water Research, Marlow.

Parr, N., Charles, A.J. and Walker, S. (Eds) (1992) *Water Resources and Reservoir Engineering*. Proceedings of the seventh conference of the British Dam Society, Telford, London.

Pearse, F. (1992) *The Dammed: Rivers, Dams and the Coming of the World Water Crisis*, Bodley Head, London.

Pearson, P.E., Whitefield, F.B. and Krasner, S.W. (1992) *Off Flavours in Drinking Water and Aquatic Organisms*, Pergamon Press, Oxford.

Peters, C.J., Young, R.J. and Perry, R. (1978) *Chemical Aspects of Aqueous Chlorination Reactions: a Literature Review*, Water Research Centre, Medmenham.

Peters, R.J.B., De Leer, E.W.B. and De Galan, L. (1990) Dihaloacetonitriles in Dutch drinking waters. *Water Research*, **24**, 797.

Peterson, L.R., Denis, D., Brown, D., Hadler, J.L. and Helgerson, S.D. (1988) Community health effects of a municipal water supply hyper-fluoridation accident. *American Journal of Public Health*, **78** (6), 711.

Pocock, S.J. (1980) Factors influencing household water lead: a British National Survey. *Archives of Environmental Health*, **35**, 45.

Pontius, F.W. (Ed.) (1990) *Water Quality and Treatment: a Handbook of Community Water Supplies*, McGraw-Hill, New York.

Powell, R., Packham, R.F., Lacey, R.F. and Russell, P.F. (1982) Water quality and cardiovascular disease in British towns. *Technical Report 178*, Water Research Centre, Medmenham.

Pressdee, J. and Hart, J. (1991) Algal/bacterial toxin removal from water. *Report FR 0223*, Foundation for Water Research, Marlow.

Price, M. (1991) *Introducing Groundwater*, Chapman and Hall, London.

Raghunath, H.M. (1987) *Groundwater*, Wiley, New York.

Reasoner, D.J., Blannon, J.C. and Geldreich, E.E. (1987) Microbiological characteristics of third faucet point-of-use devices. *Journal of the Water Works Association*, **79**, 60.

Reichard, E. *et al.* (1990) *Groundwater Contamination and Risk Assessment*. IAHS Publication 196, International Association of Hydrological Sciences Press, Wallingford, Oxford.

Richardson, A.J., Frankenberg, R.A. and Buck, A.C. (1991a) An outbreak of waterborne cryptosporidiosis in Swindon and Oxfordshire. *Epidemiology and Infection*, **107**, 509.

Richardson, K.J., Stewart, M.J. and Wolfe, R.L. (1991b) Application of gene probe technology to the water industry. *Journal of the American Water Works Association*, **83** (9), 71.

Ritter, W.F. (1990) Pesticide contamination of ground water in the United States—a review. *Journal of Environmental Science and Health*, **B25** (1), 1.

Rivett, M.O., Lerner, D.N. and Lloyd, J.W. (1990) Chlorinated solvents in UK aquifers. *Water and Environmental Management*, **4**, 242.

Rook, J.J. (1974) Formation of haloforms during chlorination of natural waters. *Water Treatment and Examination*, **23**, 234.

Rosa, P.J. De and Parkinson, R.W. (1986) Corrosion of ductile iron pipe. *Technical Report 241*, Water Research Centre, Swindon.

Rose, J.B. (1990) Emerging issues from the microbiology of drinking water. *Water Engineering and Management*, July, 23.

Royal Society (1984) *The Nitrogen Cycle of the United Kingdom. A Study Group Report*, Royal Society, London.

Sands, J.R. (1969) The control of animals in water mains. *Technical Paper 63*, Water Research Association, Medmenham.

Skirrow, M.B. (1982) *Campylobacter enteritus*: the first five years. *Journal of Hygiene, Cambridge*, **89**, 175.

Smith, A.L. and Rogers, D.V. (1990) Isle of Wight metering trial. *Water and Environmental Management*, **4**, 405.

Solbe, J.F. De L.G. (1983) Water authority regulations concerning the use of angler's bait on potable supply reservoirs. *Technical Report 188*, Water Research Centre, Medmenham.

Sollars, C.J. *et al.* (1989) Aluminium in European drinking water. *Environmental Technology Letters*, **10**, 13.

Solt, G.S. and Shirley, C.B. (1991) *An Engineer's Guide to Water Treatment*, Avebury Technical Press, Aldershot.

States, S.J., Wadowsky, R.M., Kuchta, J.M., Wolford, R.S., Conley, L.F. and Yee, R.B. (1990) *Legionella* in drinking water. In: *Drinking Water Microbiology* (Ed. G.A. McFeters), Springer-Verlag, New York, pp. 340–367.

Stout, J., Yu, V.L. and Best, M.G. (1985) Ecology of *Legionella pneumophilia* within water distribution systems. *Applied and Environmental Microbiology*, **49**, 584.

Tebbutt, T.H.Y. (1979) *Principles of Water Quality Control*, Pergamon Press, Oxford.

Thornton, I. and Culbard, E. (Eds) (1987) *Lead in the Home Environment*, Science Reviews, Northwood.

Turner, P.C., Gamme, A.J., Hollinrake, K. and Codd, G.A. (1990) Pneumonia associated with contact with cyanobacteria. *British Medical Journal*, **300**, 1440.

USEPA (1988) Handbook for special public notification for lead for public drinking water supplies. *Report 570/9-88-002*, Office of Water, US Environmental Protection Agency.

USEPA (1990) Drinking water regulations under the Safe Drinking Water Act. *Fact Sheet, May*, US Environmental Protection Agency.

USEPA (1991) *ARARs Q's and A's Compliance With New SDWA National Primary Drinking Water Regulations for Organic and Inorganic Chemicals (Phase II)*. US Environmental Protection Agency.

USEPA (1993) *Drinking Water Regulations and Health Advisories.* Health and Ecological Criteria Division, US Environmental Protection Agency, Washington DC.

USPHS (1991) Review of fluoride benefits and risks. *Report of the Ad Hoc Sub-committee on Fluoride of the Committee to Co-ordinate Environmental Health and Related Programs*, US Public Health Service, Washington.

Van der Kooij, D., Visser, D.A. and Hijnen, W.A.M. (1982) Determining the concentration of easily assimilable organic carbon in drinking water. *Journal of the American Water Works Association*, **74**, 540.

Van der Wende, Characklies, W.G. and Smith, D.B. (1989). Biofilms and bacterial drinking water quality. *Water Research*, **23**, 1313.

Vesey, G. and Slade, J. (1991) Isolation and identification of *Cryptosporidium* from water. *Water Science and Technology*, **24**, 165.

Visscher, J.T. (1988) Water treatment by slow sand filtration: considerations for design, operation and maintenance. In: *Slow Sand Filtration: Recent Developments in Water Treatment Technology* (Ed. N.J.D. Graham), Ellis Horwood, Chichester, pp. 1–10.

Vogt, R. (1986) Water quality and health study of possible relation between aluminium in drinking water and dementia. *Sosiale og Okonomiske Studier 61*, Central Bureau of Statistics of Norway, Oslo.

Waggoner, P. (Ed.) (1990) *Climate Change and US Water Resources*, Wiley, New York.

Wanielista, M.P. (1990) *Hydrology and Water Quality Control*, Wiley, New York.

Ward, R.C. and Robinson, M. (1990) *Principles of Hydrology*, McGraw-Hill, New York.

Water Authorities Association (1985) *Guide to the Microbial Implications of Emergencies in the Water Services*, Water Authorities Association, London.

Water Engineering and Development Centre (1991) *The Worth of Water: Technical Briefs on Health, Water and Sanitation*, Intermediate Technology Publications, Loughborough.

Water Research Centre (1991) *A Guide to Water Service Pipes in Scotland*, Water Research Centre, Swindon.

Water Research Centre (1992) *A Guide to Water Service Pipes*, Water Research Centre, Swindon.

Water Services Association (1990) *Waterfacts '89*, Water Services Association, London.

White, G.C. (1972) *Handbook of Chlorination*, Van Nostrand Reinhold, New York.

White, S.F. and Mays, G.D. (1989) *Water Supply Bylaws Guide*, 2nd edn, Water Research Centre and Ellis Horwood, Chichester.

Whitemore, T.N. and Carrington, E.G. (1993) Comparison of methods for recovery of *Cryptosporidium* from water. *Water Science and Technology*, **27** (3–4), 69.

Wilkins, M. (1993) The fight against *Cryptosporidium*. *Water Bulletin*, **544**, 12 February, 12.

Williams, R.J., Bird, S.C. and Clare, R.W. (1991) Simazine concentrations in a stream draining an agricultural catchment. *Water and Environmental Management*, **5**, 80.

Wood, F.C. (1987) Saline distillation. In: *Handbook of Water Purification* (Ed. W. Lorch), Ellis Horwood, Chichester, pp. 467–487.

Wolfe, R.L. (1990) Ultraviolet disinfection of potable water. *Environmental Science and Technology*, **24**, 768.

World Health Organization (1984) *Guidelines for Drinking Water Quality*, Vol. 2. *Health Criteria and Other Supporting Information*, World Health Organization, Geneva.

World Health Organization (1987) *Drinking Water Quality: Guidelines for Selected Herbicides*, World Health Organization, Copenhagen.

World Health Organization (1993) *Revision of the WHO Guidelines for Drinking Water Quality*, World Health Organization, Geneva.

World Health Organization (In press) *Guidelines for Drinking Water Quality, Vol. 3: Surveillance and Control of Community Supplies.* World Health Organization, Geneva.

Worthing, C.R. and Walker, S.B. (1983) *Pesticide Manual*, 7th edn, British Crop Protection Council, London.

Zoetman, B.C.J. and Brinkmann, F.J.J. (1976) *Hardness of Drinking Water and Public Health*, Pergamon Press, Oxford.

Index